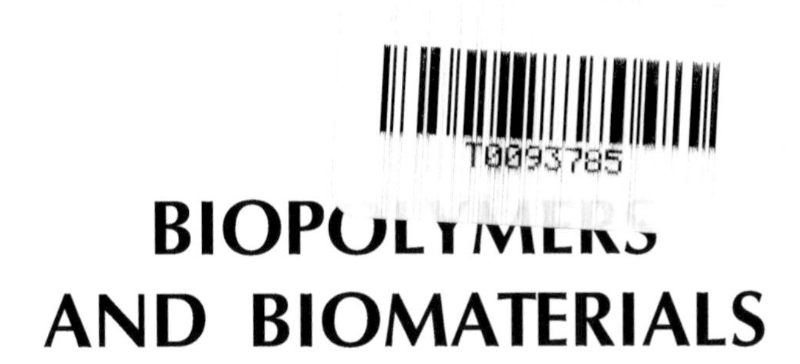

BIOPOLYMERS AND BIOMATERIALS

BIOPOLYMERS AND BIOMATERIALS

Edited by

Aneesa Padinjakkara

Aparna Thankappan

Fernando Gomes Souza, Jr.

Sabu Thomas

Apple Academic Press Inc.
3333 Mistwell Crescent
Oakville, ON L6L 0A2 Canada

Apple Academic Press Inc.
9 Spinnaker Way
Waretown, NJ 08758 USA

First issued in paperback 2021

Exclusive worldwide distribution by CRC Press, a member of Taylor & Francis Group
No claim to original U.S. Government works

ISBN 13: 978-1-77-463054-9 (pbk)
ISBN 13: 978-1-77-188615-4 (hbk)

Library and Archives Canada Cataloguing in Publication

Biopolymers and biomaterials / edited by Aneesa Padinjakkara, Aparna Thankappan, Fernando Gomes Souza, Sabu Thomas.

Includes bibliographical references and index.
Issued in print and electronic formats.
ISBN 978-1-77188-615-4 (hardcover).--ISBN 978-1-315-16198-3 (PDF)

1. Biopolymers. 2. Biomedical materials. I. Padinjakkara, Aneesa, editor

TP248.65.P62B56 2018 572 C2018-900509-2 C2018-900510-6

Library of Congress Cataloging-in-Publication Data

Names: Padinjakkara, Aneesa, editor.

Title: Biopolymers and biomaterials / [edited by] Aneesa Padinjakkara, Aparna Thankappan, Fernando Gomes Souza, Sabu Thomas.

Description: Toronto ; New Jersey : Apple Academic Press, 2018. | Includes bibliographical references and index.

Identifiers: LCCN 2018002157 (print) | LCCN 2018003010 (ebook) | ISBN 9781315161983 (ebook) | ISBN 9781771886154 (hardcover : alk. paper)

Subjects: LCSH: Biopolymers.

Classification: LCC TP248.65.P62 (ebook) | LCC TP248.65.P62 B54584 2018 (print) | DDC 572--dc23

LC record available at https://lccn.loc.gov/2018002157

Apple Academic Press also publishes its books in a variety of electronic formats. Some content that appears in print may not be available in electronic format. For information about Apple Academic Press products, visit our website at **www.appleacademicpress.com** and the CRC Press website at **www.crcpress.com**

CONTENTS

ABBREVIATIONS

ACC acid copper chrome
AFM atomic force microscopes
AgNPs silver nanoparticles
ARI abrasion resistance index
BBD Box–Behnken design
BPPM bis(3-pentadecylphenol) methane
CBS N-cyclohexyl-2-benzothiazyl sulphonamide
CC cyclic carbonate
CCA copper chrome arsenic
CCB copper chrome boron
CMC carboxymethyl cellulose
CNCs cellulose nanocrystals
CNF cellulose nanofibrils
CNSL Cashew nut shell liquid
CNTs carbon nanotubes
CP coordination polymers
CS chitosan
DFMPM di-α-formylmethoxy bis(3-pentadecylphenyl)methane
DMA dynamic mechanical analysis
DMSO dimethyl sulphoxide
DoE design of experiments
DSC differential scanning calorimetry
ECP epoxy cardanol prepolymer
EDX energy dispersive X-ray
FE SEM field emission scanning electron microscope
FRI Forest Research Institute
FTIR Fourier-transform infrared spectroscopy
FWHM full width at half maximum
HA hydroxyapatite
HDI hexamethylene diisocyanate
HPLC high-performance liquid chromatography
IPDI isophorone diisocyanate

LOSP	liquid organic solvent preservatives
MDF	medium-density fibre
MDM	monocyte-derived macrophages
MEKP	methyl ethyl ketone peroxide
MIC	minimum inhibitory concentration
MOFs	metal organic frameworks
NFC	nanofibrillated cellulose
NFCs	natural fibre composites
NIPUs	non-isocyanate polyurethanes
NR	natural rubber
NSOM	near field scanning optical microscope
OCMC	oxidized carboxymethylcellulose
P3HB	poly-(R)-3-hydroxybutyrate
PALF	pineapple leaf fibres
PBS	phosphate buffered saline
PBS	polybutylene succinate
PE	polyethylene
PEF	polyethylene furanoate
PEG	polyethylene glycol
PET	polyethylene terephthalate
PET	polyethylene terephthalate
PHA	polyhydroxyalkanoates
PHB	polyhydroxybutyrate
PLA	polylactic acid
PU	polyurethane
PVA	polyvinyl alcohol
SAXS	small-angle X-ray scattering
SBF	simulated body fluid
SEM	scanning electron microscope
SFM	scanning force microscopy
SIMS	secondary ion mass spectrometer
STI	shear thinning index
STMs	scanning tunnelling microscopes
TDI	toluene diisocyanate
TEM	transmission electron microscopy
TGA	thermal gravimetric analysis
THF	tetrahydrofuran

TPS thermoplastic starch
TSP textured soy protein
UPS ultraviolet photoelectron spectroscopy
WAXD wide-angle X-ray diffraction
WAXS wide-angle X-ray scattering
WPC wood plastic composites
XPS X-ray photoelectron spectroscopy
XRD X-ray diffraction
ZIBOC zinc, boron and copper

PREFACE

The purpose of this book entitled *Biopolymers and Biomaterials* is providing the collective knowledge in the field of biopolymers and biomaterials to the society. This book involves different studies on biopolymers and biomaterials, their results, interpretation and the conclusion arrived through investigations. Biopolymers are attracting great attention nowadays to meet the growing environmental concerns and energy demands. Development of various biomaterials creating significant advancements in the medical field and many biopolymers are used for the fabrication of biomaterials. Now, it is important to discuss both biopolymers and biomaterials together, which can be the futuristic component.

The beginning of the book explains different biopolymers used in the textile industry, their advantages, disadvantages and applications. It further describes about the importance of wood preservation, development of advanced functional materials from cashew nut shell, synthesis and characterization of cashew nut shell liquid matrix compositions, polyurethane synthesis from renewable resource and recycling of textile mill (cellulosic) waste into carboxymethyl cellulose (CMC).

The second part of the book deals with bamboo/unsaturated polyester composite–based sustainable noise control materials, properties of chemically treated *Bauhinia racemosa*/glass fibre polymer composites, plasma modification for low temperature dyeing of silk fabric, replacement of carbon black by rice hulls and soy short fibres in natural rubber composites, bioresins and bioplastics synthesized from agricultural products for novel applications, green approach in coating industries and mechanical properties of sandwich composites made by using natural fibres and glass fibre mat.

The final section of the book illustrates cross-linked polyvinyl alcohol and hydroxyapatite for tissue engineering, wound management using in situ nanosilver-immobilized Chitosan/oxidized CMC blend membranes, biodegradable polyether urethane for controlled release of antibiotics, biodegradation studies of polyurethane, particularly for therapeutic applications, and microscope and spectroscopy characterization of eco-friendly composites and nanocomposites.

This book is very useful for the researchers, students and professionals in the area of biopolymers and biomaterials. It is contributing data regarding diverse novel research work in the world based on eco-friendly, bio-based, biocompatible and biological materials for essential applications. In the book, an effort has been made for the collection and arrangement of useful information for humanity.

The editors would like to gave their heartfelt gratefulness to all the contributors of this book who extend outstanding support for the successful fulfilment of this endeavour. We are thankful to them for the commitment and the earnestness they have shown with their contribution in the book. Without their devotion and support, this work could not have been achievable. We would like to acknowledge all the reviewers who have taken their precious time to make valuable comments on each chapter. We are also obligated to the publisher, Apple Academic Press, for perceiving the need of this new book and for realizing the increasing importance of this field.

Aneesa Padinjakkara
Aparna Thankappan
Fernando Gomes
Sabu Thomas

ABOUT THE EDITORS

Aneesa Padinjakkara is a Research Fellow at the International and Inter University Center for Nanoscience and Nanotechnology, Mahatma Gandhi University, Kottayam, India. She has presented several papers at conferences focusing on advanced materials and has participated in several workshops and seminars.

Aparna Thankappan, PhD, is a Postdoctoral Research Fellow at the International and Inter University Center for Nanoscience and Nanotechnology, Mahatma Gandhi University, Kottayam, India. She has published over a dozen articles in professional journals and has presented at several conferences as well. She has co-authored several book chapters. She received her PhD in photonics from Cochin University of Science and Technology, India.

Fernando Gomes Souza, Jr., PhD, is currently Professor at the Institute of Macromolecules Professor Eloisa Mano Federal University of Rio de Janeiro (IMA/UFRJ), in the Laboratory and Biopolymers, Sensors / LIPS Technology Center University City, Brazil. He is also Professor of the Civil Engineering Program of COPPE / UFRJ and a Young Scientist of the State of Rio de Janeiro (FAPERJ-2015). His expertise is mainly focused on the use of renewable resources and nanocomposites in sensors, drug delivery, and environmental recovery. mDr. Souza has coordinated 10 research projects with financial support from government-sponsoring agencies. He has published 88 scientific articles, with several accepted in press now, as well as two chemistry books in Portuguese and three biomaterials book chapters in English. He has presented almost 200 papers at conferences and scientific meetings and has translated several computer programs related to educational chemistry.

Sabu Thomas, PhD, is a Professor of Polymer Science and Engineering at the School of Chemical Sciences and Director of the International and Inter University Centre for Nanoscience and Nanotechnology at Mahatma

Gandhi University, Kottayam, Kerala, India. The research activities of Professor Thomas include surfaces and interfaces in multiphase polymer blend and composite systems; phase separation in polymer blends; compatibilization of immiscible polymer blends; thermoplastic elastomers; phase transitions in polymers; nanostructured polymer blends; macro-, micro- and nanocomposites; polymer rheology; recycling; reactive extrusion; processing–morphology–property relationships in multiphase polymer systems; double networking of elastomers; natural fibers and green composites; rubber vulcanization; interpenetrating polymer networks; diffusion and transport; and polymer scaffolds for tissue engineering. He has supervised 68 PhD theses, 40 MPhil theses, and 45 Masters thesis. He has three patents to his credit. He also received the coveted Sukumar Maithy Award for the best polymer researcher in the country for the year 2008. Very recently, Professor Thomas received the MRSI and CRSI medals for his excellent work. With over 600 publications to his credit and over 23683 citations, with an h-index of 75, Dr. Thomas has been ranked fifth in India as one of the most productive scientists.

CHAPTER 1

BIOPOLYMERS IN THE TEXTILE INDUSTRY

ASIM KUMAR ROY CHOUDHURY

Retired Professor and HOD (Textile) Government College of Engineering & Textile Technology, Serampore, Hooghly 712201, West Bengal, India, E-mail: akrc2008@yahoo.in

CONTENTS

ABSTRACT

Textile fibres, the basic raw material for textile manufacturing, are not inherently green. Though some of them are biodegradable, more biodegradable polymers can be made by biological means. Biopolymers often have a well-defined structure. In contrast, most synthetic polymers possesses much simpler and more random (or stochastic) structure.

Biodegradable polymers have achieved a position of great interest in recent years mainly due to biomedical applications. Biodegradable polymers break down in physiological environments by macromolecular chain scission into smaller fragments, and ultimately into simple stable end products. In recent years, the use of biopolymers, i.e. fibres and plastics made from corn, sugar, starch and other renewable raw materials, has expanded.

1.1 INTRODUCTION

With advances in chemistry, technological progress and the growth of material science, a new class of synthesized or manmade materials, called polymers or plastics, has been introduced. Polymers or plastics are found everywhere in our world and used in everyday life in a wide range of applications such as textile, food packaging, automobiles, electronics, building materials and furniture on account of their remarkable performances. In terms of properties, polymers are generally lighter than glass, metals or ceramics, can be rigid or flexible and opaque or fully transparent. Most of the plastics used worldwide are still made from petroleum, a nonrenewable resource. These petroleum-based polymers are extremely resistant to natural decomposition. Consequently after using, they accumulate and damage the environment and the ecosystem. The lack of biodegradability, environmental concern and the depletion of the oil have promoted worldwide research to develop biopolymers, bio-based and biodegradable polymers, as an alternative to petroleum-based plastics.

The biopolymers have been considered since 1940s; Henry Ford used soya plastic to construct various car parts in an effort to demonstrate his belief that 'farms are the factories of the future'. Biopolymers are produced by biological systems (i.e. microorganisms, plants and animals) or are chemically synthesized from biological starting materials (e.g. sugars, starch, natural fats or oils etc.). They are more biodegradable than vegetable or animal-derived natural fibres. Biopolymers account for just over 1% of polymers by 2015.[6] However, the expected growth is 3–4 times in the coming 7–8 years.

The term 'biopolymers' is loosely defined as polymeric materials consisting, at least a significant part, of biological components, in which 'biological' means (recently) produced by living organisms, that is not produced from petroleum. Biopolymers can be thermoplastic or thermoset; they can also be composites or homogeneous and biodegradable or not.

A definition for biopolymers, sometimes found in patent literature, is based on the amount of 'modern carbon' that needs to be present in a biopolymer. 'Modern carbon' is defined in the ASTM D6866 standard, and it is the carbon that contains a specific minimum amount of the carbon-14 isotope. In this way, it can be proven that the carbon in the material is not from fossil origin.[3]

Biopolymers have been known since the dawn of civilization—leather, cotton, wool, natural rubber and cork are all biopolymers. Although these materials are still popular for specific applications, most polymeric materials those being used today are synthetic and are based on the petroleum-derived resources.

In recent years, the research and development of biopolymers have been gaining significant momentum, driven by 'green chemistry' and sustainability principles that are increasingly adapted in the industry. The increased research and development of renewable energy sources, specifically of biofuels such as bioethanol which is produced from grains or biomass and biodiesel which is produced from plant oils, also drive the development of biopolymers. By reacting plant oils with methanol, biodiesel, for example, is produced. This results in fatty acid methylesters, which is the actual biodiesel, and large amounts of glycerol as a by-product. The glycerol can be converted to di-functional compounds, which in turn can be used as monomers in biopolymer production.

Although some biopolymers, such as polylactic acid (PLA), are already becoming regular as 'green' and biodegradable packaging materials, whereas other biopolymers are more esoteric. An example is BioSteel™ which is a protein fibre produced from milk of goats that had been genetically modified with spider silk genes. The polymers are reportedly up to 10 times stronger than steel for the same weight.[3]

1.2 MERITS AND DEMERITS

Biopolymers slowly enter various polymer markets, namely textile, plastic and so forth. Their advantages are sometimes shadowed by their disadvantages, at least, at the present state of development. Before selecting a biopolymer for a particular end use, both merits and demerits are to be carefully considered.

1.2.1 ADVANTAGES OF BIOPOLYMERS

- They are fully bio-based.
- Much lower 'oil (petroleum)' is needed for production.
- Lower amount of greenhouse gases is emitted during their production. Ingeo® (PLA from NatureWorks) requires 60% less greenhouse gases and 50% less nonrenewable energy than other polymers.[5]

1.2.2 DISADVANTAGES OF BIOPOLYMERS

- The competition for biological sources for use as food and fuel
- Additional sorting during recycling to avoid contamination
- Performance still inferior to oil-based polymers—poorer heat and moisture resistance

1.3 APPLICATION OF BIOPOLYMERS

An increasing trend for biopolymer production and application is being seen due to environmental awareness in the past years and eco-friendliness of biopolymers. In the textile sector, due to insufficient mechanical properties of the biopolymers, they occupy a relatively low market share compared to conventional polymers, challenges during polymer processing and their higher price. The production of biopolymers (commonly known as bioplastics) is continuously increasing and recorded as 1.5 million tons in 2012, which is expected to reach to 6.7 million metric tons in 2018.[7]

The use of bio-based products has increased at a steady pace in the last decade. In 2005, they accounted for 7% of global sales and around US $ 77 billion (£ 49 billion) in value within the chemical sector. It is estimated that by 2020, the global market for bio-based products will reach to US $ 250 billion (£ 158 billion); and by 2030, one-third of chemicals and materials will be produced from biological sources, including biopolymers and bioplastics.

Various factors influence the production, growth of market and utilization of biopolymers worldwide, such as the convenience of their production and processability methods, properties, cost, biocompatibility and their dependence from foodstuff-based raw materials. The interest and acceptance of customers for biopolymer products are affected by these factors, and with that the decision of polymer producers to actively introduce biopolymers in the market are also influenced. Figures 1.1 and 1.2 represent the worldwide consumption (in percentage) of various biopolymers in 2013 and consumption (in 1000 t) of biopolymers in various applications in 2011, respectively.[1]

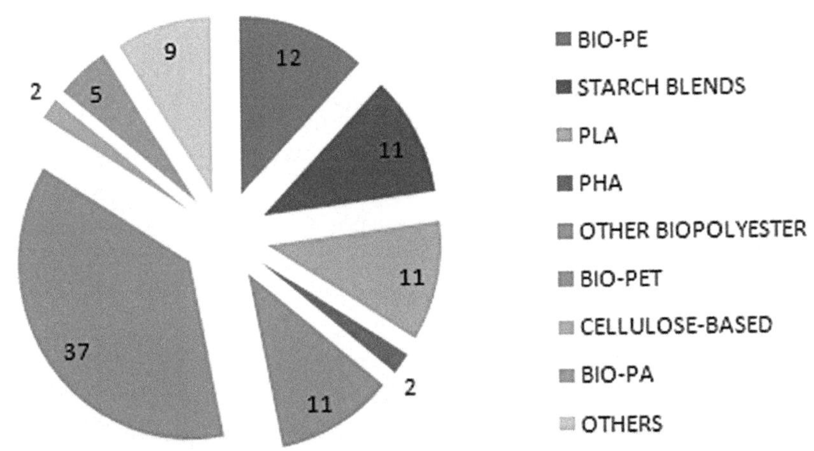

FIGURE 1.1 Worldwide consumption (%) of various biopolymers in 2013.

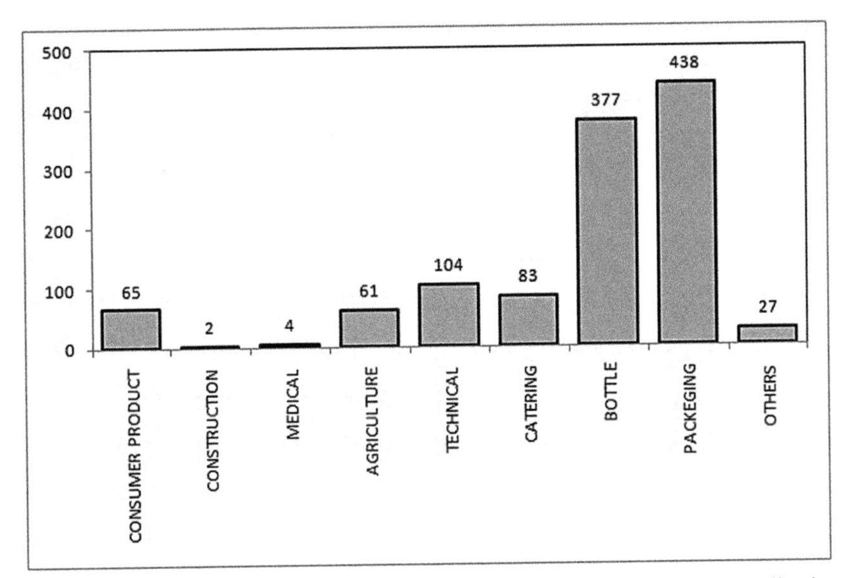

FIGURE 1.2 Worldwide consumption (in 1000 t) of biopolymers in different applications in 2011.

Biopolymers have versatile applications. A few are mentioned below:

- Drug-delivery systems (medical field)
- Wound closure and healing products (medical field)
- Surgical implant devices (medical field)
- Bioresorbable scaffolds for tissue engineering
- Food containers, soil retention sheeting, agriculture film, waste bags and packaging material in general
- Nonwoven biopolymers can also be used in agriculture, filtration, hygiene and protective clothing.

The following biopolymers have high potential for various applications:

- Starch-based polymers (packaging)
- Polylactide acid
- Polyhydroxyalkanoates (PHAs)/Polyhydroxybutyrate (PHB)
- (co)PA—(castor oil based—PA11)
- Polybutylene succinate (PBS) and biopolyester-based copolymers
- Polyethylene furanoate (PEF)—alternative for polyethylene terephthalate (PET), made from two building blocks: furandicarboxylic acid and monoethylene glycol

Although biomaterials made from proteins, polysaccharides and synthetic biopolymers are preferred, they lack the mechanical properties and stability in aqueous environments necessary for medical applications. Cross-linking improves the properties of the biomaterials, but most cross-linkers either cause undesirable changes to the functionality of the biopolymers or result in cytotoxicity. Glutaraldehyde, the most widely used cross-linking agent, is difficult to handle and contradictory views have been presented on the cytotoxicity of glutaraldehyde cross-linked materials. Recently, polycarboxylic acids that can cross-link in both dry and wet conditions have been shown to provide the desired improvements in tensile properties, increase in stability under aqueous conditions and also promote cell attachment and proliferation. In order to obtain biopolymeric materials with properties desired for medical applications, green chemicals and newer cross-linking approaches are necessary.[12]

A special type of 'application' is the biocomposites. These are mostly fibre-reinforced composites. Obviously, the well-known and popular 'composite wood products' such as oriented strand board or medium density fibreboard are biocomposites. In many cases, research about these products has mainly focused on improving the environmental properties of the binder, especially on reducing formaldehyde emissions. Mostly all types of natural fibres, such as flax, bamboo, natural wool and many others that can be bonded together to form useful composites, can be used for the production of fibre-reinforced biocomposites.

1.4 METHODS OF MANUFACTURE

In general, there are three ways to produce biopolymers that are reflected in the patent literature:

1. Polymers directly extracted or removed from biomass such as some polysaccharides and proteins. There may be partial modification of natural bio-based polymers (e.g. starch).
2. Polymers produced by microorganisms (fermentation) or genetically modified bacteria followed by polymerization or direct bacterial fermentation processes (e.g. PHAs).
3. Polymers produced by classical chemical synthesis starting from renewable bio-based monomers such as PLA.

The field of application for a polymer and its performance in technical textiles are depending on its mechanical properties. Thus, an important factor, which affects its market value, is tensile strength. Polyethylene furanoate is found to have better mechanical properties among the biopolymers. Polylactic acid is seen on second position in the terms of tensile strength. Polyethylene Furanoate is currently used for production of bottles, and further applications are restricted due to limited production of PEF. The PLA production has already reached industrial scale and shows enormous potential in many fields of application. The melt spinning of PLA was performed at extrusion temperature of 230°C followed by quench by air at 0.55 m/s and 18°C with drawing ratio of 1.95 and speed of 2500 m/min to obtain 'POY' yarn of 165 dtex. The yarn was textured (false-twist) at 205°C before the manufacturing of a PLA T-shirt. The filament extrusion was performed successfully, whereas winding appeared to be challenging. The mechanical properties of the spun PLA yarn were compared with the reference polyester yarn. The tenacity of PLA yarn is much lower than PET yarn, whereas elongation is seen in the same range.[1]

1.5 CLASSIFICATION OF BIOPOLYMERS

Biopolymers can be broadly classified into three groups:

- Polynucleotides (RNA and DNA), which are long polymers composed of 13 or more nucleotide monomers
- Polypeptides, which are short polymers of amino acids
- Polysaccharides, which are often linear, bonded polymeric carbohydrate. This group includes alginates, microbial cellulose, chitin and chitosan.

1.6 IMPORTANT BIOPOLYMERS

A large number of biopolymers are available naturally or manufactured. However, only a few of them have been used commercially. Some of the important biopolymers are discussed below.

1.6.1 SOYBEAN FIBRE

Soybean fibre is a man-made regenerated protein fibre from soybean protein blended with poly(vinyl alcohol). It is biodegradable, non-allergic and microbicidal. The clothing made from the soy fibre is less durable, but it has a soft, elastic handle. Soybean protein is a globular protein, and it has to undergo denaturation by alkali/heat/enzyme and degradation to convert the protein solution into a spinnable dope.

1.6.2 POLY (ALKYLENEDICARBOXYLATE) POLYESTERS

Monomers for aliphatic poly (alkylene dicarboxylate) polyesters (APDs) can be petroleum derived (i.e. not renewable) or biomass derived (i.e. renewable), the former being the major route. Both can be prepared to the same degree of purity, but the latter is still costlier.

Common dicarboxylic and diol monomers found in APDs are shown in Figure 1.3. They include succinic acid (SA), adipic acid (AA), ethylene glycol (EG) and 1,4-butanediol (1,4BD). Polybutylene succinate is an aliphatic polyester with similar properties to those of PET. Polybutylene succinate is produced by the condensation of SA and 1,4-butanediol.

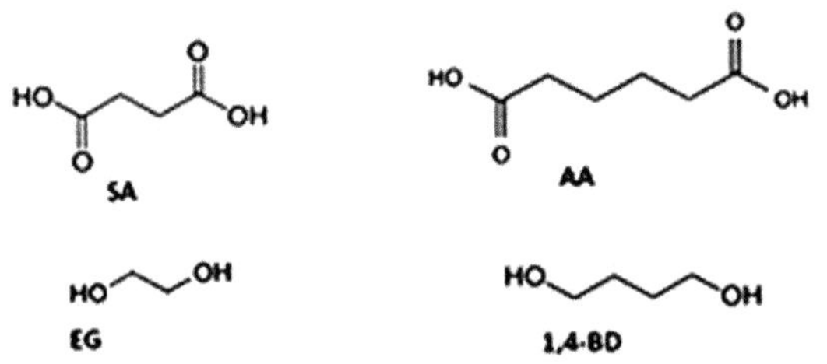

FIGURE 1.3 Dicarboxylic and diol monomers found in poly (alkylene dicarboxylate) polyesters.

Polybutylene succinate is a semicrystalline polyester with a melting point higher than that of PLA. Its mechanical and thermal properties depend on the crystal structure and the degree of crystallinity. The glass transition temperature (T_g) is approximately −32°C, and the melting temperature is approximately 115°C. In comparison with PLA, PBS is tougher in nature but with a lower rigidity and Young's modulus.[2]

Biosuccinic acid (SA) is produced directly by fermentation of bioengineered yeast and *E. coli*. Catalytic hydrogenation of biosuccinic acid produces 1,4-butanediol, which can also be produced by fermentation. Bioethylene glycol is produced from bioethylene, a product of catalytic dehydration of fermentation-derived ethanol. Bioadipic acid can be produced by a number of fermentation-based processes.

Application: aliphatic poly (alkylene dicarboxylates) are used in polyurethanes for coatings, adhesives and foams; flexible packaging; agricultural films; compostable bags and in blends and composites with other bio-based polymers to enhance properties.[8]

1.6.3 BIO-POLYAMIDE (NYLON)

Castor oil has been a non-food crop source of biopolymers. Polyamide 11 from castor oil was patented in 1944 by French scientists, and from 2004 it is marketed by Arkema as Rilsan for sportswear. Toray and Radici are now marketing another castor oil-derived polyamide, PA 6–10. Sofia launched a hybrid polyamide fibre, Greenfil by texturizing 70% synthetic PA 6 and 30% biosourced PA 10. A Greenfil sock is 5–10 times stronger, but 2–3 times costlier too.[14]

Bio-polyamides are also a subject of industrial research, for example by DuPont with co-polyamides prepared from plant oils, INVISTA with polyamide terpolymer and Rhodia.

1.6.4 BIO-POLYETHYLENE

Polyethylene (PE) is an important engineering polymer traditionally produced from fossil resources. Bio-based PE has exactly the same chemical, physical and mechanical properties as petrochemical PE. The sequence for biological method is as follows:

Fermentation of sugarcane/sugar beet/starch crop → bioethanol → distilled at high temperature over a solid catalyst → ethylene → microbial PE or green PE.

Bio-polypropylene is at least partially made from renewable resources using methanol and metathesis chemistry.[3]

1.6.5 BIODEGRADABLE POLYURETHANES

Biodegradable polyurethanes (PURs) are known for toughness, durability, biocompatibility and biostability. Unlike polyester derivatives, polyether-based PURs are quite resistant to degradation by microorganisms. Biodegradable PURs employed as thermoplastics are basically synthesized using a diisocyanate, a diol and a chain-extension agent. The reaction between a cyclic carbonate and an amine rendering the urethane bond is the first representative example avoiding diisocyanate. In particular, the polyaddition reaction between L-lysine and a bi-functional five-member cyclic carbonate in the presence of a strong base. Some have reported the enzymatic synthesis of PERs by enzymatic polyesterification.[11]

1.6.6 POLYLACTIC ACID

Polylactic acid is known since 1845 but was not commercialized until early 1990. It is the only melt-processable fibre from annually renewable natural resources such as corn starch (in the United States), tapioca products (roots, chips or starch mostly in Asia) or sugar cane (in the rest of world). It is a thermoplastic, aliphatic polyester similar to synthetic PET. The sequence of manufacture steps is as follows:

Corn → starch → unrefined dextrose → fermentation → D- and L-lactic acid à monomer production → D-, L- and meso-lactides → polymer (PLA) production → polymer modification → fibre, film, plastic, bottle manufacture.

The polymerization reaction is shown in Equation 1.1.

Lactic acid → Lactide → Polylactic acid (1.1)

Polylactic acid has high strength, good drape, wrinkle- and UV light-resistance properties. Its melting point is 170°C, and density is 1.25 g/cm^3. The limiting oxygen index (25) is higher than PET and much higher than polypropylene (PP). Polylactic acid, therefore, possesses reduced flammability, less flame retardants. Water uptake is low (0.4–0.6%), but higher than PET and PP. It possesses good durability under a range of conditions.

Not surprisingly, a lot of research is still being done on the most common biopolymer that is PLA. The research is often focused on improving polymer properties such as increasing toughness, crystallinity, impact modification and so forth. The companies such as Metabolix, PURAC, Arkema, Biovation and others are active in this field.[3]

PLA is regarded as one of the most promising bioplastics. As a result, it has raised particular attention as a potential replacement for petroleum-based polymers in many areas such as textiles, bottles, thermoformed containers, paper and cardboard coating. Polylactic acid has low heat resistance unless it can be fully crystallized. However, PLA suffers from low crystallization kinetics unless it is subjected to high orientations. Hence, increasing the crystallization rate in processing techniques, such as injection moulding, where orientation levels are relatively low, is required to improve its thermal resistance. One way to improve the crystallization kinetic is to find out a suitable bio additive. However, the understanding of nucleation and crystallization mechanisms is required to optimize the crystallization kinetics and subsequently to identify or develop the best bio-based additives.

Application: iron-able woven shirts, microwavable trays, hot-fill applications and even engineering plastics. Biomedical applications include sutures, stents, dialysis media and drug delivery devices. Polylactic acid can be used for rigid thermoforms, films, labels and bottles, but not for hot-fill containers or gaseous drinks such as beer or sodas.

1.6.7 BACTERIAL POLYESTERS

One of the most important fields of research are the PHAs which are biopolymers produced biochemically by genetically modified micro-organisms or modified plants. The resulting polymers are often copolymers from 3- and 4-hydroxybutyric acid and/or hydroxyvaleric acid. The companies such as Metabolix and Novomer have filed patents on this topic. Using the correct bacteria, PHAs can be produced from wastewater as shown in a number of patent applications or from waste fish- or palm oil or from biogas (e.g. from a landfill digesters). Polyhydroxyalkanoates can also be produced from glycerol, algae or even aromatic sources. Poly(3-hydroxybutyrate) or PHB can also be produced from transgenic plants. Examples of a plant that can be genetically engineered to produce PHB are grasses similar to switch grass. These grasses are also studied for the bioethanol production in the United States.

The bacterial polyesters, PHAs with poly-(R)-3-hydroxybutyrate (P3HB) as the first homologue (Fig. 1.4), are produced by microorganisms. Bacterial storage compound PHB copolymer named 'Biopol' is developed by Zeneca Bioproducts through fermentation of P3HB followed by copolymerization with PHV. It is a high molecular weight polyester and thermoplastic (melts at 180°C) and can be melt spun into biocompatible and biodegradable fibres that are suitable for surgical use.

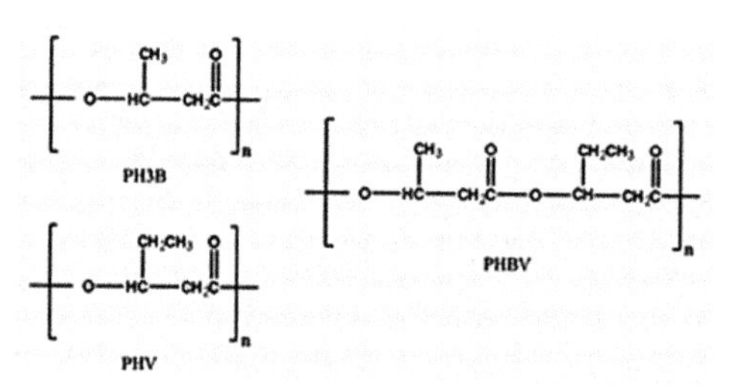

FIGURE 1.4 Chemical structures of P3HB and its copolymer PHBV.

Advantages include production from fully renewable resources, fast and complete biodegradability and excellent strength and stiffness. The disadvantages are high thermal degradability, brittleness and high price.[4]

1.6.8 SODIUM ALGINATE FIBRE

Sodium alginate is a polymeric acid, composed of two monomer units, which are as (a) L-guluronic acid (G) and (b) D-mannuronic acid (M) (Fig. 1.5).

It is nontoxic and nonirritant. Alginate fibre generates a moist healing environment and is used for wound dressing. Calcium alginate is created by adding aqueous calcium chloride to aqueous sodium alginate.

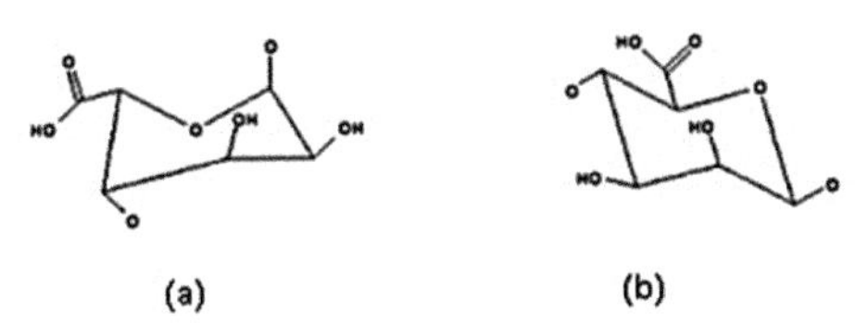

FIGURE 1.5 Two monomers of alginic acid: (a) L-guluronic acid and (b) D-mannuronic acid.

1.6.9 CHITIN AND CHITOSAN

Both polysaccharides may be regarded as derivatives of cellulose, in which chitin bears an acetamido group and chitosan bears an amino group instead of the C-2 hydroxyl group in cellulose (Fig. 1.6).

Chitosan is obtained by deacetylation of chitin. Chitosan is a linear polyamine having reactive amino and hydroxyl groups. It chelates many transitional metal ions. Currently, the commercial source of chitin is shrimp shells, but the polymer also is found in the shells of crabs and lobsters.

Derivatives of chitin have been used to impart antistatic and soil-repellent finishing to the textiles. Although chitin is used in printing and finishing preparations, chitosan is able to remove dyes from discharge water. Both have remarkable contribution to medical-related textile sutures, threads and fibres.

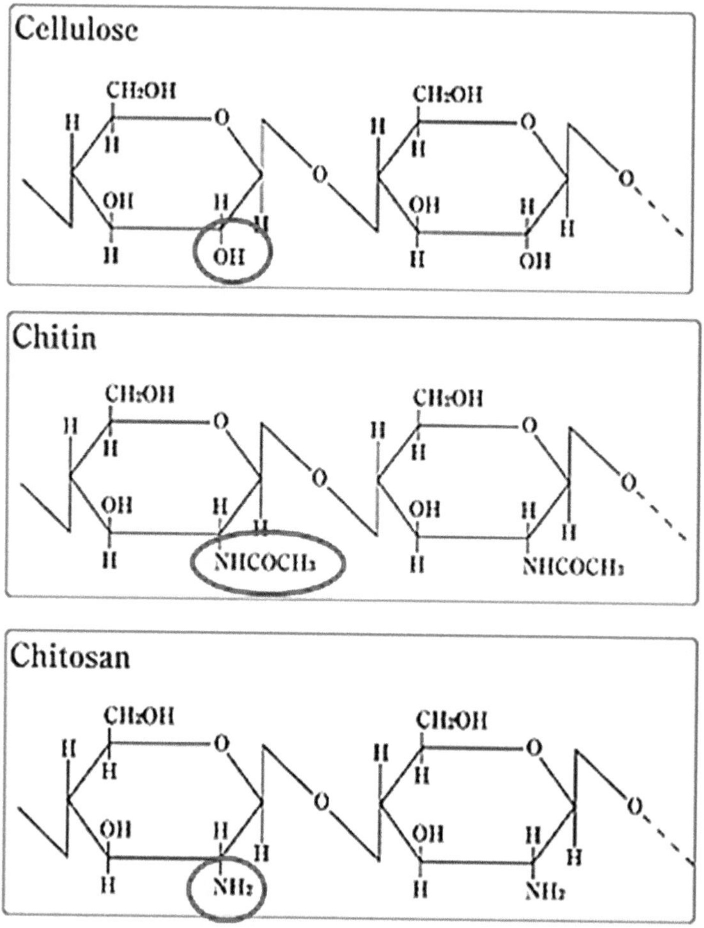

FIGURE 1.6 The chemical structure of cellulose, chitin and chitosan.

Chitosan, a Million Dollar Natural Polymer, discovered by Rouget in 1859, is a technologically important polysaccharide biopolymer. It is composed of glucosamine and N-acetylglucosamine units linked by 1–4 glucosidic bonds, chemically. Being nontoxic, biodegradable, biocompatible and microbe resistant has given it huge potential in a broad range of scientific areas such as biomedical, food, agricultural, cosmetics, textiles, pharmaceutical and other industries.[12]

For shrink-resistant finishing, enzyme-treated woollen fabric was finished with three different finishing polymers, and the performance properties of chitosan finished woollen fabric are found to be better than in other finished fabrics. Thermal properties of chitosan-finished wool are similar to synthetic polymer finish. Chitosan finish resists the denaturation of wool fibre better than that of other synthetic finishes. Surface morphology of wool fibre shows that masking of cuticle scales by chitosan finish is better than that of other synthetic finishes. The cross-sectional view infers that chitosan can also be diffused well inside the wool fibre. It is concluded that chitosan-based shrink-resistant finishing could be preferred over synthetic polymer finish for woollen materials.[10]

Some of the attributes which make chitosan, cyclodextrin, sericin protein and alginate suitable alternative agents for the functional finishing of textile materials are increased sustainability, environment friendliness, reduced pollution, green chemistry, renewability and intrinsic biological activity.[9]

Recent biopolymer-related patent literature covers different types of bio-monomers which can be turned into polymers using classical chemical synthesis. For example, quite some research is to be done on bio-based 'polymer grade' acrylic acid and methacrylic acid (US patents, US20,140,206,831, US20,130,165,690) which can, for example, be produced from bio-derived glycols and polyols, such as glycerol and sorbitol.

1.7 CONCLUSION

Biopolymers are very important for various types of medicine, surgery and healthcare products and the extension of application of the materials used is because of the versatility of biopolymer. Petroleum oil is the fuel that drives the global economy, but oil reserves are going down. Moreover, there are major concerns about the future because of our great dependency on oil and its impact on the environment[15].

The global trend towards sustainability, green chemistry and renewable energy and raw materials also has a big impact on the research and development of polymers. A large number of patent applications relating to biopolymers are being filed, covering an impressive amount of new polymers and monomers. Although real-life applications appear to be limited still, this research can be the basis for a strong growth in the near future.[3]

KEYWORDS

- **textile fibre**
- **biopolymer**
- **bacterial polyester**
- **polylactic acid (PLA)**
- **chitin and chitosan**

REFERENCES

1. Anonymous. Melt Spinning of Bio-Based Polymers: Overview on Properties and Potential of Melt Spinnable Biopolymers. *Melliand Int.* **2015,** *4*, 245–246
2. Babu, R. P. et al. *Prog. Biomater.* **2013,** *2*(8), http://www.progressbiomaterials.com/content/2/1/8 (accessed 15 Dec 2014).
3. Bleys, G. Technology Watch: Biopolymers, November, essenscia, 2015. www.PURpatents.com (accessed 10 Jan 2015).
4. Chod´ak, I. Sustainable Synthetic Fibres: The Case of Poly(hydroxyalkanoates) (PHA) and Other Fibres. In *Sustainable Textiles: Life Cycle and Environmental Impact;* Blackburn, R.S., Ed.; Woodhead: Cambridge, UK, 2009; p 88.
5. Ditty, S. 7 Biopolymer Eco Fabrics You Need to Know About, http://source.ethical-fashionforum.com/ (accessed Aug 12, 2013).
6. Doug, S. Bioplastics: Technologies and Global Markets. BCC Research Reports PLS050A, 2010.
7. Endres, H.-J.; Siebert-Raths, A. *Technische Biopolymere;* Carl Hanser Verlag: Munich/Germany, 2009. ISBN: 978-3-446-41683-3.
8. Gotro, J. Aliphatic Poly(alkylene dicarboxylate) Polyesters: Organic or Not Organic? Polymer Innovation Blog. http://polymerinnovationblog.com/ (accessed April 15, 2013).
9. Islam, Shahid-ul-, Shahid M.; Mohammad, F. Green Chemistry Approaches to Develop Antimicrobial Textiles Based on Sustainable Biopolymers-A Review, *Ind. Eng. Chem. Res. (Am. Chem. Soc.)* **2013,** *52,* 5245–5260.DOI: org/10.1021/ie303627x.

10. Lakshmanan, A.; Lee, J. S.; Jeyaraj, J. M.; Kochadai, A. S. Comparison of Biopolymer Finishing with Functional Finishing on Wool Fibre. *Indian J. Fibre Text. Res.* December **2015,** *40,* 447–452.
11. Lendlein, A.; Sisson, A. *Handbook of Biodegradable Polymers;* Wiley-VSH: Weinheim, Germany, 2011.
12. Sofia. New Yields in Castor Oil Polyamides. WSA: the international magazine for performance and Sports Materials, May–June, 2012. file:///H:/biopolymer%20conf/.
13. Raafat, D.; Sahl, H. Chitosan and its Antimicrobial Potential—A Critical Literature Survey. *Microbiol. Biotechnol.* **2009,** *2,* 186–201.
14. Reddy, N.; Reddy, R.; Jiang, Q. Crosslinking Biopolymers for Biomedical Applications, Trends *Biotechnol.* Jun, **2015,** *33*(6), 3629. DOI: 10.1016/j.tibtech.2015.03.008. (accessed April 14, 2015).
15. Roy Choudhury A.K.,Green Chemistry and the Textile industry, Textile Progress, 2013, Volume 45, Issue 1, (Textile Institute, UK). https://doi.org/10.1080/00405167.2013.807601.

CHAPTER 2

IMPORTANCE OF WOOD PRESERVATION: AN INDIAN OVERVIEW

APARNA KALAWATE

Zoological Survey of India, Western Regional Centre, Vidyanagar, Sector-29, P.C.N.T. Post, Rawet Road, Akurdi, Pune 411044, India, E-mail: aparna_ent@yahoo.co.in

CONTENTS

ABSTRACT

Due to the stringent forest policy, wood is not available from natural forests to the wood-based industry. Short-rotation plantation wood that grows fast has emerged as a major raw material for the future development of industry. Plantation timbers are susceptible to the attack of various wood-destroying agencies. Only a preservative treatment gives adequate and economic service life to a plantation timber. Due to existence of numerous species of insects and fungi that destroy wood, climatic and biological conditions in tropical countries, like India, make wood preservation more important. In addition, hot and humid climate speeds up decomposition of wood. India is one of the major wood-consuming countries of the world. Preservative treatment of timber has become an indispensable need in India. Nevertheless, the quantity of timber which is given preservative treatment is rather limited. This subject has started gaining importance recently. The current chapter addresses the importance of wood preservation in India.

2.1 INTRODUCTION

Wood preservation is a process of impregnation of chemical into wood for ensuring long-term resistance against the attack of various wood-destroying agencies, although material will not last forever—even steel and stone will corrode and crumble with time. Wood is used for making tools, utensils, shelter, ships and vehicles. India, with developing economy, needs very large resources of timbers for diverse purposes. The quantity of available timbers in the country falls short compared with the requirement of various wood-based industries, and the shortage is expected to increase with the rapid pace of industrialization. Due to stringent forest policy, wood is not available to the wood-based industry from natural forest. Short-rotation plantation wood is emerging fast as a major raw material for the future development of industries. Most plantation-grown timbers need preservative treatment to give adequate and economic service life due to its low durability. Prophylactic treatment is needed for timber awaiting conversion during storage. Preservation treatment can reduce timber requirement by enhancing wood service life, which can help in managing forests on a sustainable basis to serve the dual purpose of providing a stable raw material supply and maintaining ecological balance. India is a major wood-importing country in the world. Hence, it may wish to preserve the wood in order to conserve foreign currency by reducing import of wood.

The heartwood and sapwood of nondurable species and only the sapwood of durable species need protection against wood-destroying agencies. In 1908, wood preservation was introduced in India by Sir Ralph

Pearson of the Indian Forest Service. Pearson, in India, is called the father of wood preservation for his valuable work.[19] In 1854, the first wood preservation plant of India was established at Bally (Howrah). India has introduced two of the most effective and widely used wood preservatives, namely copper chrome arsenic (CCA) and copper chrome boron (CCB) to the world. The current level of wood treatment in the country is very low with a negligible contribution to the wood economy.[7] Different institutions have made significant research contributions and generated tremendous data on wood preservation, but because of the lack of general awareness for enabling legislative support, usage of wood preservation has not taken off yet in the country.[7]

The preservation industry has developed with the development of railroad system at par with most other developed countries. The most popular wood preservatives are CCA, CCB, acid copper chrome (ACC), creosote, and recently, liquid organic solvent preservatives (LOSP) have also appeared in the market. The major user of CCA is the cooling tower industry that uses more than 50% of the current CCA produced in the country.[18] Now, due to the carcinogenicity of CCA, its use is slowly diminishing from the industry. Copper chrome arsenic has been replaced by CCB. Borax treatment (boric acid) is the main preservative treatment, being used in the furniture industry. Liquid organic solvent preservatives are being used as brush-in applications and in the remedial treatments. The overall picture of preservative used (around 1350 t CCA equivalent) is quite disappointing, compared with the volume of nondurable woods that is 22.5 million m^3 used, annually.[18]

In India, the Bureau of Indian Standard is the apex bodies that formulates the codes. The major standard for wood preservation is IS: 401-1967 (Indian Standard—Code of Practice for Preservation of Timber), which can also be called the Bible of Wood Preservation. It covers types of preservatives, their brief descriptions, methods of treatment and the type and choice of treatment for different species of timbers for a number of uses. This standard includes the preservatives and methods of treatment that has given satisfactory results under Indian condition of service.

The process for treating refractory timbers, such as eucalyptus, has also been covered in it. In the current chapter, an attempt has been made to highlight the importance of wood preservation.

2.2 WOOD PRESERVATION IN SUSTAINABLE DEVELOPMENT

In India, where extreme climatic conditions prevail, using plantation timber without treatment with preservative chemicals is very difficult. India is one of the major wood-consuming countries in the world. Currently, there are more than 150 treatment plants with an annual treating capacity of over 2.5 million m^3, although the existing level of treatment is only about 1.5 million m^3.[4] Approximately 400,000 m^2 of plywood and particle board are treated annually by using about 1000 t of various preservatives.[20]

In recent years, great changes have taken place in the Indian panel industry. Production of medium-density fibre (MDF) board and particle boards have increased dramatically, and new plants are being planned to be installed. MDF is used extensively in factory-assembled and ready-to-assemble furniture, cabinets, underlayment, drawer fronts, moulding and counter tops. MDF is also replacing thin plywood and wet-process hardboard in the production of moulded and flush-door skins,[17] but these products are susceptible to biological attack. Hence, preservative treatments with suitable chemicals are required to make it resistant to the attack of wood-destroying organisms. Wood–plastic composite is gaining more acceptances in the market. In this product, wood comprises about 40–50% of the composition and so the chemical preservative treatment is necessary. It has become necessary to manufacture products with a minimal environmental impact as people are aware of the hazardous impact of chemical preservatives on the environment and mankind. Wood is a biologically derived renewable raw material and has a low processing energy demand. It is recyclable and biodegradable.

In earlier days, people could procure abundant quantities of high-quality wood from natural forests at their discretion. But now, the availability of these wood raw materials has become scarce; even no significant supply can be expected from the natural forests in the course of time. The dependence of wood industries on fast grown plantation wood as the main source of raw material is predicted to significantly increase in the coming years.[3,5] Under ideal conditions, wood can serve for years. Protection must be provided during processing and use to get rid of the condition that permits the development of wood-degrading organisms. The biological organisms that can degrade wood are mainly fungi and insects. Wood preservation not only increases the life of products but also decreases the chances of replacement of wood. Thus, it helps in sustainable utilization of wood. By increasing the lifespan of wooden material used in various applications, the carbon inside the wood is locked and sequestrated. Carbon sequestration in wooden materials is used as a means for mitigating climate change.[8] Hence, by treating the wood and wood-based panel products with small amount of preservative, can last for a longer duration than untreated ones. Wood preservation can help in maintaining the sustainable use of forest resources, increasing the life of wood and wood-based panel material and carbon sequestration.

2.3 THE DEMAND AND SUPPLY OF WOOD IN INDIA

The total forest cover of India is 697,898 km^2 which measures 21.23% of the total geographic area of the country.[9] The total growing stock of wood in the country is estimated to be 5658 m^3—comprising 4173 m^3 inside forest areas and 1485 m^3 outside recorded forest areas.[9] The average per hectare in the forests is 54 m^3.[9]

The government owns more than 58 million ha of the total forest cover 68.4 million ha in the country. As per the Forest Sector Report 2010 by Indian Council of Forestry Research and Education, Dehradun, it produces just 2.38 million m^3 (cum) of timber a year. The remaining 18.7 million ha of private forests, plantations as well as trees cover yield 44.3 million m^3 of timber annually. This is 20 times the government production.

The present demand of wood in India is about 29 million m^3, whereas the estimated production is only about 16 million m^3.[21] Productivity of India's forests' wood is 0.7 cum/ha/year against the global productivity of 2.1 cum/ha/year.[10] The sustainable use of resources may help in bridging the huge gap between demand and supply of wood in India. As per the report of Bob Flynn,[2] the share of global imports of industrial roundwood for India is 5%. India's wood fibre deficit was estimated to be 12.5 million m^3 in 2012. India's imports of logs doubled in volume from 3.2 million m^3 in 2006 to an estimated 6.4 million m^3 in 2012. Approximately 74% of India's log imports were hardwood and 26% softwood. Cultivation of plantation of timbers by the farmers may minimize this gap to some extent. In addition, treating wood with proper preservative chemical with requisite retention level may help in increasing the life of wood, thus, reducing the gap in demand and supply of wood.

2.4 MAJOR DEVELOPMENTS IN WOOD PRESERVATION

In the current chapter, the history of wood preservation is mentioned in brief. Wood preservation is an old science known to world. Use of coal-tar creosote was the major development in the history of wood preservation. It was patented in 1836 by Moll. Creosote was used in pressure impregnation process, known as Bethel or full-cell process, which was patented by John Bethell in 1838. This was the first major use of pressure for treating wood and remaining the basis of most of the modern wood-treating operations.[6] In 1839, the Boucherie method was developed. In 1884, Boulton published his work on 'the Antiseptic Treatment of Timber', which provided the basis for the boultonizing process for seasoning wood. The 1850–1900 era was dominated by the use of creosote for preservation. The 1920–1940 period was the CCA-dominated era. The year 1950 was dominated by the use of inorganic borates–organotin compounds. LOSPs were majorly used in 1960. Alkyl ammonium compounds were used more during 1980. The 1990 era was dominated by the use of chrome- or arsenic-free preservatives. In 1995, use of vapour phase (methyl borate, ethyl borate, butyl borate and propyl borate) treatment was mostly utilized for preservation.

In India, scientific study on wood preservation was started by Ralph Pearson in 1908 at Forest Research Institute (FRI), Dehradun. Kamesam, together with Falk of Germany, developed the preservative Falkamesam. Copper chrome arsenic and CCB known preservatives were first formulated by an Indian scientist, Sonti Kamesam, of FRI, Dehradun in 1930.

Work on the development of arsenic-free wood preservative is under development in many countries including India. Recently, Indian Plywood Industries Research and Training Institute, Bangalore has developed copper ethanolamine boron as chromium- and arsenic-free wood preservative.[13] In this composition, ethanolamine acts as fixative for copper and boron in the wood. Several biocides, for example Chitin synthesis inhibitors such as Lufenuron and Diflubenzuron, were also tested for their efficacy as wood preservatives.[11,12] ZIBOC (zinc, boron and copper) has been developed by FRI, Dehradun as an alternative to CCA.[22] Work has also been initiated in the direction of developing and utilizing nanobiocides for wood preservation.[14]

2.5 SOME IMPORTANT WOOD-DESTROYING ORGANISMS

One should be aware of the organisms decaying wood and wood-based panel products before undertaking any treatment with preservative chemical. In India, approximately 33% of the wood is lost due to biodegradation. Some of the wood-destroying organisms are explained here in brief.

2.5.1 FUNGI

Fungi causing damage to timber are mainly mould, sapstain and decay.

2.5.1.1 MOULD

The common mould fungi belong to ascomycetes and deuteromycetes (Fungi imperfecti). They are superficial surface growths and indicate localized damp conditions conducive to fungal decay. They occur on the surface of timber with woolly or powdery appearance and can be removed by brushing. As mould fungi cannot utilize lignin and cellulose, loss of timber strength does not occur. The damaged wood reduces the aesthetic value.

2.5.1.2 SAPSTAIN FUNGI

Sapstains are mainly confined to sapwood in ray cells in which they get food material. The fungi causing sapstain have coloured hyphae. They secrete soluble pigments which cause the formation of coloured deposits in ray cells. Fungi take food from the stored sugar and starch in the wood and are incapable of utilizing lignin and cellulose. They can cause discolouration but cannot affect the strength of the wood.

2.5.1.3 WOOD ROT OR DECAY FUNGI

2.5.1.3.1 Brown-rot Fungi

Many brown-rot fungi produce bracket-shaped fruit bodies on the trunks of dead trees. The brown-rot fungi only utilize cellulose, changes the structure of the lignin and turns it brown, hence, called brown-rot fungi. Brown rots are predominantly members of the Basidiomycetes.

2.5.1.3.2 White-rot Fungi

It attacks all the components of wood that is cellulose, hemicelluloses and lignin. Removal of these components from the wood gives bleached appearance and hence called white-rot fungi. The lightening in colour may be general or it may occur in patches causing a white 'Pocket' rot. They include both Ascomycetes and Basidiomycetes.

2.5.1.3.3 Soft Rot

They grow on wood in damp environment. These are mostly seen on wooden window frames, the timbers of cooling towers and wood in marine environment and are characterized by the wet wood being softened progressively from the surface. The attacked surface is eroded and weathered away. Hard woods are more prone to attack than softwoods. These are usually more tolerant to preservatives.

2.5.2 INSECTS

Different groups of insects which attack timber are as follows:

Wood-destroying insects		
Coleoptera	**Isoptera**	**Hymenoptera**
i. Ambrosia beetle	i. Ground-dwelling termite	i. Carpenter ants
ii. Flat-headed beetle and long horn beetle	ii. Wood-dwelling termite	
iii. Powderpost beetle		

2.5.2.1 AMBROSIA BEETLE (PIN-HOLE BORERS/ SHOT-HOLE BORERS)

These are first to appear on the scene and inflict severe damage. These borers are called as bark beetle (Scolytidae family) or ambrosia beetle.

There are some beetles that bore directly into wood and feed on wood in both larval and adult stage. Ambrosia beetles, both in larval and adult stage, get nourishment from a mould type of fungus known as ambrosia. The fungus is introduced in the tunnels in sapwood and sometimes in the heartwood, where it grows on walls. The degradation of round timber is enormous and is caused by borer holes as well as the black stains caused by the fungus. They damage timber by making pinhole or shot hole. As the ambrosia fungus cannot grow in dry wood, damage can occur in green timber. The ambrosia fungus and the insect die because of deprivation of their food on conversion and drying of timber.

2.5.2.2 FLAT-HEADED BORERS AND LONG-HORN BEETLES

As natural seasoning advances, there is a decline in the pin-hole borer attack. They are unable to thrive on wood with moisture content less than 50%. Flat headed beetle belong to the family Buprestidae and long horn beetle are from Cerambycidae. The larvae feed under the bark in the sapwood or heartwood. The attack can be recognized by the presence of galleries that are tightly packed with dust. Long-horn beetles or round-headed borers attack felled logs and wood work in village huts and buildings. They attack the wood on which the bark is present and is fresh and green. Drying of wood for 4–6 months makes it immune to such attack.

2.5.2.3 POWDERPOST BEETLES

The beetles belonging to the families of Lyctidae (true powderpost beetle) and Bostrychidae (false powderpost beetle) are commonly called powder-post beetle. The term ‘powderpost’ is derived from the fact that interior of the attacked wood is reduced to a flour-like powder. Damage caused by lyctus beetles is second after termites. The larvae make small tunnels up to 6 mm in diameter by boring into the sapwood of seasoned hardwoods. Starch is a main component of larvae diet. The wood of ring-porous species is more susceptible to such attack and the extent of infestation is almost proportional to the starch content. An advanced stage is indicated by multiple perforations in the form of round pinholes. Piles of wood floor beneath the infested timber also indicate the presence of these borers.

2.5.3 TERMITES

2.5.3.1 DRY WOOD TERMITES

In certain areas, such as in moist coastal regions and north-east India, dry wood termite causes considerable damage to wooden structures and constructional timbers. They damage even dry and sound-seasoned wood and do not maintain any contact with ground for moisture.

2.5.3.2 GROUND DWELLING

These are also called as subterranean termites. Ground dwelling are the most important and major destroyers of wood and wood-based products. They operate from a well-organized, usually underground, colony and span large distances through tunnels to reach the food source. The tunnels serve both as protection from sunlight, enemy and also prevent loss of moisture from their bodies. They work in concealment, eating away from the inside and leaving the exterior untouched. The galleries in the wood are often filled with earth or deposits of excrement.

2.5.3.3 CARPENTER ANTS

Carpenter ants may be black or red. Although they usually live in stumps, trees or logs, they often damage poles or structural timbers. Carpenter ants use wood for shelter and not for food. They usually prefer wood that is naturally soft or has been softened by decay. The galleries are large and smooth, unlike those of termites. Mounds of saw dust indicate their presence.

Protection of plantation wood from above-mentioned wood-destroying agencies will go a long way in conserving our natural forests and making wood available in future to meet the ever-increasing demand of wood by wood-based industries in India.

The selection of treatment process is very important for proper preservation of the product. The commonly used treatment methods are mentioned below.

2.5.3.4 BRUSHING/SPRAYING

Water-soluble or oil-soluble preservatives can be brushed or sprayed on solid timber or finished components of furniture, joinery and so on in house construction and for packing cases. The treatments will form an envelope around the timber and complete penetration will not be possible. The preservative chemicals used for this are as: boric acid: Borax (1:1.5); sodium penta chlorophenate 2% water soluble; CCA, CCB, ACC 3–4%; zinc napthanate 8%; copper napathnate 6%; trichorophenol 4%; bifenthrin 0.025%; chlorpyrifos 1%.

After application of water-soluble preservatives, the woods have to be dried to required moisture content. Therefore, as there is no requirement of drying the wood after treatment, with oil-soluble preservatives are more suitable for brushing.

2.5.3.5 *DIFFUSION TREATMENTS*

This is the most commonly used treatment method in the country. It is ideally suited for large-scale applications. Timber is dipped in water/solution of boric acid: borax; sodium pentachlorophenate (1:1.5:1) or CCA, CCB, ACC 3–4% for certain period of time and the timber is stacked closely under slow-drying conditions for certain time, before drying it to required moisture content. Depending on the thickness, dipping time and stacking time are to be altered.

2.5.3.6 *VACUUM CUM PRESSURE IMPREGNATION*

Vacuum cum pressure impregnation is also known as pressure process. It is carried out to get maximum penetration with adequate retention and hence, more service life. Vacuum cum pressure impregnation requires elaborate treatment plants. Both water-soluble preservatives and oil-soluble preservatives can be used in this system. For vacuum cum pressure impregnation, timber has to be dried to 12% m.c.

The preservatives used for this treatment are:

Water-soluble preservatives: boric Acid; borax (1:1.5); CCA, CCB, ACC.

Oil-soluble preservatives: zinc naphthanate, copper naphthanates trichorophenol, lindane 20 EC (emulsifiable concentrate), chloropyrifos 20 EC/50 EC, bifenthrin 2.5 EC.

2.5.3.7 GLUELINE TREATMENTS

This method is widely adopted by the Indian wood industry. This method is commonly called glue line poisoning. Some preservative chemicals can be incorporated into the adhesive formulation for enhancing durability of plywood made from woods that are susceptible to borer attack. This is a simple process in which the preservative chemical along with the adhesive penetrate into the veneer layers during the process of manufacture of plywood. The chemicals used for this treatment are provided in IS 12120-1992 as follows: boric acid 1% (for urea/melamine resin), chlorpyrifos 205 g ai/m^3 (for urea/melamine/phenolic resin). Synthetic pyrithroid that is bifenthrin at 5 g ai/m^3 was found to be effective.

Wood-preservative treatments are necessary at three stages in the service life of wood products. First, immediately after felling to prevent staining fungi and pin-hole borers, second, to enhance service life by inhibiting attack of decay fungi and insects and third, to use a remedial treatment for an observed attack.

2.6 ENVIRONMENTAL ISSUES OF TREATED WOOD-BASED PANEL PRODUCTS

Nowadays, concerns have been developed for minimizing the usage of chromium- and arsenic-based preservatives. The preservative-treated material, especially with hazardous chemicals such as chromium and arsenic, poses health risk. Till now, CCA- and CCB-treated products were widely used in residential application. The treated wood are sometimes burnt for fuel after its service life. The burning of toxic-treated wood may emit toxic chemicals. The disposal of treated wood should not be near water source as it may contaminate the water. The treated wood can be used as a fuel in the boiler (closed chamber). The research on proper disposal of treated wood is under process in world laboratories.

Since the last few years, environmental concern has immensely influenced the wood preservation research and development in India. Many chemicals that were being used indiscriminately are now banned. The wood preservation industry with its use of some broad-spectrum chemicals such as pentachlorophenate, inorganic arsenicals and creosote came under early scrutiny. Now, the wonderful chemical such as CCA is also under scrutiny for its use in wood preservation industry. In addition to restrictions on biocide usage, the disposal of treatment-associated process and product waste is becoming an ever-increasing problem.

Currently, the driving forces in wood preservation include both regulatory and market or cost-based issues. In addition, environmental concerns, disposal issues and public perceptions are directing researchers and the industry to develop and use environmentally acceptable preservative chemicals and treatment methods to protect wood. The primary objective of the preservative treatment of wood is to increase the life of the material in service, thus decreasing the ultimate cost of the product and avoiding the need for frequent replacements. Therefore, intense research is going on in the laboratories of world to develop environmentally acceptable solution for copper fixation in wood. The dangers posed to wood-treatment workers by most of the conventional proprietary wood preservatives, in addition to environmental degradation, are becoming a matter of major concern worldwide.[1]

2.7 CONCLUSION

The judicious use of wood can help in minimizing the load on forests. In India, negligible importance was given to wood preservation. As the resources are limited, the wood needs to be used judiciously. A country like India, where the gap between demand and supply of wood is more, sustainable use of resources is the only option to bridge the gap. India is the major wood consumer in the world. Earlier, less importance was given to the treatment of wood and wood-based panel products. Now, slowly, the perception of the industry is changing as the consumers are demanding insect-proof wood-based panel products. Hence, wood preservation is gaining importance in India. Treating wood with preservative chemicals not only increases the service life of wood but also locks the carbon inside the wood. The wood-destroying organisms decay the wood products and add carbon dioxide to the atmosphere.

REFERENCES

1. Barnes, H. M. Wood Protecting Chemicals from the 21st century, Proceedings of International Research Group on wood preservation, 24th Annual Conference Meeting, Orlando, Florida, USA, 16–20 May 1992, IRG/WP 93-30018.

2. Flynn, B. RISI Viewpoint: India's Demand for Log Imports Set to Double Over the Next 10 Years. RISI Wood Biomass Market, 2013. http://www.risiinfo.com/risi-store/do/product/detail/2013indias-forest-products-industry.html (accessed March 2, 2015).
3. Carle J.; Holmgren, P. Wood from Planted Forests: A Global Outlook 2005–2030. *Forest Prod. J.* **2008,** *58,* 6.
4. Dev, I.; Bhojvaid, P. P. Status of R and D in Wood Preservation in India. In *Proceedings of National Workshop on Wood Preservation in India: Challenges, Opportunities and Strategies*; Rao, K. S., Gairola, S., Aggarwal, P. K., Eds.; IWST: Bangalore, 2003; p 144.
5. FAO. Status of the World's Forests Report 2009. (Food and Agriculture Organization of the United Nations Rome). 2009.
6. Freeman, M. H.; Shupe, T. F.; Vlosky, R. P.; Barnes, H. M. Past, Present, and Future of Preservative Treated Wood. *Forest Prod. J.* **2003,** *53*(10), 8–15.
7. Gairola, S. C.; Aggarwal, P. K. Status of Wood Preservation in India. 2005—IRG/WP 05-30386. 2005.
8. IPCC. *Fourth Assessment Report. Inter-governmental Panel on Climate Change;* Cambridge University Press: Cambridge. UK, 2007.
9. ISFR. State of Forest Report 2013. Forest Survey of India. Dehradun. India. p 252.
10. Joshi, S. C.; Aggarwal, P. K. *Dynamics of Wood Utilization in India. Proceedings of International Conference on Future of Panel Industry—Challenges and Key Issues;* Sujatha, D., Nandanwar, A., Prakash, V., Lalithamba, C. S., Ezhil, K. V., Eds.; IPIRTI: Bangalore, 2012; p 203.
11. Kalawate, A. Bioefficay of Diflubenzuron Against Subterranean Termite. *Int. J. Fundamental Appl. Sci.* **2012a,** *1*(4), 67–70.
12. Kalawate, A. Lufenuron a New Breakthrough in Control of Wood Destroying Insects. *Indian J. Forestry* **2012b,** *35*(4), 447–451.
13. Kalawate, A. Evaluation of Copper Ethanolamine Boron Based Wood Preservative to Control Wood Destroying Insects. *Mole. Entom.* **2013,** *4*(2), 6–12.
14. Kalawate, A. Exploratory Studies on Nanobiocide for Wood Preservation. *Indian J. Forestry* **2014,** *37*(1), 9–16.
15. Kumar, S. Status of Wood Preservation Industry in India. 2005—IRG/WP 05-30388.
16. Negi, J. D. S. Biological Productivity and Cycling of Nutrients in Managed and Man-made Ecosystems. Ph.D. Thesis, Garhwal University, Srinagar. India. 1994.
17. Padmanabhan, S.; Kamal, S. Z. Recent Development in Treatment of Wood and Wood Based Panel Products. In *Proceedings of National Workshop on Wood Preservation in India: Challenges, Opportunities and Strategies;* Rao, K. S., Gairola, S., Aggarwal, P. K., Eds.; IWST: Bangalore, 2003; p 144.
18. Pandey, C. N.; Sujatha, D. Crop Residues, the Alternate Raw Materials of Tomorrow for the Preparation of Composite Board. https://www.icac.org/projects/CommonFund/20_ucbvp/papers/06_pandey.pdf (accessed March 3, 2015).
19. Tripathi, S.; Bagga, J. K.; Jain, V. K. Preliminary Studies on ZiBOC—A Potential Eco-friendly Wood Preservatives. The International Research Group on Wood Protection, IRG Doc. No. IRG/WP 05-30372, 2005.

CHAPTER 3

CASHEW NUT SHELL LIQUID-BASED ADVANCED FUNCTIONAL MATERIALS

SHABNAM KHAN[1,1], LAXMI[1], HINA ZAFAR[2], ERAM SHARMIN[3], FAHMINA ZAFAR[1,*,#], and NAHID NISHAT[1,*]

[1]*Inorganic Materials Research Laboratory, Department of Chemistry, Jamia Millia Islamia, New Delhi 110025, India,*
[*]E-mail: fahmzafar@gmail.com (F. Zafar); nishat_nchem08@yahoo.com (N. Nishat)

[2]*Division of Inorganic Chemistry, Department of Chemistry, Aligarh Muslim University, Aligarh 202002, India*

[3]*Department of Pharmaceutical Chemistry, College of Pharmacy, Umm Al-Qura University, PO Box 715, 21955 Makkah Al-Mukarramah, Saudi Arabia*

[#]Authors have equal contribution

CONTENTS

[1] Authors have equal contribution

ABSTRACT

As the days are proceeding, there is growing interest in the development of advanced materials from renewable resources rather than petroleum-based feedstocks. This is on the account of depletion of petroleum resources leading to increase in the cost of their products. This reflects a global requirement for sustainability without resource exhaustion. Cashew nut shell liquid (CNSL) is an agro byproduct of cashew nut processing industry. It contains anacardic acid, cardol, 2-methyl cardol and cardanol. All these constituents have versatile applications. So, CNSL is regarded as a versatile and valuable raw material for wide applications such as additives for lubricants, diesel engine fuels, pour point depressants, antioxidants, stabilizers, flame retardants, resins, inks, hydro-repellents, fine chemicals, intermediates as well as in biomedical applications. In the development of new green advanced functional materials, CNSL plays an important role due to its availability, low cost and biodegradability along with its unique structural features. The aim of the present chapter is to focus on the development of CNSL-based advanced functional materials such as soft materials, nanocomposites and coordination polymers and their applications.

3.1 INTRODUCTION

In the field of renewable resources, the knowledge and chemistry applied to the fabrication of commercial products has a marked importance in capturing the interest of both academic and industrial world researchers. During past few years, on the account of the day by day alarming rise in the cost of petrochemical products, the use of renewable resources for the synthesis of advance functional materials emerged as an area of keen interest. Renewable resources, commonly known as biomass, refer to material with recent biological origin, including plant materials, animal manure and agricultural crops.[1] The most widely used renewable raw materials include: wood, proteins, cellulose, lignin, tannins, starch, vegetable oils, chitin and chitosan and so on.[2,3] These biomass-derived materials have been utilized in the preparation of a variety of chemicals.[4–6] Vegetable oils such as soya bean, tung, linseed, rapeseed, castor and cashew nut shell or cashew nut shell liquid (CNSL) emerged as one of the most promising renewable resource with a wide array of applications.[7] In this regard, CNSL, an agricultural by-product extensively available in many parts of the world, come into view as a unique starting precursor as it contains a phenolic moiety with an unsaturated 15-carbon long alkyl chain.[8] The use of CNSL, which is also a non-edible oil, as a chemical raw material does not have a marked influence on the food supply chain unlike other renewable chemical feedstocks such as edible oils. Greater utilization of CNSL as a starting material is witnessed for the industrial polymer products in the view of its low cost, ample availability (450,000 metric tons per year) and chemically reactive nature.[9] Cashew nut shell liquid, a cost effective resource, gives most of the reactions of phenol; thus, it can be used as a substitute for phenol in many applications with equivalent or better performances. Hence, the use of CNSL as an alternate to phenol serves the purpose of conservation of synthetically derived substance and utilization of economical by-product. Literature indicated several reviews summarizing the composition, reactions or applications of CNSL.[10–18] In the present era, CNSL seems to hold a position of immense interest due to its renewable nature[19] as well as economic benefits to mankind. In the current chapter, considering the properties and enhanced applications of CNSL and its derivatives, we have focused on the advanced functional materials derived from CNSL—a 'functionally blessed' renewable resource. In the proceeding sections, there will be brief discussions about CNSL, its constituents, derivatives or polymers obtained from different chemical reactions and their applications, with special emphasis on CNSL-based advance functional materials and their applications.

3.2 CASHEW NUT SHELL LIQUID

Cashew nut shell liquid, a greenish-yellow liquid in the soft honey comb of the cashew nut shell, is a product of the cashew tree, *Anacardium occidentale.* It constitutes about 20–25% of the total weight of cashew nut. Industrial grade commercial CNSL occurs as reddish-brown viscous oil with a distinctive smell, and this is quite different from other vegetable oils. CNSL has inherent antimicrobial and anti-termite properties. It is immiscible in water but miscible in most of the organic solvents. Various kinds of methods are usually employed in the extraction of CNSL from nuts. Hot oil process and roasting process are the two most common methods used in the extraction of CNSL in which CNSL oozes out from the shell.[20,21] The other reported techniques for the extraction of CNSL includes extraction using solvents such as benzene, toluene, petroleum hydrocarbon solvents or alcohol, or supercritical extraction of oil using a mixture of CO_2 and isopropyl alcohol.[22]

Cashew nut shell liquid is also known as aromatic oil consisting of a mixture of natural phenols with around 90% anacardic acid (AA, a salicylic acid derivative with linear 15-carbon alkyl chain) as a major component and 10% allied compounds such as cardanol, cardol and 2-methyl cardol. The components of CNSL are themselves mixtures of four constituents differing in side chain unsaturation, such as saturated (~5–8%), monoene (~48–49%), diene (~16–17%) and triene (~29–30%). Cashew nut shell liquid which occurs naturally contains mainly four components, that is AA, cardol, cardanol and 2-methyl cardol (Fig. 3.1). Technical grade CNSL is obtained through decarboxylation of naturally occurring CNSL that converts AA to cardanol due to which it hardly contains any AA.

- **Anacardic acid** are phenolic lipids consisting of salicylic acid substituted with 15 carbon atoms hydrocarbon chain having one, two, three or no double bonds at meta position.
- **Cardanol** has around 30–41% phenols with C_{15} hydrocarbon chains with three double bonds (C_8–C_9, C_{11}–C_{12} and C_{14}–C_{15}), 16–22% phenols with hydrocarbon chains with two double bonds (C_8–C_9 and C_{11}–C_{12}), 25–36% phenols with hydrocarbon chains with one double bond (C_8–C_9).
- **Cardol** is a phenol having two hydroxyls at position 1 and 5 and a hydrocarbon chain of 15 carbon atoms at position 3.

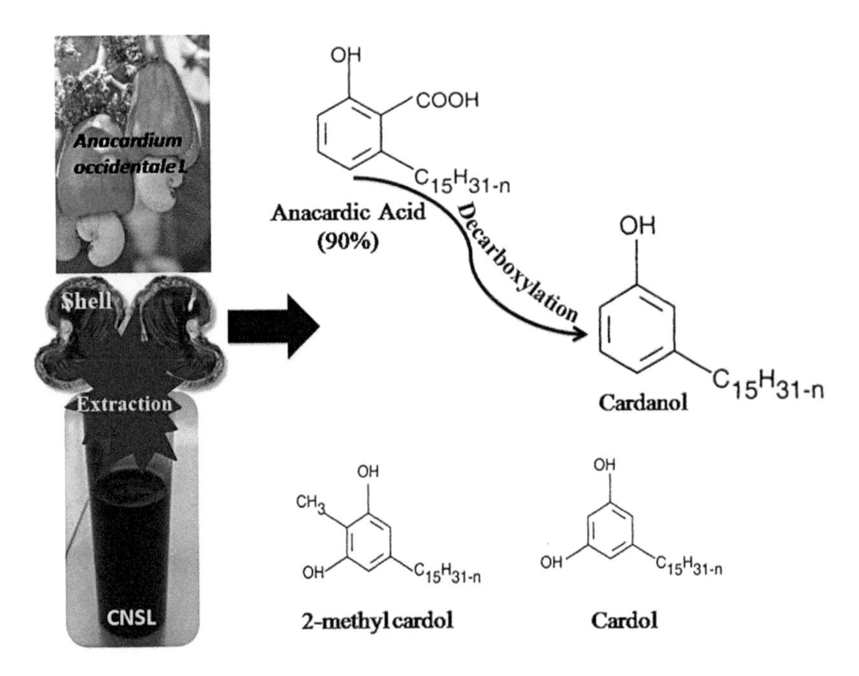

FIGURE 3.1 Cashew nut shell liquid (CNSL) and chemical structure of its constituents.

3.3 CHEMICAL TRANSFORMATION OF CASHEW NUT SHELL LIQUID/CARDANOL INTO DIFFERENT MATERIALS

Cashew nut shell liquid/cardanol is associated with three types of functional sites such as phenolic hydroxyl, aromatic ring and long alkyl side chain with unsaturation. The reactive phenolic hydroxyl group imparting synthetic flexibility, the long alkyl chain with non-isoprenoic *cis* double bonds offering it amphiphilic and lipidic character and the aromatic ring allows for p-p stacking and functionalization. CNSL is made an attractive precursor by all these functional sites collectively and is used for chemical modifications to produce a wide range of functional materials (Fig. 3.2).

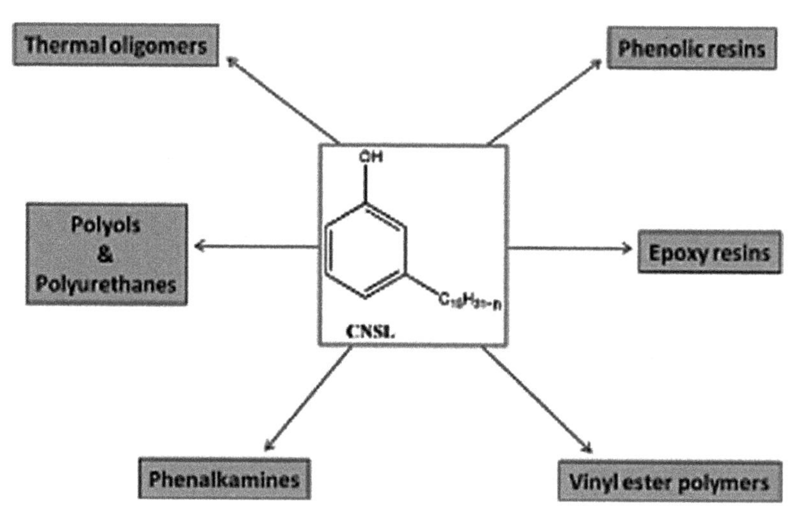

FIGURE 3.2 Reactions of CNSL in polymer chemistry.

3.3.1 OLIGOMERS OR POLYMERS VIA REACTION AT DOUBLE BONDS OF ALKYL SIDE CHAIN

The aliphatic side chain of CNSL provides hydrophobicity to the materials, which in turn proves to be an important property for many applications. The basis for addition polymerization can be the side chain unsaturation [23,24] using free radicals or ionic initiators. Acid catalysts (H_2SO_4, HCl, para toluene sulfonic acid etc.) can also be used to obtain polymeric materials.

The acid catalysed polymer of CNSL is rubber like which is quite less susceptible to oxidation. Earlier, there have been many studies on the kinetics and mechanism using catalysts such as H_2SO_4 and H_3PO_4 and Lewis acids (boron trifluoride etherate). The studies revealed that cardanol undergoes oligomerization under acidic conditions through side chain unsaturation, and the oligomerization involves initiation with side chain unsaturation followed by cationic chain growth. The polymerization of CNSL can also be achieved with the help of salts, namely zinc chloride, stannic chloride, ferrous sulphate, aluminium sulphate, boron trifluoride and salts of iron, cobalt, nickel, boron, chromium, lead, silver, mercury, manganese and so on up to the extent of 1–6%.

Thermal oligomerization can be considered the simplest heating method to polymerize CNSL via side chain unsaturation. Cashew nut shell liquid undergoes polymerization when heated to 200°C or above. Several agents such as cationic, anionic or oxidizing agents can be used to accelerate the process. If the reaction proceeds only with the aid of heat, the salts present in the CNSL catalyse the polymerization. Alkalis can also be employed as catalysts for the polymerization of CNSL. The first step proceeds through the reaction of phenolic hydroxyl with the alkali resulting in an adduct that further acts as a catalyst for the polymerization process. The kinetics and mechanism involved in the oligomerization of cardanol over acid catalysts was studied by Manjula et al.[25] Rodrigues et al.[9] studied the thermal oligomerization of cardanol without any kind of catalyst at 140°C for different times (5–40 h). The reaction initiated at unsaturated side chain, both from internal double bond (monoene, diene) and vinyl bond (triene). It was pointed out by the rheology studies that dimer was the main oligomer obtained from it.

3.3.2 PHENOLIC RESINS

The synthesis of condensation polymers by reaction with formaldehyde, furfural, hexamethylene tetramine (HMTA) and others is quite possible due to the phenolic properties of CNSL/cardanol.[26] These polymers are similar to the phenolic resins but show lesser reactivity during synthesis and later cross-linking. Thus, CNSL can be utilized for the synthesis of resole and novolac type resins due to its comparable similarity with phenolic resins.

- **Resole** is formed in the presence of alkaline catalysts and with more formaldehyde content with respect to CNSL, in which the methylol phenols condense either through methylene linkages or ether linkages. In later cases, subsequent loss of formaldehyde resulting in methylene link formation is observed.
- **Novolac** is synthesized in the presence of acid catalyst and with less amount of formaldehyde with respect to CNSL, with no reactive methylol groups and hence incapable of condensing with other molecules on heating in the absence of hardening agents.

Cardanol-formaldehyde resins showed improvement in flexibility (due to the 'internal plasticization' effect of C_{15} chain) that led to a better processability in comparison to conventional phenolic resins. The long alkyl side chain is responsible for the hydrophobic nature of the polymer, making it water repellent and resistant to weathering. Cashew nut shell liquid based resins have remarkable resistance to the softening action of mineral oils and are resistant to acids and alkalis.[27] In the field of structural applications, cardanol-formaldehyde resins revealed much lower tensile strength as compared with phenol-formaldehyde resins due to the stearic hindrance and eventually reduced molecular interactions provided by the C-15 side chain.[16] Dileep Tiwari et al.[29] synthesized cardanol-based phenolic resins in the presence of tricarboxylic acid catalyst with the help of microwave (MW) irradiation. The main advantage of this method is twofold reduction in the reaction time as compared with conventional method.

It is also revealed by literature survey that amine-based catalyst is used at different molar ratios for the curing of cardanol-formaldehyde resins. The inherent release of water during the curing of cardanol-formaldehyde resins leads to the formation of porous materials, which is also common in phenolic resins. Sometimes, an epoxy resin, diglycidal ether of bisphenol-A is usually mixed with cardanol-formaldehyde resins to reduce the amount of water released during the curing process. Riya and Deepak[28] synthesized novolac resins based on renewable resources cardanol and furfural in four different mole ratios, namely 1.0:0.3, 1.0:0.4, 1.0:0.5 and 1.0:0.6, respectively, with the use of dicarboxylic acid catalyst such as succinic acid. The most suitable curing agent that is HMTA was used for their curing. The modified resins have the potential to reduce the dependency on petrochemical derived phenolic resins. In addition, furfural, being a product of vegetable origin and its abundance in unlimited quantities, proves to be much more cost-effective as compared with formaldehyde.

The synthesized resin system finds diverse applications in composite matrix, surface coatings, brake linings, lamination industry, pesticides, azodyes and others.

3.3.3 EPOXY RESINS

There have been many reports on the synthesis and characterization of different types of epoxy resins based on CNSL and its derivatives.[30] The card-bisphenol (a reaction product of cardanol and phenol) on reaction with epichlorohydrin resulted in the production of epoxy resins that displayed superior performance over conventional epoxy resins. The epoxy resins have found applications as protective coating materials, auto primers, linings for cans, drums, pipes, potting and encapsulation of electrical and electronic components, and in electrical laminates as well. The epoxy coatings based on CNSL and its derivatives can be air dried in the absence of any expensive hardener, as in the case of conventional epoxy resins. Epoxy resins have also been tailored to obtain solvent-less adhesives that are curable at room temperature with better impact strength and shear resistance by synthesizing liquid resin from bisphenol-A, epichlorohydrin and a diglycidyl ether of CNSL as a plasticizer.[13] Shukla et al.[31] synthesized cardanol-based self-curable epoxide resins that can be used as surface coatings. Due to the inherent self-curable characteristic of the developed epoxide resins, the coatings based on cardanol do not require additional/external cross-linker to be integrated in the surface coatings. These coatings can be applied as primer coat or top coat on metallic substrates. Kanehashi et al.[32] synthesized cardanol-based epoxy resin from amine and epoxy cardanol prepolymer (ECP) by thermal polymerization for coating at room temperature. Thermal analysis disclosed that ECP coating exists at room temperature due to the flexible side chains of ECP. The ECP coating also showed better chemical stability as compared with the commercial cashew coating.

3.3.4 VINYL ESTER POLYMERS

The production of eco-friendly vinyl ester polymers can be achieved by the reaction of unsaturated end of the epoxy resin with acid functional acrylic/methacrylic monomer.[33–35] Sultania et al.[36–39] have synthesized cardanol-based epoxidized novolac through esterification of epoxy resin with methacrylic acid, in the presence of triphenyl phosphine as a catalyst, at 80–100°C. The reaction proceeds through a transition phase that results in the formation of acid catalyst complex through the interaction of acid and triphenylphosphine. It was found that the reaction was spontaneous, irreversible and produced highly ordered activated complex as confirmed by kinetic studies. Cardanol-based epoxidized novolac resin cured at 120°C by mixing it with benzoyl peroxide and styrene, and the kinetic studies also revealed that the curing reaction was slightly faster than the poly (vinyl ester) resin. The synthesized cardanol vinyl ester resins were found to be thermally stable up to 260–285°C.[37]

3.3.5 PHENALKAMINES

Cashew nut shell liquid based phenalkamines is a class of materials that have characteristic properties of both aliphatic amines (rapid cure, good chemical resistance) and polyamides (low toxicity, good flexibility and good chemical resistance).[40] These phenalkamines have a combination of attractive properties of both amines and aromatic ring due to the aliphatic polyamine substituents attached to the aromatic ring. The aromatic backbone imparts high chemical resistance; hydrophobic side chains make them water-resistant, and the high cross-link density is due to the amine groups. With the help of phenalkamines along a wide range of temperatures, the epoxy resins can be cured. For the low temperature cure of epoxy resins, different Mannich bases can be produced by the reaction of cardanol with formaldehyde and aliphatic amines, such as ethylene diamine, diethylene triamine and triethylene tetramine.[41] Phenalkamine curing agents based paint formulations prove to be better protective coatings with excellent resistance towards corrosion. These coatings appeared to be very surface tolerant and could be applied on non-well prepared metallic surfaces even under humid conditions. Phenalkamine-based coatings are particularly applied in heavy duty industrial and marine applications.

3.3.6 POLYOLS AND POLYURETHANES

Polyurethanes are formed by the addition polymerization reaction between diisocyanates with di or polyfunctional hydroxyl compounds, or other compounds with a large number of active hydrogen atoms. A wide range of polyols and polyurethanes from cardanol derivatives with enhanced thermal, mechanical and chemical characteristics have been reported many times.[42–45]

S. Gopalakrishnan et al.[46] reported the synthesis of polyurethanes from cardanol-fufural resin as a polyol by the condensation reaction between cardanol and furfural in the presence of adipic acid as a catalyst. The resin was further reacted with toluene diisocyanate (TDI) to yield hard segment polyurethanes. Synthesis of soft segment polyurethanes also occurred by the reaction of commercial polyol, polypropylene glycol-1200 with TDI and cardanol-furfural resin. The thermal studies revealed greater stability of hard segment polyurethanes than the soft segment polyurethanes.

By applying suitable methodologies, Suresh et al.[47] demonstrated the preparation of cardanol-based low viscosity polyols with a range of hydroxyl values. The synthesis of polyol was carried out first by preparing monoglycidyl ether of cardanol followed by the ring opening reaction to yield the diol or reaction with diethanolamine to form a triol. Alternatively, triol can also be obtained by reacting glycerol monochlorohydrin with cardanol. In comparison to triglyceride oil-based polyols, the polyols based on cardanol possess enhanced hydrolytic stability, and the polyurethanes prepared from them showed improved thermal stability. Cardanol-based polyols results in tough, rigid polyurethanes. The semi-rigid and rigid polyurethanes obtained were based on diol and triol based polyurethanes, respectively. However, it was revealed that diol and glycard based polyurethanes exhibit higher thermal stability.

The Mannich polyols synthesized from cardanol proved to be an excellent replacement for Mannich polyols based on petro-based derivatives (e.g. nonylphenol). The Mannich polyols based on cardanol have found applications in the preparation of rigid polyurethane foams with better physico-mechanical and fireproofing properties. Particularly for 'spray' rigid polyurethane foams, moulded rigid polyurethane foams and as cross-linking agent in many other polyurethane applications (e.g. coatings), these cardanol-derived Mannich polyols are suitable.[9]

3.4 APPLICATIONS OF CASHEW NUT SHELL LIQUID BASED POLYMERIC MATERIALS[11]

Cashew nut shell liquid based polymers have generally excellent properties and compatibility with a wide range of polymers such as alkyds, melamines, polyesters and so on which make them unique for many applications (Fig. 3.3). Cashew nut shell liquid proves to be the most attractive starting precursor for the synthesis of value-added products by the virtue of its low cost. The presence of long side chain imparts flexibility and water resistance to the polymers; its low fade characteristic makes it a desirable constituent of brake lining formulations, whereas the inherent antimicrobial and anti-termite properties make it suitable material for antimicrobial films and coatings. Cashew nut shell liquid is used in a number of surface coating applications due to its solubility in a number of common solvents. A serious drawback associated with CNSL-based polymeric materials is the colour imparted by the oil. Some of the important applications of CNSL-based materials will be discussed further.

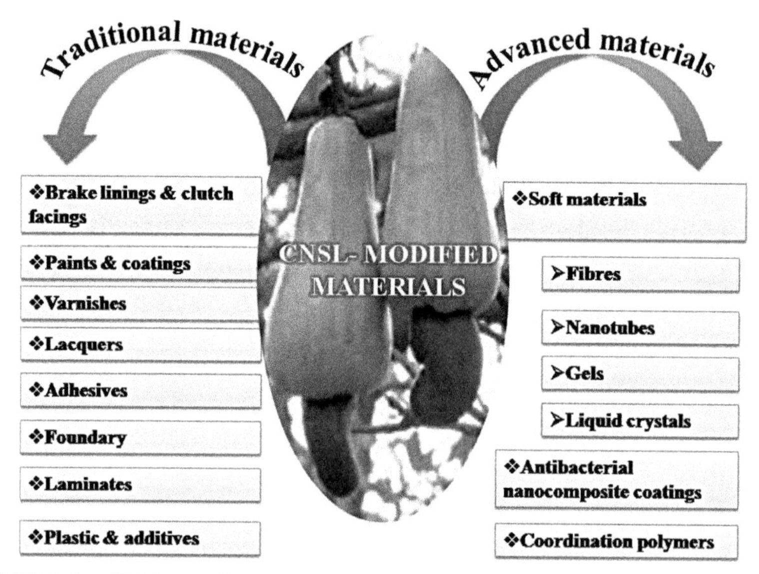

FIGURE 3.3 CNSL-modified materials.

3.4.1 BRAKE LININGS AND CLUTCH FACINGS

Brake linings and clutch facings industry is one of the largest consumers of CNSL. The outstanding blend properties of CNSL such as heat absorption generated due to friction, enhanced retention of braking efficiency, high impact strength and modulus, low fade characteristics, and low cold wear make it to be a natural choice for this application. Resins based on CNSL have good miscibility with rubber and other natural and synthetic materials. The CNSL-based frictional materials proved to be suitable for low speed automobiles in which the temperature does not exceed 250°C. The resins used in brake linings are usually prepared by the acid catalysed side chain polymerization followed by conventional condensation with formaldehyde. CNSL-furfural reaction products have very even frictional characteristics over wide temperature ranges. CNSL-modified phenolic resin with carbon fibre, powdered cast iron, and $BaSO_4$ possess excellent fade and wear characteristics. The blends of CNSL-modified phenolic resins with rubber and certain additives have been used for preparation of friction material having low brake noise. The fatty acids from oils on reaction with CNSL are converted to hard infusible particles that can be useful as friction dust. The fatty acid-CNSL reactions are usually catalysed by potassium acetate, pyridine and so on at 200–300°C, and the products obtained can be used for friction dust formulations. Phenol-cardanol aldehyde resins have been used as binders for extrudable brake linings. Moulding compositions were based on CNSL resins which helped in obtaining filled compositions useful for brake and clutch linings as well as floor coverings.[11]

3.4.2 PAINTS AND COATINGS

Cashew nut shell liquid based polymers have been found to be highly suitable for surface coating applications due to chemically stable nature of CNSL, resin solubility in various solvents, inherent hydrophobicity and chemical resistance, film forming ability and high degree of unsaturation. The condensation product of CNSL, phenol and formaldehyde on treatment with chloroacetic or bromoacetic acid under alkaline conditions and neutralization with ammonium hydroxide or an amine is a water-based coating material that has good hot water and chemical resistance properties. Paint formulations of water-based stable emulsions of oil-in-water type can be prepared by the use of CNSL-formaldehyde resins. To prepare anti-fouling paints, the incorporation of copper aceto-arsenite or linseed oil rosin condensate into CNSL-aldehyde was adopted. The flexibility and drying characteristics can be improved by the addition of unsaturated oils to CNSL. Styrenation of CNSL-based phenolic resin improves the paint performance. Polymerized CNSL on mixing with oils and resins such as linseed, china-wood and coumarone gives coatings for cloth, fabrics, papers and metal sheets. The wires after dipping in these liquids carries a liquid film which, on drying, is flexible but hard and tough. It was depicted by the literature survey that polyester resins based on CNSL and its derivatives can be used as good coating materials. Cardanol-based vinyl resins provide films with excellent adhesion and flexibility, resistant to oils and greases, water and chemicals. Phosphated CNSL-formaldehyde reaction products yield coatings with improved heat resistance, adhesion and flexibility. CNSL oil modified alkyd resin finds application in surface coatings industry and printing inks. Coatings with inherent insecticidal properties were obtained by adding dichlorodiphenyltrichloroethane, gamexane and so on to CNSL or chlorinated CNSL after treatment with formaldehyde. Surface coatings have been developed from CNSL for protecting concrete against attack by chemical fertilizers. CNSL-hexamine resins with lead and cobalt naphthenates after mixing it with calcined mud, dehydrated castor oil and solvents are used for development of corrosion-resistant primers. CNSL-glycerin reaction products based coatings establish tough elastic films.[11] Emilie Darroman et al.[48] synthesized bio-based epoxy blends exhibiting interesting properties for coating applications. Commercial epoxidized cardanol (from CNSL) was cured with two different amines, isophorone diamine and Jeffamine T403, to produce materials exhibiting good properties. Sorbitol and isosorbide are two kinds of sucrose epoxy derivatives that were used as blends for enhancing the properties of epoxy cardanol-derived materials. Mukesh Kathalewar et al.[49] synthesized a new class of non-isocyanate polyurethanes (NIPUs) via eco-friendly approach which can be used in coating application. The cyclic carbonate (CC) based on CNSL was allowed to cure with diamines, hexamethylene diamine and isophorone diamine, to produce NIPU. The study vividly explained that CNSL-based CC can act as component for coating formulation and the coating properties can be enhanced through appropriate selection of amine cross-linker.

3.4.3 VARNISHES

Esters of acrylic acid and cardanol give materials that are useful for manufacturing varnish that are resistant to water and dilute Na_2CO_3 solutions. Electrical insulating varnishes were prepared by the treatment of CNSL with formaldehyde and further compounding the resultant material with pure phenolic or alkyd resin in suitable proportions. Films of these materials show resistance towards water and chemicals and can be used as insulating varnishes with high electrical resistance and also as laboratory table tops. Quick drying varnishes have also been prepared by the use of mixtures of polymerized CNSL and bodied linseed oil.[11]

3.4.4 LACQUERS

Cashew nut shell liquid based lacquers could be used for insulation and also for protective or decorative purposes for furniture, buildings, automobiles and others. The films based on them showed optimum toughness and elasticity, excellent gloss and excellent adhesive qualities. The obtained dried films were found to be superior to ordinary oil paints in terms of resistance towards oils, greases, moisture and chemicals. Lacquers for special applications such as screen lacquering are prepared by the condensation of CNSL and formaline in the presence of alkalis, and for gold lacquering CNSL-formaldehyde condensate is treated with $Ti(OBu)_4$.[11]

3.4.5 ADHESIVES

Cashew nut shell liquid emerged as a desirable starting precursor for adhesive formulations due to the properties such as high polarity, innate tackiness of phenolic materials and the ease in the liquid to solid conversion. Applications as adhesives for plywood have been found by CNSL-formaldehyde resins. Polymerized CNSL after treatment with phenol and formaldehyde in the presence of acids can be used to make plywood adhesives. Plywood adhesives can be made fireproof by incorporating ammonium phosphate and ammonium dihydrogen phosphate or diammonium hydrogen phosphate to CNSL-formaldehyde. Cashew nut shell liquid adhesive formulations based on CNSL and formaldehyde can replace phenol-formaldehyde adhesives used in the manufacture of moisture-proof adhesives. Adhesives can also be synthesized by using CNSL with gelatinous materials such as animal glue, blood and egg albumin and so on.[11]

3.4.6 FOUNDARY

In foundary industry, the resins based on CNSL utilizing the phenolic nature of CNSL prove to be an excellent replacement to conventional phenolic materials. By using CNSL-modified phenolic resins mixed with amines and ammonium salts of inorganic acids and water-soluble surfactants, the synthesis of low-viscosity caking agents for sand moulding was carried out. Resin prepared from phenol and CNSL ethers by reaction with aldehydes has also been used in moulding powders or as a binder.[11]

3.4.7 LAMINATES

Cashew nut shell liquid based resins can be used for improved heat resistance, the bonding of the reinforcement materials, and the flexibility of laminates, while reducing brittleness and age hardening. Cashew nut shell liquid and phenol on reaction with aldehydes give resins for laminated products. CNSL-modified phenolic resins are employed for making laminates with improved dimensional stability. Epoxidized CNSL increases flexural and tensile strength of laminates. Thermally stable and fire-resistant laminates based on CNSL have also been reported. Metal polymer laminates requiring high separation forces have been developed from CNSL.[11]

3.4.8 PLASTIC MATERIALS AND ADDITIVES

High performance and speciality polymers can be obtained from CNSL by various kinds of methods.[50] CNSL-formaldehyde resins can be used for the manufacture of lightweight, sandwich-type plastic, composite panels suitable for partitions, flush door and so on. Cashew nut shell liquid based foam plastics tend to become lighter upon ageing for a period of 12 days at 100°C. Polymerized CNSL-furfural resins are resistant to acids and alkalis and thus found application in insulation parts of storage batteries. The increased heat resistance of phenolic moulding composition can be achieved by the addition of CNSL. Cashew nut shell liquid resin based mouldings have good antistatic and electromagnetic properties and is proved to be a good coating material for buildings.[11]

3.5 ADVANCED FUNCTIONAL MATERIALS AND APPLICATIONS OF CASHEW NUT SHELL LIQUID

Innumerable opportunities are provided by the nature to synthesize functional materials from a wide variety of raw materials such as carbohydrates, lipids, nucleotides and many more.[51] Formulating processes for the conversion of bio- or industrial waste into a wide range of value-added chemicals and functional materials have been proven to be an alternate approach towards the concept of biorefinery.[52] Several methods are now known to convert raw materials into molecular precursors and applied end-materials.[52–54] Cashew nut shell liquid/cardanol can be turned into a number of value added materials due to the in-built functionalities present in the molecule. In this section, we are focusing briefly on the role of CNSL/cardanol in generating a variety of advanced functional materials.

3.5.1 *SOFT MATERIALS*

Soft materials are basically the self-assembled nano- or meso-structures which are obtained from small molecules but sometimes include polymer-based materials as well. Such assembled nanostructures can be achieved by self-assembly; it has emerged as an important tool in the field of nanotechnology.[55] Some of the cardanol-based soft nanomaterials are briefly discussed below that may find diverse applications in various fields such as biomaterials, templated syntheses, drug carriers, medicines and biosensors.

3.5.1.1 *FIBRES AND NANOTUBES*

Self-assembly emerged to be a unique process that is quite sensitive to the changes in the chemical structure of molecules. By having a control over small structural changes, one can synthesize a wide variety of self-assembled nanomaterials from simple building blocks. Cardanol possessing the inherent flexible functionality allows simple chemical modifications for the generation of distinctive soft materials. By simply attaching glucopyranose moiety to cardanol, George John et al.[52] synthesized cardanol-based amphiphilic glycolipids (Fig. 3.4), GlyLip-1 and GlyLip-2, and also studied their self-assembly properties in water. A wide variety of soft materials, such as twisted fibres, helical coils and nanotubes, were synthesized through the self-assembly of such amphiphilic glycolipids.[56] In this system, the sugar moiety acts as a non-ionic polar head group that imparts chirality to the molecule. The nanotubes obtained from cardanyl glycolipids were 40–60 nm wide, 10–100 μm long and 8–15 nm thick. Energy-filtered transmission electron microscopy studies revealed that thinnest nanostructures of diameter 30–35 nm were obtained through self-assembly of GlyLip-1 in aqueous media. These nanostructures gradually transformed into structures of length up to hundreds of micrometres and about 10–15 nm of internal diameter. The mixture of four different components with differing side chain unsaturation, GlyLip-1, formed nanotubes, whereas the fully saturated component, GlyLip-2, formed only twisted fibres. Another most important issue to understand is the role of *cis*-double bond unsaturation in the self-assembly process [57–59] and to outline the nanotube formation containing inner hollowness from typical twisted ribbons. The methodical study of binary mixtures of glycolipids with different ratios demonstrated the influence of side chain unsaturation on the self-assembled morphology. The monoene component of GlyLip-1 upon self-assembly in water resulted in the formation of nanotubes including somewhat coiled nanofibres. These coiled nanofibres sometimes can be considered the precursor in the formation of nanotubes.[60] The self-assembly of a typical composition of 90:10 and 80:20 (saturated:monoene) mixture forms twisted ribbons in water and the doping of monoene component up to 30–40% does not have a profound effect on the twisted morphology. However, an equimolar composition of 50:50 gets assembled to form loosely coiled-ribbon morphology, in between the twisted and tight helical coil.

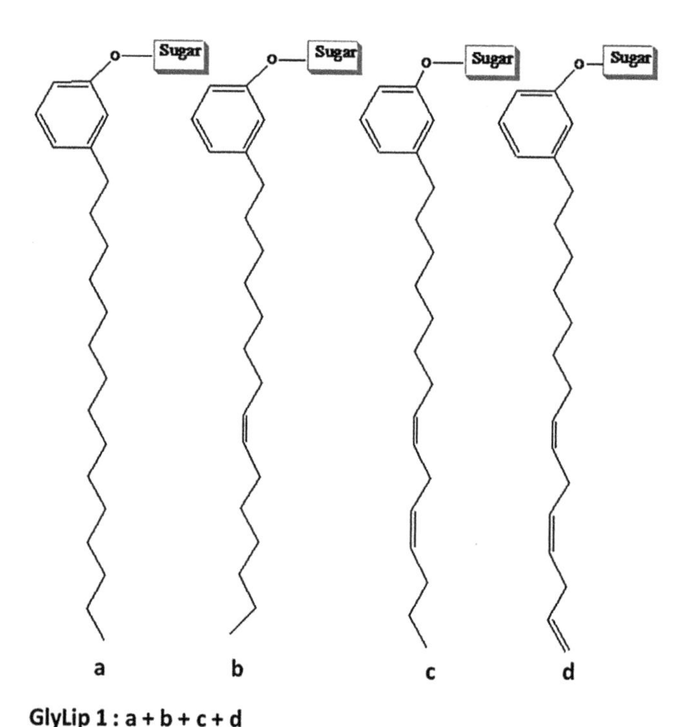

FIGURE 3.4 Chemical structures of cardanol glycolipids.

3.5.1.2 GELS

Gels are the soft nanomaterials whose mechanical properties lie between elastic solids and viscous liquids. A gel-state can be defined as the balanced state between precipitation and solubilization of the molecules in a solvent.[61] In case of gels, the self-assembly of molecules results in the formation of fibrous network entrapping the solvent through capillary forces which resists the flow of the medium. The gels with strong inter-molecular interactions can be obtained by careful controlling of both the concentration and solvent conditions. Further, the orientation and directional behaviour of self-assembled structures can be controlled by both the solvent polarity and solubility of the amphiphiles. It was revealed that solvent polarity influenced the orientation of amphiphiles in the formation of self-assembled stacks.[62,63] The degree of alkyl side chain unsaturation of cardanyl glycolipids had a critical role in regulating the self-assembly behaviour.[57] The gelation capability of the cardanyl glycolipids was tested in a number of organic solvents as well as in water-alcohol mixtures. The saturated derivative of cardanyl glycolipids, GlyLip-2 get assembled to form gel in the common organic solvents such as cyclohexane, toluene and xylene as well as in aqueous solutions with water miscible solvents such as dimethyl sulphoxide (DMSO) and tetrahydrofuran (THF). However, GlyLip-1, mixture of four different glycolipids differing in side-chain unsaturation, form gels only in organic solvents and doesn't show gelation behaviour in either aqueous solution or mixture of water–water-miscible solvents such as water–THF or water–acetone.

To understand the self-assembly phenomenon of glycolipids better, the gelation capability of the individual mono-, di- and saturated components were tested in a wide range of solvents. The gelation ability for the cardanol glycolipids tends to decrease with increasing degree of side-chain unsaturation. The monoene component, GlyLip-1b, forms only partial gel in many organic solvents such as cyclohexane and toluene. The polarity and solvent nature also had a pronounced effect on the packing mode of the glycolipids in case of gel formation. The aqueous gel from GlyLip-1 results in intertwined and long-range ordered bilayer structures, whereas the corresponding organogel results in extended bilayer structures. Thus, the structure of a wide range of self-assembled materials from cardanyl glycolipids can be tuned up by proper selection of degree of unsaturation in side chain and the solvent.

3.5.1.3 LIQUID CRYSTALS

Liquid crystals exhibit the properties of both solids and liquids as they flow like a liquid and form solid crystals with positional and/or orientational order. Liquid crystals also tend to show thermotropic or lyotropic properties that is response to the changes in heat and concentration, respectively. The cardanol-based glycolipids, GlyLip-1 and GlyLip-2, were found to show both thermotropic and lyotropic behaviour in the formation of liquid crystalline materials.[64] The degree of unsaturation in alkyl side chain and the hydration effect markedly influences the phase behaviour of cardanol-derived glycolipids. The glycolipid with saturated analogue, GlyLip-2, usually show characteristic smectic phase at 142°C, whereas the unsaturated analogue (monoene derivative) shows the same smectic phase at slightly lower temperature of 131°C. The hydrated cardanol-based glycolipids exhibit lyotropic mesomorphism. The fully hydrated saturated glycolipid, GlyLip-2, forms lamellar phases at 77.8°C by showing a gel-to-liquid transition. However, the transition temperature for mono-, di- or triene-glycolipids is about 45°C and the glycolipids retained their liquid crystalline state at close to room temperature 33°C.

3.5.2 ANTIBACTERIAL NANO-COMPOSITE COATING MATERIALS

The development of bactericidal coatings through green chemical methods is gaining widespread attention as an environmental-friendly application. Cardanol, due to the presence of side chain unsaturation, undergo air-assisted autoxidation as in case of drying oil to form polymeric films; hence it is used in the preparation of coatings.[65,66]

Being a phenolic lipid, cardanol is able to form nanoparticles by in situ reduction of the corresponding metal ions. They can also be used in the formation of optically transparent, scratch free and wrinkle-free polymeric films through cross-linking at ambient conditions. The autoxidation process of the non-isoprenoic double bonds of cardanol yields numerous free-radical species during the cross-linking process that can reduce metal ions. The free radicals generated during the autoxidation of polyunsaturated hydrocarbon chains are capable of reducing metal ions, such as Au(II) and Ag(II), to form zero oxidation state nanoparticles. The elimination of external toxic reducing and stabilizing agent is the main advantage of this method for the preparation of nanoparticles. The cross-linked polymeric network matrix itself acts as the capping agents for the metal nanoparticles resulting in nanoparticles-embedded films. A smooth and scratch-free film was produced from poly(cardanyl acrylate) polymer doped with metal ion through the autoxidation drying process. On the other hand, a poly(pentadecylphenyl acrylate) polymer with a saturated hydrocarbon chain failed to undergo oxidative drying owing to the absence of essential side chain unsaturation, thus preventing nanoparticle formation in the matrix. Efficient antimicrobial activity was exhibited by silver nanoparticles (AgNPs) against a broad range of bacteria. Literature suggests that the mechanism of AgNP-mediated antibacterial activity involves the interaction of AgNPs with the outer membrane constituents of bacteria causing structural changes and degradation, eventually resulting in cell death. AgNPs less than 15 nm in size are proved to have better antibacterial activity. The coatings embedded with AgNP were found to be effective against airborne gram-positive *S. aureus* and Gram-negative *E. coli.*

3.5.3 COORDINATION POLYMERS

Over the past few years, a number of coordination polymers (CP), also called metal–organic frameworks, have attracted great attention and have successfully become a remarkable hot spot in the field of all sciences due to their inherent features such as porosity, large inner surface area, tunable pore size and topologies. These features lead to their versatile architectures and promising applications (Fig. 3.5), such as in ion exchange, gas adsorption and storage, in particular of hydrogen and methane, gas and liquid separation processes, drug delivery, sensor technology, heterogeneous catalysis, hosts for metal colloids or nanoparticles or polymerization reactions, pollutant sequestration, microelectronics, luminescence, non-linear optics and magnetism.[67–76]

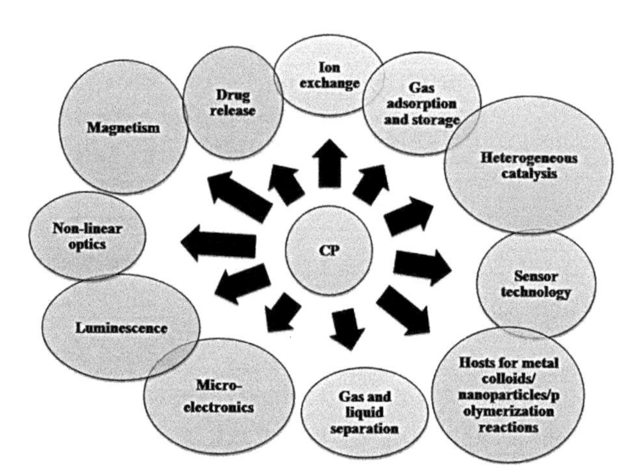

FIGURE 3.5 Application of coordination polymers (CP).

CP are usually synthesized from self-assembly of metal ion nodes with organic ligands (Fig. 3.6). The tailoring of CP or metal organic frameworks (MOFs) is controlled by the choice of coordinated ligand(s) as well as the choice of metal(s), as the structure of the resultant MOF_s is governed by the geometry and connectivity of an organic linker (or bridging ligand). By the proper adjustments of linker geometry, length, ratio and functional group, properties such as size, shape and internal surface of MOFs for a targeted application can be tuned.[76]

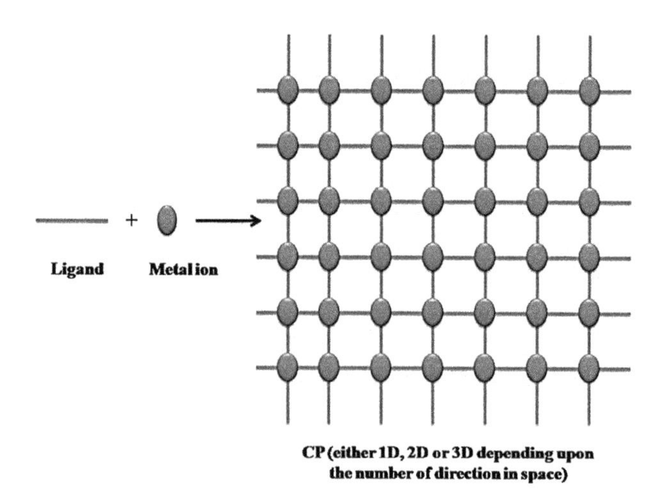

FIGURE 3.6 Schematic representation of the formation of CP.

Ligands and their coordination complexes/polymers are generally petro-based and synthesized by multi-step processes in the presence of volatile organic solvents; along with this, their limited solubility forces the scientists and industrialists to develop these coordination materials through alternative route with desirable properties. The antibacterial activity and metal ion intake of cardanol-based polymeric Schiff base transition metal complexes were studied by C. Isac Sobana Raj et al.[78] The preparation involved (a) the synthesis of bis(3-pentadecylphenol) methane (BPPM) using cardanol and formaldehyde, (b) BPPM on treatment with epichlorohydrin followed by the action of sodium periodate forms di-α-formylmethoxy bis(3-pentadecylphenyl)methane (DFMPM), (c) DFMPM into (2+2) macrocyclic Schiff base ligand on reaction with ethylene diamine and finally (d) Schiff base complexes with transition metal salts. The ligands and complexes were studied using techniques such as IR, UV-Visible, ^{1}H-NMR and elemental analysis, melting point, conductivity, metal ion intake and antibacterial activity were also studied. B. Ravindranath et al. reported lipophilic metal derivatives of AA with an unusual high degree of selectivity.[79] AA–metal (Mg^{2+}, Ca^{2+}, Ba^{2+}, Fe^{2+}, Cu^{2+}, Zn^{2+}, Co^{2+}, Mn^{2+} and Ni^{2+}) derivatives in ratios of both 1:1 and 2:1 were prepared and characterized. The complexes were characterized by using UV. The selectivity and permeability of the metal ions across a liquid membrane were also investigated. Kazuo Tsujimotoa et al. reported AA chelate iron interaction with the active site of the enzyme in which the hydrophobic tail portion slowly begins to interact with the hydrophobic domain close to the active site. Formation of AA-ferric ion complex in the ratio of 2:1 was detected as the base peak in the negative ion electrospray ionization mass spectrometry.[80] Lin Jinhuo and HU Binghun synthesized cardanol-aldehyde condensation polymeric coating material containing boron-nitrogen coordinating bond and characterized by IR, X-ray photoelectron spectroscopy, high-performance liquid chromatography and DTA-TG. The coatings showed excellent physico-mechanical properties, better corrosion resistance and were found to be stable at high temperatures.[81] F. B. Hamad et al. prepared novel heterogeneous copper (II) Schiff base catalysts by using a CNSL templating agent.[82]

Our research group has developed coordination complexes/polymers from CNSL by following the principles of green chemistry.[83–95] It has emerged as an alternate approach for the development of polymers via eco-friendly route, with the minimum use of toxic chemicals, particularly organic solvents that are hazardous to health and environment. Green technologies involve bio-based materials along with their synthesis through green chemistry that is via solvent free, facile MW assisted routes. In our unpublished work,[91–93] we have developed coordination complexes/polymer networks from CNSL/cardanol and its derivatives as organic ligands and divalent transition metals [Cu(II), Co(II), Mn(II) and Zn(II)] ions as metal nodes. These complexes were synthesized from technical CNSL (containing 62.86% cardanol)/pure cardanol and divalent metal salts via conventional heating and facile MW irradiation methods via solvent-free method. The structure and morphology of these materials were confirmed by Fourier-transform infrared spectroscopy (FTIR), UV, scanning electron microscope (SEM) and X-ray diffraction (XRD). Solubility of these materials was tested in various polar and non-polar solvents (e.g. xylene, benxene, DMSO, ethanol, diethylether and THF). They showed very good solubility in these solvents due to presence of polar groups and long alkyl chain. CNSL-formaldehyde–metal complexes/polymers with divalent transition metals, CNSL-F-M(II), were synthesized by in situ solvent less method. Polymers, CNSL-F-M(II)PU, were prepared by the reaction of CNSL-F-M(II) with calculated amount of TDI in presence of dibutyl tin dilaurate as a catalyst with minimal amount of solvent at room temperature. Films of these mixtures were cured at room temperature.[91] In another (unpublished) work, our group has synthesized CP from CNSL and furfuraldehyde (CNSL-Fur)-based ligand and divalent transition metal ions as metal nodes. The synthesized system is completely derived from agro byproduct. The polymer of CNSL-Fur-.M(II) were synthesized by the same methods as Col-F-M(II).[92] The synthesized systems are CNSL-Fur-M(II)PU. Films of these polymers were cured at ambient temperature. Optical microscopy was also used to investigate the morphology of some samples. It shows uniform defined pattern of arrangement of metal ions with organic ligands (Fig. 3.7). In our recently published work, we have developed nanostructured CP based on cardanol-formaldehyde polyurethanes with 'd_5' Mn(II) and 'd_{10}' Zn(II) as metal nodes following the green and sustainable chemistry principles. Films of these CP were allowed to cure under ambient conditions.[93] The structure and morphology was confirmed by FTIR, UV-Vis, field emission scanning electron microscope (FE SEM), EDX, transmission electron microscopy (TEM). The thermal studies indicating better thermal stability were carried out by TGA/DTA/DTG and DSC. Through preliminary dye adsorption technique, the use of the developed coordination polyurethanes in the toxic dye removal from waste water was also determined.

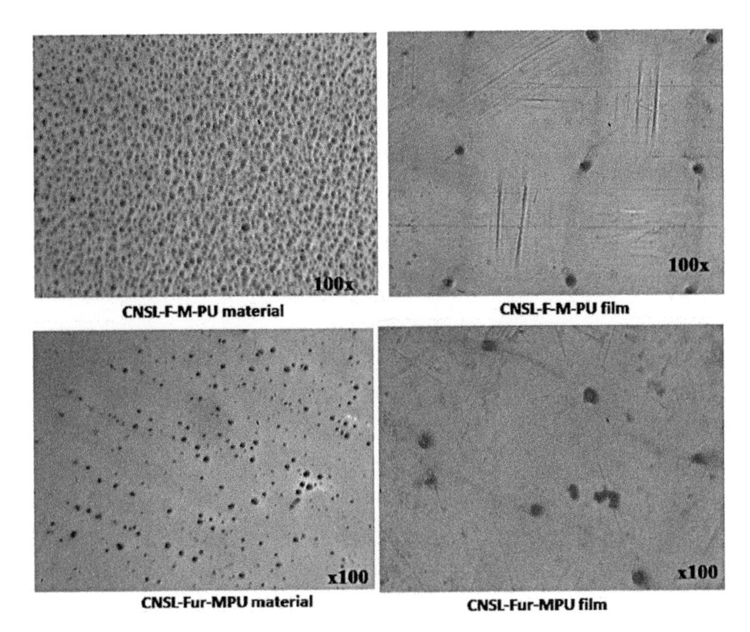

FIGURE 3.7 Optical micrographs of CP at magnification 100×.

The polyurethanes films of CNSL-F-Mn(II)PU and CNSL-Fur-Mn(II) PU also show porous nature by FE SEM (Figs. 3.8 and 3.9). Brunauer–Emmett–Teller (BET) technique confirms the mesoporosity of the system. The synthesized systems show nanosized particles (5–20 nm size) by TEM (Fig. 3.10). XRD analysis of these systems showed amorphous behaviour. Some samples were tested for their antibacterial activity, and we found positive results.

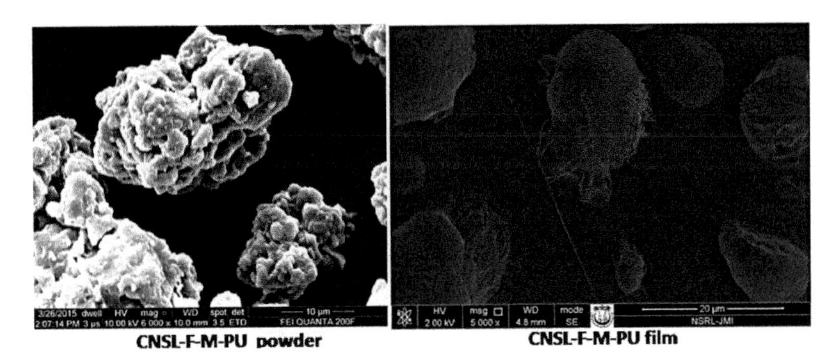

FIGURE 3.8 FE SEM of Col-F-M(II)-PU.

FIGURE 3.9 FE SEM of CNSL-Fur-M(II)PU powder sample.

Recently our research group has also attempted to cure CNSL-Fur-M(II) and CNSL-For-M(II) system (where M=Mg(II)) with HMTA via solvent less method.[94,95] The free standing films of these systems were formed at 170–180°C for 30 min to 1 h. The cross-linked CNSL-Fur-Mg(II) system showed semi-crystalline behaviour, and the calculated average crystallite size were found in the range of 38–48 and 2–47 nm in case of cross-linked CNSL-Fur-Mg(II) and CNSL-For-Mg(II), respectively. Optical morphology of the films showed well-defined pattern of Mg (II) within polymeric matrix. The film of CNSL-Fur-Mg(II) showed excellent chemical (acid, salt and alkali) resistance and thermal stability (5–10 wt% loss at 280–300°C).[95] For versatile applications such as protective coatings, adhesives, drug carrier and adsorption and others, these systems can be used.

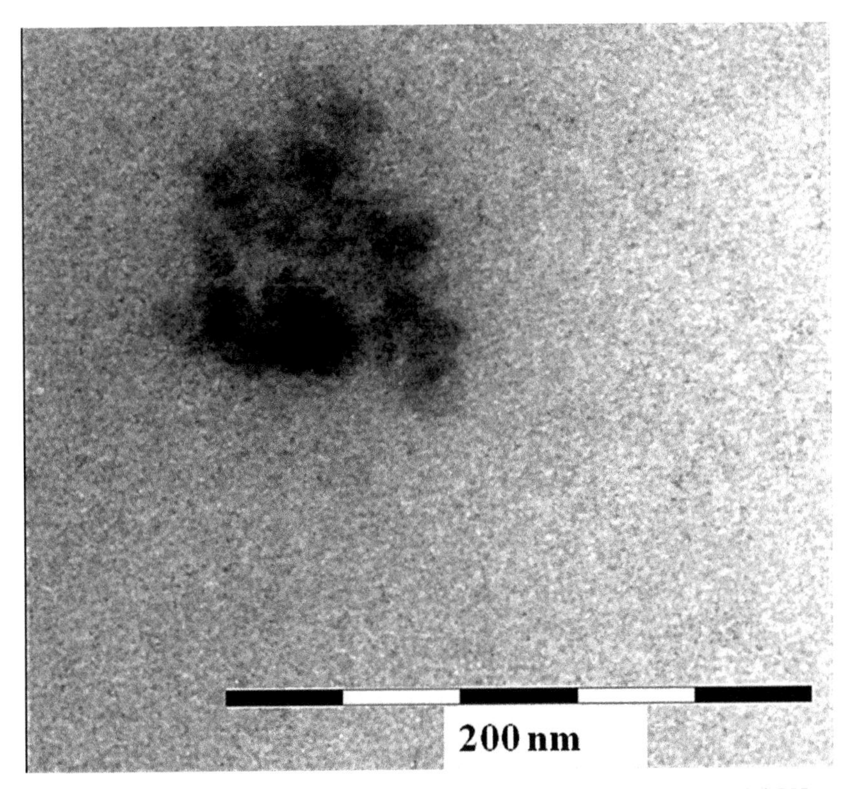

FIGURE 3.10 Transmission electron microscopy micrograph of CNSL-F-M(II) PU.

3.6 CONCLUSION

CNSL/cardanol is a renewable resource and by-product of the cashew industry. It is inherently associated with reactive phenolic hydroxyl group that provides synthetic flexibility, long alkyl chain with saturation and unsaturation attributing amphiphilic and lipidic character, and the aromatic ring allowing for p–p stacking and functionalization. Through these functional sites, CNSL/cardanol can be converted to valuable polymeric materials that may find versatile application such as paints, coatings, varnishes, additives, adhesives, lacquers. Among them, various forms of advanced nanomaterials have also been generated from CNSL/cardanol such as soft materials (lipid nanotubes, twisted/helical nanofibres, low-molecular-weight hydro/organogels and liquid crystals), antibacterial nanocomposites and CP which find applications in versatile field. CNSL/cardanol can be considered as a tenable precursor for chemical modifications to generate a library of amphiphiles and functional monomers. Thus, in a nutshell, many investigations are already conducted and a lot still remains unexplored in the field of 'the liquid from the nut shell'—Mother Nature's bountiful gift to us.

3.7 ACKNOWLEDGEMENTS

Shabnam Khan and Laxmi thanks to UGC (New Delhi, India) for Maulana Azad National Fellowship (MANF), ref # F1-17.1/2014-15/MANF-2014-15-MUS-UTT-36965/(SA-III/Website) and non-NET fellowship respectively. Dr Fahmina Zafar is thankful to UGC for Dr. D. S. Kothari postdoctoral fellowship [Ref. #F.4/2006(BSR)/13-986/2013(BSR), 2013-2016] and the Department of Science & Technology, New Delhi, India for the award of fellowship under the Women Scientists Scheme (WOS) for Research in Basic/Applied Sciences (Ref. No. SR/WOS-A/CS-97/2016), and the Head, Dept of Chemistry, Jamia Millia Islamia, for providing facilities to carry out their research work.

KEYWORDS

- **CNSL**
- **cardanol**
- **advanced functional materials**

REFERENCES

1. Crocker, M.; Crofcheck, C. Biomass Conversion to Liquid Fuels and Chemicals. *Energeia* **2006,** *17*(6), 1–3.
2. de Espinosa, L. M.; Meier, M. A. R. Plant Oils: The Perfect Renewable Resource for Polymer Science?!. *Eur. Polym. J.* **2011,** *47*, 837–852.

3. Zakzeski, J.; Bruijnincx, P. C. A.; Jongerius, A. L.; Weckhuysen, B. M. The Catalytic Valorization of Lignin for the Production of Renewable Chemicals. *Chem.Rev.* **2010,** *110*, 3552–3599.
4. Huber, G. W.; Iborra, S.; Corma. Synthesis of Transportation Fuels from Biomass: Chemistry, Catalysts, and Engineering. A. *Chem. Rev.* **2006,** *106*, 4044–4098.
5. Demirbas, A. Biorefineries: Current Activities and Future Developments. *Energy Convers. Manage.* **2009,** *50*, 2782–2801.
6. Olah, G. A. Beyond Oil and Gas: The Methanol Economy. *Angew. Chem. Int. Ed.* **2005,** *44*, 2636–2639.
7. Biermann, U.; Bornscheuer, U.; Meier, M. A. R.; Metzger, J. O.; Schafer, H. J. Oils and Fats as Renewable Raw Materials in Chemistry. *Angew. Chem. Int. Ed.* **2011,** *50*, 3854–3871.
8. Pillai, C. K. S. Challenges for Natural Monomers and Polymers: Novel Design Strategies and Engineering to Develop Advanced Polymer. *Des. Monomers Polym.* **2010,** *13*, 87–121.
9. Voirin, C.; Caillol, S.; Sadavarte, N. V.; Tawade, B. V.; Boutevin, B.; Wadagaonkar, P. P. Functionalization of Cardanol: Towards Biobased Polymers and Additives. *Polym. Chem.* **2013.** DOI: 10.1039/c3py01194a.
10. Balachandran, V. S.; Jadhav, S. R.; Vemula, P. K.; John, G. Recent Advances in Cardanol Chemistry in A Nutshell: from a Nut to Nanomaterials. *Chem. Soc. Rev.* **2013,** *42*, 427–438.
11. Lubi, M. C.; Thachil, E. T. Cashew Nut Shell Liquid (CNSL)—A Versatile Monomer for Polymer Synthesis. *Des. Monomers Polym.* **2000,** *3*, 123–153.
12. Pillai, C. K. S. Liquid Crystalline Polymers: The Effects of Chain Disruptors. *Pure Appl. Chem.* **1998,** *70*, 1249–1252.
13. Gedam, P. H.; Sampathkumaran, P. S. Cashew Nut Shell Liquid: Extraction, Chemistry and Applications. *Prog. Org. Coat.* **1986,** *14*, 115–157.
14. John, G.; Vijai Shankar, B.; Jadhav, S. R.; Vemula, P. K. Biorefinery: A Design Tool for Molecular Gelators. *Langmuir* **2010,** *26*, 17843–17851.
15. Kozubek, A.; Tyman, J. H. P.; Atta-Ur, R. Bioactive Phenolic Lipids. *Stud. Nat. Prod. Chem.* **2005,** *30*, 111–190.
16. Raquez, J. M.; Deleglise, M.; Lacrampe, M. F.; Krawczak, P. Thermosetting (Bio) Materials Derived from Renewable Resources: A Critical Review. *Prog. Polym. Sci.* **2010,** *35*, 487–509.
17. Vasapollo, G.; Mele, G.; Del Sole, R. Cardanol-Based Materials as Natural Precursors for Olefin Metathesis. *Molecules* **2011,** *16*, 6871–6882.
18. Vemula, P. K.; John, G. Crops: A Green Approach toward Self-Assembled Soft Materials. *Acc. Chem. Res.* **2008,** *41*, 769–782.
19. Manjula, S.; Sudha, J. D.; Bera, S. C.; Pillai, C. K. S. Polymeric Resin from Renewable Resources: Studies on Polymerization of The Phenolic Component of Coconut Shell Tar. *J. Appl. Polym. Sci.* **1985,** *30*, 1767–1771.
20. Tyman, J. H. P. *Synthetic and Natural Phenols*; Elsevier: Amsterdam, 2008; Ch.13.
21. Manjula, S.; Pillai, C. K. S. Thermal Characterization of Cardanol-Formaldehyde Resins and Cardanol-Formaldehyde/Poly(Methyl Methacrylate) Semi-Interpenetrating Polymer Networks. *Thermochim. Acta* **1990,** *159*, 255–266.

22. Jain, R. K.; Sivala, K. Development of a Cashew Nut Sheller. *J. Food Eng.* **1997,** *32*, 339–345.
23. Barron, H. Use of CNSL resins in modern plastics, 2nd Eed.itionedn.; Chapman and Hall: London, 1949; pp 84
24. Antony, R., Pillai, C. K. S.; Scariah, K. J. GPC studies on The Cationic Polymerization of Cardanol Initiated by Borontrifluoridediethyletherate. *J. Appl. Polym. Sci.* **1990,** *41*, 1765–1775.
25. Manjula, S.; Kumar, V. G.; Pillai, C. K. S. Kinetics and Mechanism of Oligomerization of Cardanol Using Acid Catalysts. *J. Appl. Polym. Sci.* **1992,** *45*, 309–315.
26. Srivastava, R.; Srivastava, D. Studies on the Synthesis and Curing of Thermosetting Novolac Resin Using Renewable Resource Material. *Int. J. Chem. Tech. Res.* **2013,** *5*, 2575–2581.
27. Yadav, R.; Srivastava, D. Kinetics of the Acid-Catalyzed Cardanol–Formaldehyde Reactions. *Mater. Chem. Phys.* **2007,** *106*, 74–81.
28. Srivastava, R.; Srivastava, D. Utilization of Renewable Resources in the Synthesis of Novolac Polymers: Studies on its Structural and Curing Characteristics. *Int. J. Res. Rev. Eng. Sci. Technol.* **2013,** *2*, 22–25.
29. Tiwari, D.; Devi, A.; Chandra, R. Synthesis of Cardanol Based Phenolic Resin with Aid of Microwaves. *Int. J. Drug Dev. Res.* **2011,** *3*, 171–175.
30. Aggarwal, L. K.; Thapliyal, P. C.; Karade, S. R. Anticorrosive Properties of the Epoxy–Cardanol Resin Based Paints. *Prog. Org. Coat.* **2007,** *59*, 76–80.
31. Shukla, R.; Kumar, P. Self-Curable Epoxide Resins Based on Cardanol for Use in Surface Coatings. *Pigm. Resin Technol.* **2011,** *40*, 311–333.
32. Kanehashi, S.; Yokoyama, K.; Masuda, R.; Kidesaki, T.; Nagai, K.; Miyakoshi, T. Preparation and Characterization of Cardanol-Based Epoxy Resin for Coating at Room Temperature Curing. *J. Appl. Polym. Sci.* **2013,** *130*, 2468–2478.
33. Pal, N.; Srivastava, A.; Agrawal, S.; Rai, J. S. P. Kinetics and Mechanism of Esterification of Monoepoxies. *Mater. Manuf. Processes.* **2005,** *20*, 317–327.
34. Agarwal, S.; Mishra, A.; Rai, J. S. P. Effect of Diluents on the Curing Behavior of Vinyl Ester Resin. *J. Appl. Polym. Sci.* **2003,** *87*, 1948–1951.
35. Gaur, B.; Rai, J. S. P. Rheological and Thermal Behaviour of Vinyl Ester Resin. *Eur. Polym. J.* **1993,** *29*, 1149–1153.
36. Sultania, M.; Rai, J. S. P.; Srivastava, D. Kinetic Modeling of Esterification of Cardanol-Based Epoxy Resin in The Presence of Triphenylphosphine for Producing Vinyl Ester Resin: Mechanistic Rate Equation. *J. Appl. Polym. Sci.* **2010,** *118*, 1979–1989.
37. Sultania, M.; Rai, J. S. P.; Srivastava, D. Studies on the Synthesis and Curing of Epoxidized Novolac Vinyl Ester Resin from Renewable Resource Material. *Eur. Polym. J.* **2010,** *46*, 2019–2032.
38. Sultania, M.; Rai, J. S. P.; Srivastava, D. Modeling and Simulation of Curing Kinetics for The Cardanol-Based Vinyl Ester Resin by Means of Non-Isothermal Dsc Measurements. *Mater. Chem. Phys.* **2012,** *132*, 180–186.
39. Sultania, M.; Rai, J. S. P.; Srivastava, D. Process Modeling, Optimization and Analysis of Esterification Reaction of Cashew Nut Shell Liquid (CNSL)-Derived

Epoxy Resin using Response Surface Methodology. *J. Hazard. Mater.* **2011,** *185*, 1198–1204.
40. Huang, K.; Zhang, Y.; Li, M.; Lian, J.; Yang, X.; Xia, J. Preparation of A Light Color Cardanol-Based Curing Agent and Epoxy Resin Composite: Cure-Induced Phase Separation and its Effect on Properties. *Prog. Org. Coat.* **2012,** *74*, 240–247.
41. Pathak, S. K.; Rao, B. S. Structural Effect of Phenalkamines on Adhesive Viscoelastic and Thermal Properties of Epoxy Networks. *J. Appl. Polym. Sci.* **2006,** *102*, 4741–4748.
42. Bhunia, H. P.; Nando, G. B.; Basak, A.; Lenka, S.; Nayak, P. L. Synthesis and Characterization of Polymers from Cashew Nut Shell Liquid (CNSL), A Renewable Resource III. Synthesis of a Polyether. *Eur. Polym. J.* **1999,** *35*, 1713–1722.
43. Tan, T. T. M. Cardanol–Glycols and Cardanol–Glycol-Based Polyurethane Films. *J. Appl. Polym. Sci.* **1997,** *65*, 507–510.
44. Sathyalekshmi, K.; Gopalakrishnan, S. Synthesis and Characterisation of Rigid Polyurethanes Based on Hydroxyalkylated Cardanol Formaldehyde Resin. *Plast. Rubber Compos.* **2000,** *29*, 63–69.
45. Mythili, C. V.; Malar Retna, A.; Gopalakrishnan, S. Physical, Mechanical, and Thermal Properties of Polyurethanes Based on Hydroxyalkylated Cardanol–Formaldehyde Resins. *J. Appl. Polym. Sci.* **2005,** *98*, 284–288.
46. Gopalakrishnan, S.; Sujathaa, R. Synthesis and thermal properties of polyurethanes from Cardanol-Furfural Resin. *J. Chem. Pharm. Res.* **2010,** *2*(3), 193–205.
47. Suresh, K. I.; Kishanprasad, V. S. Synthesis, Structure, and Properties of Novel Polyols from Cardanol and Developed Polyurethanes. *Ind. Eng. Chem. Res.* **2005,** *44*, 4504–4512.
48. Darroman, E.; Durand, N.; Boutevin, B.; Caillol, S. New Cardanol/Sucrose Epoxy Blends for Biobased Coatings. *Prog. Org. Coat.* **2015,** *83*, 47–54.
49. Kathalewar, M.; Sabnis, A.; D'Mello, D. Isocyanate Free Polyurethanes from New Cnsl Based Bis-Cyclic Carbonate and its Application in Coatings. *Eur. Polym. J.* **2014,** *57*, 99–108.
50. Pillai, C. K. S.; Prasad, V. C.; Sudha, J. P.; Menon, A. R. R. Polymeric Resins from Renewable Resources. II. Synthesis and Characterization of Flame-Retardant Prepolymers from Cardanol. *J. Appl. Polym. Sci.* **1990,** *41*, 2487–2501.
51. Ragauskas, A. J.; Williams, C. K.; Davison, B. H.; Britovsek, G.; Cairney, J.; Eckert, C. A.; Murphy, R.; Templer, R.; Tschaplinski, T. The Path Forward for Biofuels and Biomaterials. *Science* **2006,** *3*(11), 484–489.
52. John, G.; Vemula, P. K. Design and Development of Soft Nanomaterials from Biobased Amphiphiles. *Soft Matter* **2006,** *2*, 909–914.
53. Corma, A.; Iborra, S.; Velty A. Chemical Routes for the Transformation of Biomass into Chemicals. *Chem. Rev.* **2007,** *107*, 2411–2502.
54. Huber, G. W.; Cheda, J. N.; Barret, C. J.; Dumesic, J. A. Production of Liquid Alkanes by Aqueous-Phase Processing of Biomass-Derived Carbohydrates. *Science* **2005,** *308*, 1446.
55. *Self-assembled Nanostructures*, Zhang. J. S., Wang, Z. L., Liu, J., Chen, S., Liu, G.-Y., Eds.; Kluwer Academic Publishers: New York, USA, 2003.

56. John, G.; Masuda, M.; Okada, Y.; Yase, K.; Shimizu, T. Nanotube Formation from Renewable Resources via Coiled Nanofibers. *Adv. Mater.* **2001,** *13*, 715–718.
57. John, G.; Jung, J. H.; Masuda, M.; Shimizu, T. Unsaturation Effect on Gelation Behavior of Aryl Glycolipids. *Langmuir* **2004,** *20*, 2060–2065.
58. Kamiya, S.; Minamikawa, H.; Jung, J. W.; Yang, B.; Masuda, M.; Shimizu T. Molecular Structure of Glucopyranosylamide Lipid and Nanotube Morphology. *Langmuir 21*, 743–750.
59. John, G.; Mason, M.; Ajayan, P. M.; Dordick, J. S. Lipid-Based Nanotubes as Functional Architectures with Embedded Fluorescence and Recognition Capabilities. *J. Am. Chem. Soc.* **2004,** *126*, 15012–15013.
60. John, G.; Jung, J. H.; Minamikawa, H.; Yoshida, K.; Shimizu, T. Morphological Control of Helical Solid Bilayers in High-Axial-Ratio Nanostructures through Binary Self-Assembly. *Chem. Eur. J.* **2002,** *8*, 5494–5500.
61. *Molecular Gels: Materials with Self-Assembled Fibrillar Networks*; Weiss, R. G., Terech, P., Eds.; Springer; Dordrecht: The Netherlands, 2006.
62. Sahoo, P.; Adarsh, N. N.; Chacko, G. E.; Raghavan, S. R.; Puranik, V. G.; Dastidar, P. Combinatorial Library of Primaryalkylammonium Dicarboxylate Gelators: A Supramolecular Synthon Approach. *Langmuir* **2009,** *25*, 8742–8750.
63. Dastidar, P.; Okabe, S.; Nakano, K.; Iida, K.; Miyata, M.; Tohnai, N.; Shibayama, M. Facile Syntheses of A Class Of Supramolecular Gelator Following A Combinatorial Library Approach: Dynamic Light Scattering and Small-Angle Neutron Scattering Studies. *Chem. Mater.* **2005,** *17*, 741–748.
64. John, G.; Minamikawa, H.; Masuda, M.; Shimizu, T. Liquid Crystalline Cardanyl β-D-Glucopyranosides. *Liq. Cryst.* **2003,** *30*, 747–749.
65. Reich, L.; Stivala, S. *Autoxidation of Hydrocarbons and Polyolefins*; Marcel Dekker: New York, 1969.
66. John, G.; Pillai, C. K. S. Self-Crosslinkable Monomer from Cardanol: Crosslinked Beads of Poly(Cardanyl Acrylate) by Suspension Polymerization. *Makromol. Chem. Rapid Commun.* **1992,** *13*, 255–259.
67. Barea, E.; Montoro, C.; Navarro, J. A. R. Toxic Gas Removal-Metal-Organic Frameworks and Vapours. *Chem. Soc. Rev.* **2014,** *43*, 5419–5430.
68. Bradshaw, D.; El-hankari, S.; Lupica-spagnolo, L. Porous Metal-Organic Frameworks. *Chem. Soc. Rev.* **2014,** *43*, 5431–5443.
69. Bradshaw, D.; Garai, A.; Huo. Metal-Organic Framework Growth at Functional Interfaces: Thin Films and Composites for Diverse Applications. *J. Chem. Soc. Rev.* **2012,** *41*, 2344–2381.
70. Furukawa, S.; Reboul, J. Structuring of Metal-Organic Frameworks at the Mesoscopic/Macroscopic Scale. *Chem. Soc. Rev.* **2014,** *43*(16), 5700–5734.
71. Imaz, I.; Rubio-Martnez, M.; Garcia-Fernandez, L.; Garcia, F.; Ruiz-Molina, D.; Hernando, J.; Puntes, V.; Maspoch, D. Coordination Polymer Particles as Potential Drug Delievery Systems. **2010,** *46*, 4737–4739.
72. Li, J. Coordination Polymer Particles for New Generation of Drug Delievery, 5-7, Seminar Lecture. Nov 5, 2013.

73. Chughtai, A. H.; Ahmad, N.; Younus, H. A.; Laypkov, A.; Verpoort, F. Metal-Organic Frameworks: Versatile Heterogeneous Catalysts for Efficient Catalytic Organic Transformations. *Chem. Soc. Rev.* **2015.** DOI: 10.1039/C4CS00395K.
74. Xia, W.; Mahmood, A.; Zou, R.; Xu, Q. Metal-Organic Frameworks and Their Derived Nanostructures for Electrochemical Energy Storage and Conversion. *Energy Environ. Sci.* **2015,** *8*, 1837–1866.
75. Li, S.; Huo, F.; Metal-Organic Framework Composites: from Fundamentals to Applications. *Nanoscale* **2015,** *7*, 7482–7501.
76. Hu, X.; Zhu, Z.; Cheng, F.; Tao. Z.; Chen, J. Micro-Nano Structured Ni-MOF_s as High-Performance Cathode Catalyst for Rechargeable Li-O_2 Batteries. *Nanoscale* **2015,** *7*, 11833–11840.
77. Chneemann, A.; Bon, V.; Schwedler, I.; Senkovska, I.; Kaskel, S.; Fischer, R. A. Flexible Metal-Organic Frameworks. *Chem. Soc. Rev.* **2014,** *43*, 6062–6096.
78. Sobana, C. I.; Christudhas, M.; Gnana, G. A. Synthesis, Characterization, Metal ion intake and Antibacterial Activity of Cardanol Based Polymeric Schiff Base Transition Metal Complexes using Ethylene Diamine. *J. Chem. Pharm. Res.* **2011,** *3*, 127–135.
79. Nagabhush, K. S.; Shobha, S. V.; Ravindranath, B. Selective Ionophoric Properties of Anacardic Acid. *J. Natural Prod.* **1995,** *58*, 807.
80. Tsujimotoa, K.; Hayashia, A.; Hab, T. J.; Kubob, I. Anacardic Acids and Ferric Ion Chelation. *Z. Naturforsch. C.* **2007,** *62*, 710–716.
81. Jiuhua, L.; Binghuan, B. H. Study on the Cardanol-Aldehyde Condensation Polymer Containing Boron-Nitrogen Coordinate Bond. *Chem. J. Polym. Sci.* **1998,** *16*, 219–225.
82. Hamad, F. B.; Mobofu, E. B.; Makame, Y. M. M. Wet Oxidation of Maleic Acid by Copper (II) Schiff Base Catalysts Prepared Using Cashew Nut Shell Liquid Templates. *Catal. Sci. Technol.* **2011,** *1*, 444–452.
83. Tang, S. Y.; Bourne, R. A.; Smith, R. L.; Poliakoff, M. The 24 Principles of Green Engineering and Green Chemistry: Improvements productively. *Green Chem.* **2008,** *10*, 268–269.
84. Zafar, F.; Zafar, H.; Sharmin, E.; Nishat, N. *Development of Agro Byproduct Based Coordination Polymers for Biomedical Application.* International symposium on Advances in Biological & Material Sciences, organized by Humboldt Academy Lucknow & University of Lucknow, 15th July, 2014.
85. Zafar, F.; Khan, S.; Zafar, H.; Sharmin, E.; Nishat, N. Studies on Bio-Resource Derived Ligand and Transition Metal Ions-Based Coordination Complexes/Polymers for Biomedical Application. APA 2014, International Conference organized by Asian Polymer Association (APA), New Delhi, India, February 19–21, 2014.
86. Zafar, F.; Zafar, H.; Sharmin, E.; Nishat, N. *Synthesis and Characterization of Biologically Active Coordination Complexes/Polymers from Agro- Byproduct.* POLYCHAR 22, World Forum on Advanced, Materials, organized by STIAS Research Centre, Stellenbosch, South Africa, April 7–11, 2014.
87. Zafar, F.; Zafar, H.; Sharmin, E.; Nishat, N. *CNSL Based Coordination Polymers as Drug-Carrier System.* 1st International conference on emerging trends of nanotechnology in drug discovery, organized by Sri Venkateswara College & Department of

Biochemistry, University of Delhi, India in association with Centro de Quimica da Madeira, University of Madeira, Portugal, India, May 26–27, 2014.

88. Zafar, F.; Zafar, H.; Sharmin, E.; Nishat, N. *Synthesis and Characterization of Porous Coordination Complex/Polymer from Cashew Nut Shell Liquid (CNSL): A Sustainable Development.* 3rd International Conference on Innovative Approach in Applied Physical, Mathematical/Statistical, Chemical Sciences and Emerging Energy Technology for Sustainable Development (APMSCSET-2014) organized by Social Welfare Foundatio in association with Krishi Sanskriti, JNU, Delhi, India, September 27–28, 2014.
89. Zafar, F.; Zafar, H.; Sharmin, E.; Nishat, N. *Synthesis, Characterization and Antimicrobial Activity of Cardanol-Metal Ions Coordination Complexes/Polymers.* Jointly organized by Indian Institute of Technology (IIT) Delhi, ENEA, Rome, Italy and National Research Council of Italy under the auspices of APA, India, October 27–30, 2014.
90. Zafar, F.; Zafar, H.; Sharmin, E.; Nishat, N. *Cashew Nut Shell Liquid Based Advanced Functional Materials.* International Conference on Natural Polymers, Bio-polymers, Bio-materials, Their Composites, Nanaocomposites, Blends, Ipns, Polyelectrolytes and Gels: Macro to Nanoscales (ICNP–2015) Organized by International Unit on Macromolecular Science and Engineering (IUMSE) Kottayam, Kerala, India & Wroclaw Uniwersity of Technology, Faculty of Electrical Engineering 27 Wybrzeze Wyspianskiego St 50-370 Wroclaw, Poland & Laboratório de Biopolímeros e Sensores/LaBioS, Centro de Tecnologia [UTF-8] Cidade Universitária, AV Horácio Macedo 2030, Bloco J, Brazil, at Kottayam, Kerala, India, April 10–12, 2015.
91. Zafar, F.; Kaur, B.; Khan, S.; Zafar, H.; Sharmin, E.; Nishat, N. Development of mesoporous nanocoordinated polyurethanes from CNSL: Formaldehyde and divalent transition metal ions. Communicated 2018.
92. Zafar, F.; Khan, S.; Prashant, Zafar, H.; Sharmin, E.; Nishat, N. Synthesis and Characterization of Nanocoordinated Polyurethanes from Agrobyproducts (CNSL & Furfuraldehyde) Derived Ligand and Divalent Transition Metal Ions. To be communicated 2018.
93. Khan, S.; Laxmi; Zafar, F.; Nishat, N. Development of Bio-derived Nanostructured Coordination Polymers Based on Cardanol-Formaldehyde Polyurethanes with 'd^5' Mn(II) and 'd^{10}' Zn(II) Metal Nodes: Synthesis, Characterization and Adsorption Behavior. *RSC Adv.* **2016,** *6*(55), 50070–50082.
94. Zafar, F.; Khan, S. Zafar, H.; Begum, M.; Zafar, H.; Sharmin, E.; Nishat, N. Synthesis and Characterization of Mg (II) Containing Nanocoordination Polymer Film Derived from Agrobyproduct (Cardanol and Furfuraldehyde). To be communicated 2018.
95. Zafar, F.; Khan, S.; Shabnam; Zafar, H.; Sharmin, E.; Nishat, N. Development of Mg (II) Containing Nanocoordination Polymer Film Based on CNSL-Formaldehyde and HMTA. To be communicated 2018.

CHAPTER 4

SYNTHESIS AND CHARACTERIZATION OF CASHEW NUT SHELL LIQUID MATRIX COMPOSITIONS FOR COMPOSITES APPLICATIONS

A. V. SAWANT[1], A. R. TAKALKAR[1,2], and K. PADMANABHAN[1,3]

[1]*Centre for Excellence in Nano-composites, School of Mechanical and Building Sciences, VIT University, Vellore 632014, India, E-mail: ajinkyavsawant7@gmail.com*

[2]*E-mail:takalkaradi@gmail.com*

[3]*Email: padmanabhan.k@vit.ac.in*

CONTENTS

ABSTRACT

Among all the other sources, cashew nut shell liquid (CNSL) is an abundantly available natural source for synthesizing phenolic compounds. Cardanol is an excellent monomer, which is isolated from CNSL for producing polymer. These are polymerized with aldehydes and acids in a particular mole fraction in the presence of catalysts so that it can be converted into resins. In the current work, formaldehyde with catalysts such as NaOH or NH_4OH and acids such as HNO_3 or H_2SO_4 were combined for resin preparation. For the thermal characterization of the synthesized CNSL resins, studies on differential scanning calorimetric and thermal gravimetric analysis were carried out. Characterization studies were also carried out with respect to the mechanical properties such as hardness and rigidity. Scientists are optimistic that the resin formulations with various tunable properties could be used for composites applications when acid- and alkali-based synthesis of the CNSL resins yielded many significant compositions with varied properties.

4.1 INTRODUCTION

Cashew nut shell liquid (CNSL), the byproduct of cashew industry, is a unique resource of unsaturated long chain phenolic distillate, which is cardanol.[1] The natural meta-substituted alkyl phenol can yield a series of phenolic resins when catalysed with aldehydes or acids (Fig. 4.1). Due to the phenolic structure of cardanol, it can be polymerized and suitably modified for various applications. Menon et al. studied and characterized the polymerization of cardanol.[2] Papadopoulou et al. successfully synthesized CNSL phenol-formaldehyde and studied its thermal properties and stability and also compared it with standard phenol formaldehydes.[3] Resins derived from CNSL are widely used in industries as friction materials, laminates, adhesives, surface coatings, flame retardants, anticorrosive paints and so forth.[4]

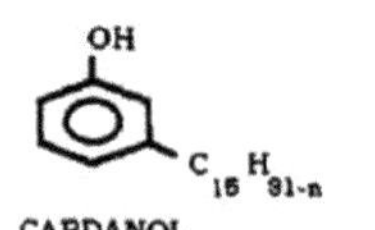

FIGURE 4.1 (a) Cardanol, major constituent of cashew nut shell liquid (CNSL), (b) structure of CNSL constituents.

Our main goal in the present investigation is to synthesize a superior CNSL matrix material for composite applications with the aid of thermal characterization of CNSL matrices, obtained with alkali and acid catalysis.

4.2 EXPERIMENTAL DETAILS

Cashew nut shell liquid was obtained from KS Chemicals, Cuddalore, India. At a specific temperature, the samples were heated for 2 h. By using four different alkali catalysis procedures, combinations of CNSL resin, listed in Table 4.1, were synthesized.

TABLE 4.1 Alkali Catalysed Cashew Nut Shell Liquid (CNSL) Resin.

Sample no	Material composition			Temp	Properties observed
	CNSL (g)	Formaldehyde (30% by weight) (g)	Catalyst (5% by weight)	Heat (°C)	
1	30	8	–	120	Tough, rigid, hard, good surface finish
2	30	8	NaOH	120	Flexible, moderate surface finish
3	30	8	NH_4OH	120	Moderate surface finish, higher hardness
4	30	8	15% NaOH	120	Moderate surface finish

In the acid catalysis route, 89 g of CNSL was mixed with 7 g of catalyst. This was stirred well and preheated to 180°C for an hour followed by 2-h curing at that temperature. Table 4.2 lists the details of acid catalysis.

TABLE 4.2 Acid Catalysed CNSL Resin.

Sample no	Material composition		Temp	Properties observed
	CNSL (grams by weight)	Catalyst (7 grams by weight)	Heat (°C)	
5	89	HNO_3	180	Foamy, soft and flexible
6	89	Conc. H_2SO_4	180	Tough and semirigid

4.3 RESULTS AND DISCUSSION

Endothermic and exothermic processes were presented by differential scanning calorimetry (DSC) curves. The DSC endothermic peaks near 250°C correspond to the mass loss that can be observed in thermal gravimetric analysis (TGA) curves as shown in Figures 4.2 and 4.3. Due to the increased heat capacity of the samples, glass transition causes endothermic shifts in the initial baseline. For all the six samples which are related to the thermal decomposition and degradation of the resin, exothermic peaks are observed around temperatures ranging from 400 to 430°C, as shown in the DSC and TGA graphs. Normally, DSC peaks are directly related to enthalpy changes in the samples. The parameters studied with the aid of TGA instrument are (a) CNSL decomposition, (b) peak temperature (T_{max}) for significant degradation and (c) residual mass at 850°C. These parameters give the thermal stability of CNSL resin.

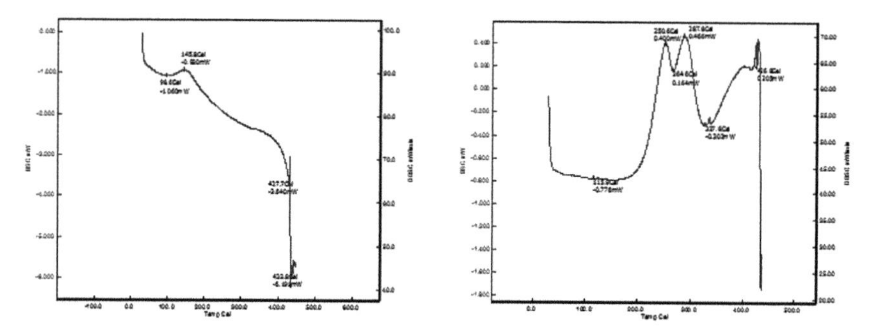

FIGURE 4.2 (a) Differential scanning calorimetry (DSC) for sample 1, (b) DSC for sample 2.

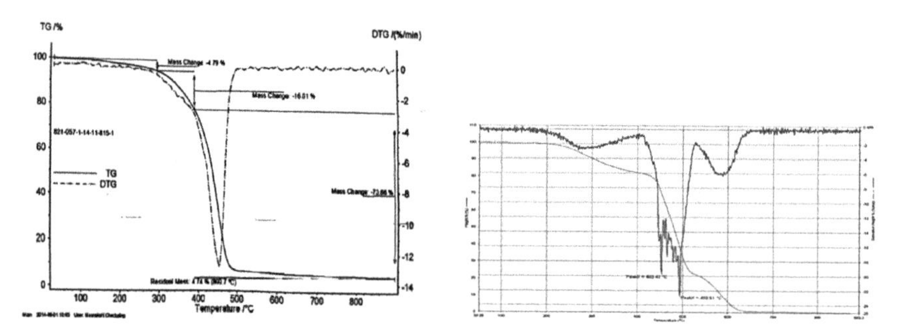

FIGURE 4.3 (a) Thermal gravimetric analysis (TGA) for Sample 1, (b) TGA for sample 5.

Table 4.3 and the curves of TGA analysis (Fig. 4.3a,b) reveal that CNSL resin decomposition is performed in three steps. In the first step, at 0–300°C, up to 5% mass loss was observed in the first 4 samples, whereas in samples 5 and 6, it is the maximum, that is 10%. This may be because of moisture removal that was retained in CNSL. In the second step, gradual weight loss occurs in the temperature, ranging from 300 to 450°C. The main cause behind this may be the degradation of the side chain and small fragments such as CH_3 and OH radicals. Cashew nut shell liquid is thermally stable at a temperature up to 450°C. In the third and last stage of thermal degradation, a weight loss of around 70% can be observed. Depolymerization and degradation of the CNSL matrix leads to this weight loss.

TABLE 4.3 Mass Change Versus Temperature Range for all Samples.

Sample no.	Mass change in % at			T_{max} in (°C)	Residual mass % at 850°C
	0–300°C	300–450°C	450–900°C		
1	4.79	16.81	73.66	452.3	4.74
2	3.61	17.14	79.24	490.06	2.3
3	5.92	21.85	72.18	460.13	1.32
4	3.58	23.93	69.22	450.9	3
5	9.34	18.13	72.51	493.51	1
6	9.92	17.44	72.62	488.13	4.3

Different physical properties are exhibited by different combinations of CNSL resin. Samples 1 and 2 can be cured with the aid of formaldehyde; this results in exhibiting the highest hardness values in both shore A and D durometer tests. Increasing the percentage of sodium hydroxide from 5 to 15 wt% causes a decrease in the hardness of the cured CNSL resin as shown in Table 4.4. Sample 5 is HNO_3 cured and is foamy, soft and flexible in appearance possessing the lowest hardness, whereas the HCl-cured sample exhibits considerable hardness. The resin becomes foamy and softer due to the formation of amines/amides with a high rate of reaction. On the other hand, chlorides seem to stabilize the chain. These resins may be chosen for composites applications depending on the strength, stiffness or toughness requirements. The DSC analysis was also carried out to measure the glass transition temperature, softening and degradation points and so forth. The glass transition temperatures are shown in Table 4.5. No noticeable difference is observed by the resins in glass transition temperatures. The resins synthesized are thus for low-end composite applications.

TABLE 4.4 Hardness Values.

Sample no.	Shore-A	Shore-D
Sample 1	97	50
Sample 2	85	43
Sample 3	60	18.4
Sample 4	71.5	38
Sample 5	28	Not possible
Sample 6	83	40

TABLE 4.5 Glass Transition Temperatures.

Sample no.	T_g
Sample 1	38.04
Sample 2	37.27
Sample 3	36.48
Sample 4	37.72
Sample 5	37.64
Sample 6	38.38

4.4 CONCLUSION

In this investigation, we could vividly draw a conclusion that the CNSL with formaldehyde curing have highest hardness (97), and the hardness can be varied with the aid of many catalysts. From the results of DSC and TGA, it is observed that the CNSL resins show good thermal stability up to 450°C. The glass transition temperature for the CNSL resins is around 38°C. The alkali-catalysed resins were found to be harder than the acid-catalysed ones. Thus, the thermal and physical properties were studied and reported.

4.5 ACKNOWLEDGEMENTS

The authors wish to acknowledge the DSC and TGA tests conducted at CIPET, Chennai.

KEYWORDS

- **cashew nut shell liquid**
- **hardness**
- **differential scanning calorimetry**
- **thermal gravimetric analysis**
- **catalysts**
- **CNSL resin**

REFERENCES

1. Gedam, P. H.; Sampathkumaran, P. S. Cashew Nut Shell Liquid: Extraction, Chemistry and Applications. *Prog. Org. Coat.* **1986,** *14,* 115–157.
2. Menon, A. R. R.; Sudha, J. D.; Pillai, C. K. S.; Mathew, A. G. *J. Sci. Ind. Res.* **1985,** *44,* 324.
3. Papadopoulou, E.; Chrysalis, K. Thermal Study of Phenol–Formaldehyde Resin Modified with Cashew Nut Shell Liquid. *Thermochim. Acta* **2011,** *512,* 105–109.
4. Lubi, M. C.; Thachill, E. B. *Des. Monomers Polym.* **2000,** *3,* 123.

CHAPTER 5

SYNTHETIC APPROACH FOR POLYURETHANE FROM RENEWABLE MATERIAL (CASHEW NUT HUSK TANNIN)

A. J. SUNIJA[1,*] and S. SIVA ILANGO[2]

[1]*Department of Chemistry, University College of Engineering Nagercoil, Anna University, Chennai, India, *E-mail: sunija_aj@yahoo.com*

[2]*Department of Chemistry, Thiagaraja College of Engineering, Madurai, India*

CONTENTS

5.1 INTRODUCTION

Cashew nut husk is a biomaterial waste from cashew industry and is an important source of tannin. In India, *Anacardium occidentale* (cashew) trees grow wild on the drier sandy soils and are cultivated in many parts of the world. India is the largest producer and processor of cashew in the world, with the total area in India under cashew cultivation being about

868,000 ha with annual production of 665,000 t, giving an average productivity of 860 kg/ha. India processes about 1,138,000 t of raw cashew nut seeds.[1] The husk of cashew is a rich source of tannins, a group of plant chemicals with documented biological activity. Estimates of tannin content in cashew nut husk run from 45 to 50%.[2] The global consumption of polyurethane (PU), a versatile polymer, had been growing at an average rate of over 7% annually for the past 15 years. Polyurethane has a world production of 14 Mt and is the sixth most widely sold plastic in the world. As per research and markets, the global market for PUs was estimated at 13,650.00 kt in 2010 and is expected to reach 17,946.20 kt by 2016.[3]

In the past few decades, the utilization of renewable resources in the synthesis of polymers has received considerable attention on account of their potential substitute for their petrochemical counterpart.[4] Polyurethanes are very versatile polymers that find applications in many fields due to their excellent properties. They are extremely versatile and are available in a variety of forms—ranging from flexible or rigid foams to elastomers, coatings, adhesives and sealants. Many researchers have reported the use of natural polymers having more than two hydroxyl groups per molecule either as a polyol or cross linker in the preparation of PU by allowing them to react efficiently with diisocyanates. Plant oil,[5] liquefied benzoylated wood,[6] lignin-based polyols,[7] soy flour,[8] *Acacia mearnsii* bark, pecan nut pith (*Carya illinoensis*), gambier leaf (*Uncaria gambir*) and various Pinus species barks[9] were successfully used as polyols in the preparation of PU. Jin-Jie synthesized rigid PUs by utilization of wattle tannin and *Acacia meamsii* bark as a partial replacement for synthetic polyols.[10] The current chapter focuses on preparation of PU from cashew nut husk and on their physical and chemical properties.

5.2 EXPERIMENTAL DETAILS

Tannin extracted from cashew nut husk is treated with hexamethylene diisocyanate (HDI) in the presence and absence of extender, 1,4-butanediol in a three-necked and round-bottomed flask to prepare PU.

5.2.1 EXTRACTION AND ANALYSIS OF TANNIN

The thin husk separated from the edible portion of cashew nut, obtained from the native plants of Kanyakumari District, Tamil Nadu, India, is used as a source for tannins. Cashew nut husk is dried in shade, powdered and sieved through 0.5 mm mesh.[2] 30 g of the powder is extracted by distillation with 200 ml of 70% aq. acetone at 60°C for 30 min.[11,12] The extract is then evaporated to become dryness and analysed by high performance liquid chromatography (HPLC) and Fourier-transform infrared (FTIR) spectroscopy. About 8 g of tannin is obtained from 30 g of husk. Figure 5.1 shows the images of cashew fruit with nut and cashew nut husk and tannin.

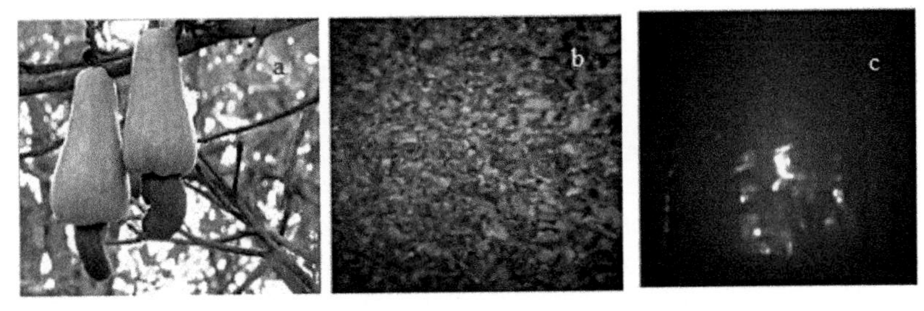

FIGURE 5.1 (a), (b) and (c) Images of cashew fruit with nut and cashew nut husk and tannin.

5.2.2 BIO-POLYURETHANE FROM TANNIN

The tannin extracted from cashew nut husk is used as a source for polyol. HDI (98+%, Alfa Aesar), 1,4-butanediol (99%, Spectrochem) and 1,4-Diazobicyclo (2,2,2) octane (98%, Alfa Aesar) are used as obtained.

5.2.2.1 SYNTHESIS OF POLYURETHANE, PU1

One part by weight of tannin is dissolved in six parts by weight of water and taken in a three-necked flask. HDI and catalyst 1,4-Diazobicyclo (2,2,2) octane (2 ml of 1% solution) are added simultaneously dropwise and the reaction mixture is stirred at a rate of 2500 rpm at 60°C and allowed to react for 8 min. Tannin and HDI are used at different proportions and temperature, but tannin/HDI ratio of 1:3 at 60°C is found to be exact for PU preparation with good consistency and reproducibility. All the other proportions resulted in products which are brittle in nature. The product obtained is quickly transferred to different moulds and allowed to cure at room temperature for 3–10 h. The yield of the product is 2 g.

5.2.2.2 SYNTHESIS OF POLYURETHANE, PU2

One part by weight of tannin is dissolved in six parts by weight of water and taken in a three-necked flask. Hexamethylene diisocyanate and catalyst 1,4-Diazobicyclo (2,2,2) octane (2 ml of 1% solution) are added simultaneously in the same ratio as in the preparation of PU1 and the reaction mixture is stirred at a rate of 2500 rpm at 60°C for 5 min. About 0.05 parts by weight of extender 1,4-butanediol is added drop wise to this reaction mixture and the reaction is allowed to continue for 8 min at 60°C. The PU so obtained is quickly transferred to a mould and allowed to cure at room temperature. The curing period is about 3–10 h. The yield of the product is 2.044 g.

5.2.3 MATERIALS AND METHODS

Infrared spectral analysis is carried out by using sixteen principal component FTIR spectrophotometer from PerkinElmer, controlled by Digital Equipment Corporation's 316-SX personal computer running infrared data manager software. The HPLC of the tannin sample is carried out by using Waters 600 series pump, C18 column, 4.6 × 150 mm, 725 Rheodyne injector; the mobile phase is 0.1% tetra fluoro acetic acid in distilled water 2487 UV detector at 220 and 254 nm. Approximately, 7 mg of sample is treated with 1.0 ml mobile phase. Injection volume is 20.00 μl and run time is 40 min. X-ray diffraction patterns are recorded by using a PANalytical X'Pert Pro Multipurpose diffractometer. The measurement conditions include 25 kV voltage and 30 mA current, angular scan speed 0.01330/s, step size 0.020 and count time 1.5 s per step. Scanning electron microscopy (SEM) is recorded by using S2400, HITACHI instrument with a secondary electron detector and an acceleration voltage of 80 KV.

According to the ASTM D570, water absorption test is conducted on PU sample by immersing the specimen in distilled water at room temperature for 25 days. The weight gained, M_t, as a result of water absorption is determined using Equation 5.1.

$$M_t = \frac{W_w - W_d}{W_d} \times 100 \tag{5.1}$$

where W_d and W_w represent the material dry weight and weight of material after being exposed to the water absorption at a period of time t, respectively.[13]

Swelling experiment is carried out by immersing the test samples PU1 and PU2 in solvents such as water, 1 N HCl, 1 N NaOH, ethylacetate, tetrahydrofuran (THF), dimethylsulphonoxide, benzene, toluene, ethanol and diesel. Periodically, these specimens are removed from the solvent and weighed after removal of the excess of solvent from the sample with the help of filter paper. Until constant weight (equilibrium swelling) is achieved, this is repeated. The average value of three parallel swelling experiments in every solvent for each sample is used. The swelling degree, q, is calculated using gravimetric method using the Equation 5.2.

$$q = \frac{m - m_0}{m_0} \quad (5.2)$$

where m and m_0 are the sample weights after and before swelling, respectively.[14]

Density is measured by 'floatation method' by using relative density bottle (Borosil R) by ASTM D1505 method. Density and specific gravity were measured by the following Equations (5.3) and (5.4):

$$\text{Density} = \frac{W_3 - W_1}{W_2 - W_1} \ \text{g/cm}^2 \quad (5.3)$$

Where
W_1 = Weight of empty bottle
W_2 = Weight of empty bottle + water
W_3 = Weight of empty bottle + solvent

$$\text{Specific gravity} = \text{Density of object/Density of water} \quad (5.4)$$

5.3 RESULTS AND DISCUSSION

5.3.1 ANALYSIS OF TANNIN

5.3.1.1 ANALYSIS BY HIGH PERFORMANCE LIQUID CHROMATOGRAPHY

The tannin obtained is analysed by HPLC. The chromatogram obtained by HPLC is as shown in Figure 5.2.

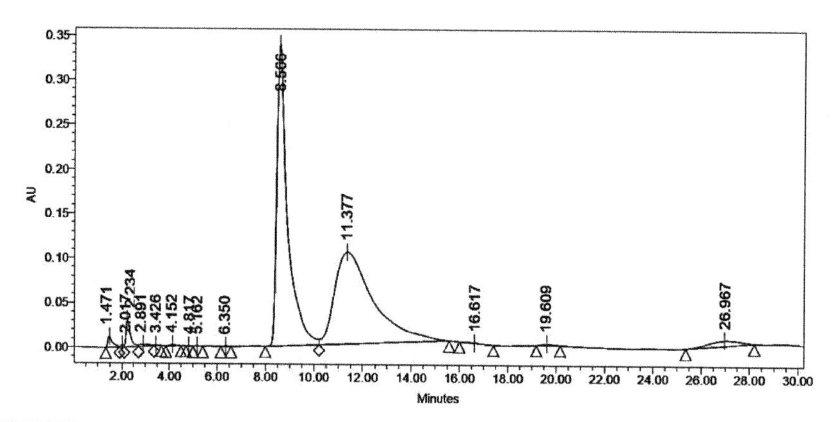

FIGURE 5.2 Chromatogram of tannin by high performance liquid chromatography.

The data when compared with the standard HPLC data[15] show that the most prominent component is catechin with retention time of 8.566 min, area 10,561,041 AU and height 338,863 mm. The other component has a retention time 11.377 min with area 11,429,480 AU and height 102,795 mm corresponding to epicatechin.

In addition, literature proved that cashew nut husk extract contains 80% tannin.[16]

5.3.1.2 FOURIER-TRANSFORM INFRARED SPECTRA OF TANNIN

Fourier-transform infrared spectra (Fig. 5.3) show characteristic absorption band at 3400 cm^{-1} which correspond to –O–H group (broad band due to H bonding). Bands at 1610 and 1518 cm^{-1} correspond to asymmetric stretching of single and double bond of aromatic compound at 1296 cm^{-1} for –C–H stretching (–C–H– linkage), at 1284 cm^{-1} due to –C–O–C– asymmetric stretching, 1448 cm^{-1} for aromatic ring stretching, at 1356 cm^{-1} correspond to –C–H deformation (–C–H– linkage), at 1236 cm^{-1} correspond to –C–H in plane bending, at 1143 cm^{-1} correspond to ring stretching (aromatic linkage) and at 1035 cm^{-1} due to –C–H bending (aromatic in-plane H bonding).[17,18]

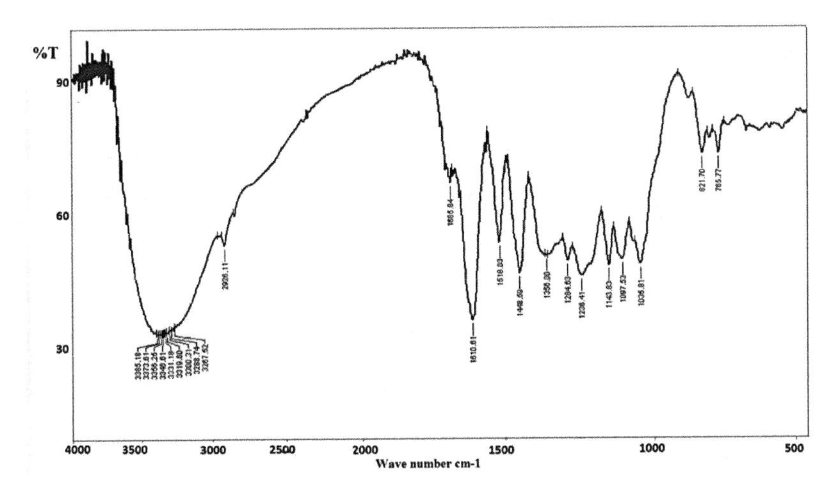

FIGURE 5.3 Fourier-transform infrared (FTIR) spectra of Tannin.

In the present study and literature, the HPLC and FTIR data[16,18] suggest that the phenolic compounds in extract of cashew nut husk were clearly identified as (+)-catechin and (−)-epicatechin. Thus, the structure of tannin in cashew nut husk could contain catechin B ring.

5.3.2 CHARACTERIZATION OF POLYURETHANE

5.3.2.1 FOURIER-TRANSFORM INFRARED SPECTRAL STUDIES OF POLYURETHANE

Fourier-transform infrared spectrum of samples PU1 and PU2 in Figure 5.4 shows characteristic stretching broad absorption band at 3300 cm^{-1} due to aromatic hydroxyl group (O–H with H bonds) and also to –N–H stretching frequency of secondary amide of urethane linkage. Bands at 2853 cm^{-1} for PU1 and 2841 cm^{-1} for PU2 correspond to N=C=O (asymmetric stretching) of urethane. Bands at 1625 cm^{-1} for PU1 and 1637 cm^{-1} for PU2 correspond to C=O (stretching) in urethane. Bands at 1330 and 1323 cm^{-1} correspond to –C–N (stretching) for PU1 and PU2, respectively. Weak band at 734 cm^{-1} occurs for both PU1 and PU2 and corresponds to methylene rocking vibration for –C–H in $(CH_2)_6$ groups. There is not much variation in the spectra of two samples except in their intensity of the bands, which is due to the presence of additional methylene and hydroxyl group of extender in PU2.[10,18]

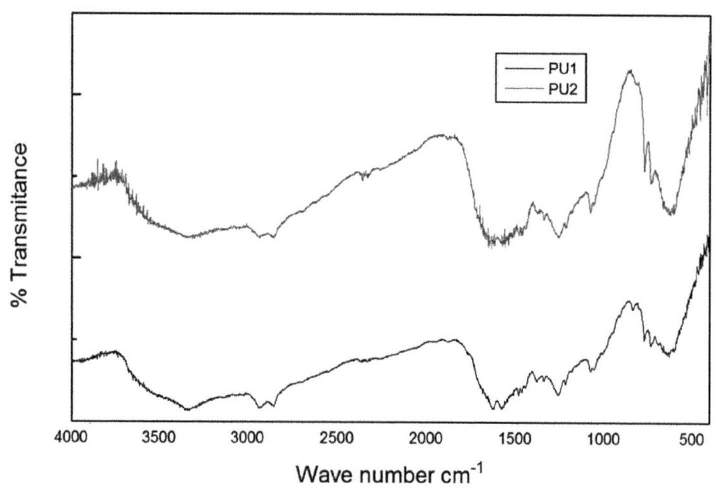

FIGURE 5.4 FTIR of PU1 and PU2.

From the spectral data and previous studies,[10,16] the expected structure for the prepared PU PU1 and PU2 are shown in Schemes 5.1 and 5.2.[19]Polyurethane formed may exist in two tautomeric forms I and II. Isocyanate substitution does not take place at the A ring as this substitution may result in highly unstable product due to steric hindrance and also due to lower electron density at the A ring than in B ring.[10]

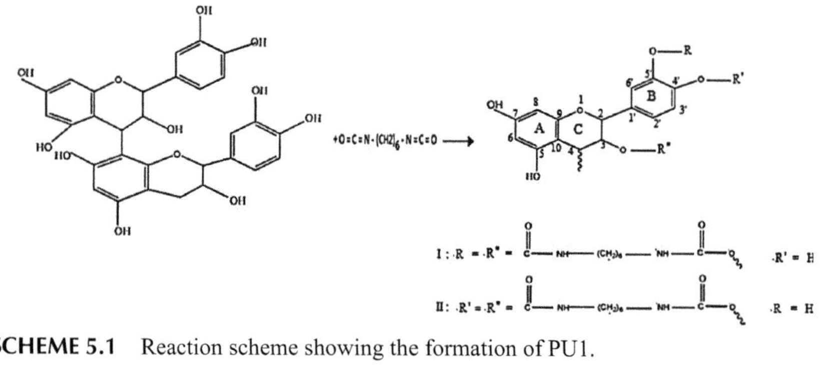

SCHEME 5.1 Reaction scheme showing the formation of PU1.

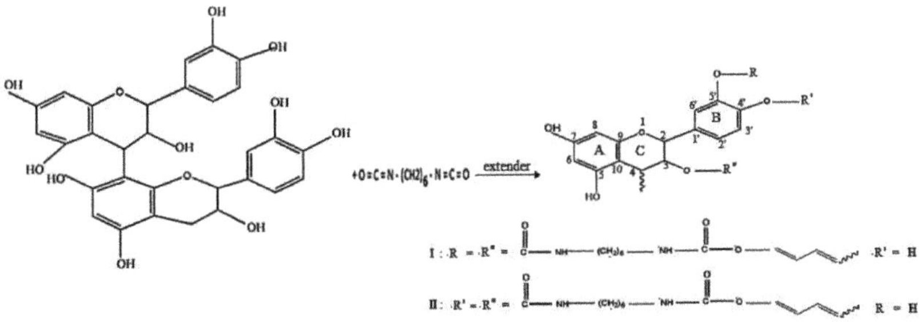

SCHEME 5.2 Reaction scheme showing the formation of PU2.

5.3.2.2 X-RAY DIFFRACTION STUDIES

It can be considered that a polymer is partly crystalline and partly amorphous. The crystalline domains act as a reinforcing grid, such as a composite material, and improve the performance over a wide range of temperature. In amorphous polymers, there is lack of periodicity in structure, and the resulting X-ray scattering curve shows one or two broad maxima called hump. The X-ray powder diffraction images of PU1 and PU2 are shown in the Figure 5.5 which shows the presence of broad humps implying that the polymer samples are more amorphous in nature.

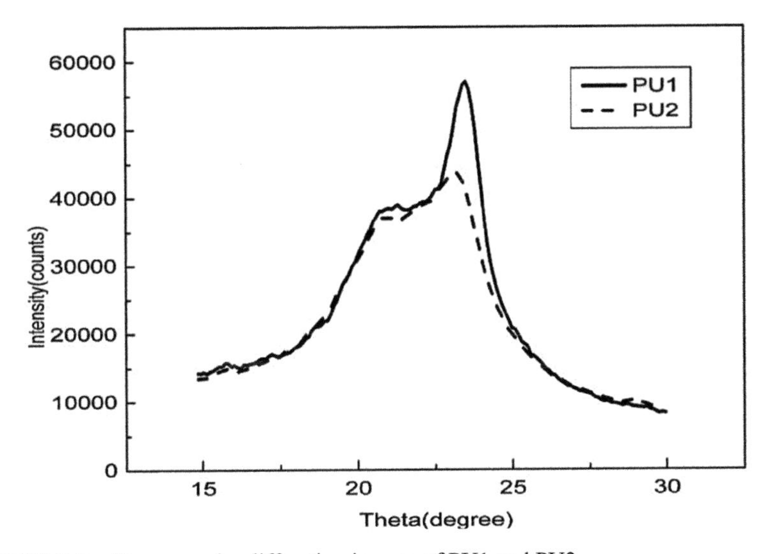

FIGURE 5.5 X-ray powder diffraction images of PU1 and PU2.

5.3.2.3 MORPHOLOGICAL STUDIES

The SEM photographs of representative areas of the samples, PU1 and PU2, are taken at different magnifications and the best images are represented in Figure 5.6.

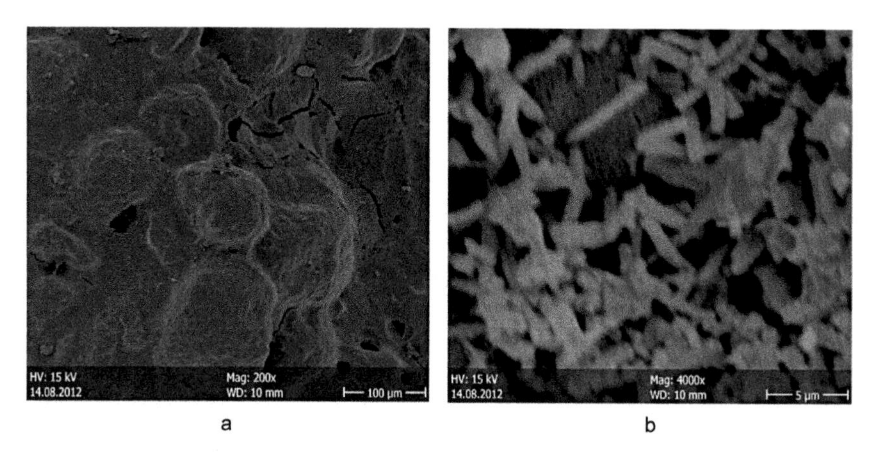

FIGURE 5.6 Scanning electron microscopy images of (a) PU1 and (b) PU2.

Typically, the morphology of PUs consists of many hard segments with size of domains ranging from ten to hundreds of angstroms, dispersed in the matrix of soft segments.[20] The primary driving force for hard domain formation is the strong intermolecular interaction between the urethane units that are capable of forming inter-urethane hydrogen bonds. In phase separation, the hard structure plays an important role. The symmetrical HDI and the methylene group of 1,4-butanediol act as soft segment.[21] Scanning electron microscopy image of PU1 (Fig. 5.6(a)) shows that no distinct hard domains which is due to the presence of less flexible soft segments (–CH_2– linkage).

Chain extender components and their properties also determine the intermolecular interactions as well as the morphology of PUs. Figure 5.6(b) shows that in PU2, the fibrillar segments of hard domains are very distinct and percolate through the soft segment matrix of the soft domains which include the methylene linkages of 1,4-butanediol and HDI.

5.3.2.4 WATER ABSORPTION STUDIES

It is revealed by water absorption studies that the samples are sensitive to moisture. It is observed that the water uptake in PU1 and PU2 samples reach the saturation stage in 25 days when equilibrium is achieved. When all the micro-voids in the 3-dimensional cross-linked network structure were filled with water molecules, equilibrium was attained.[22] Water uptake is attributed to the presence of free volume and hydrophilic functional groups such as hydroxyl groups. The water molecules absorbed may exist as bound water or unbounded clusters. Previous studies demonstrated that polymeric systems with higher cross-link density possessed lower free volume and lower water uptake.[23] The percentage water absorption is shown in Figure 5.7, which reveals that the presence of extender in PU2 decreases the cross-linking density which increases the percentage of absorption.

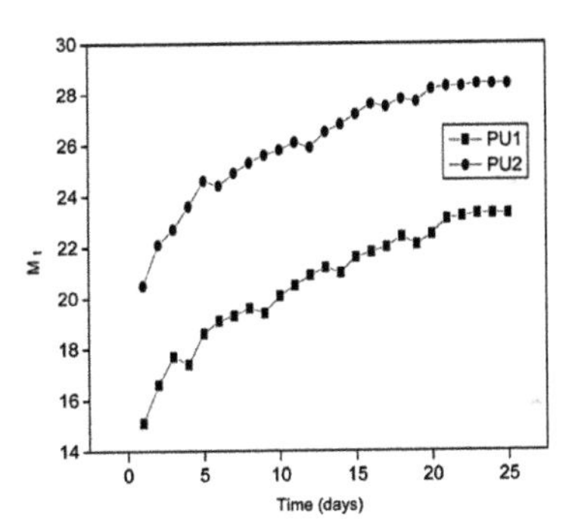

FIGURE 5.7 Plot of variation of percentage water absorption for PU1 and PU2 with time.

5.3.2.5 SOLUBILITY STUDIES

By dissolving known weight of previously dried and weighed specimen in different solvents, solubility studies were carried out. Visual examination on the solubility of PU1 and PU2 shows that the samples had good chemical resistance and are insoluble in most solvents used in the study. Very partial solubility was observed with 1 N NaOH and THF. The solubility of PU1 and PU2 are shown in Table 5.1.

TABLE 5.1 Solubility of PU1 and PU2 in Various Solvents after 25 Days of Immersion.

Sample code	Solvent									
	Water	1 N HCl	1 N NaOH	Tetrahydrofuran	Dimethylsulphon-oxide	Ethyl acetate	Benzene	Toluene	Ethanol	Diesel
PU1	–	–	+–	+–	–	–	–	–	–	–+
PU2	–	–	+–	+–	–	–	–	–	–	–+

Soluble (+); partially soluble (+–); insoluble (–)

The free energy change, during dissolution of a polymer in a solvent, is related to the change in enthalpy and change in entropy by the following thermodynamic Equation 5.5:

$$\Delta G = \Delta H - T\Delta S \tag{5.5}$$

Where ΔG = free energy change
ΔH = change in enthalpy
ΔS = change in entropy
T = temperature

If ΔH is positive and greater than $T\Delta S$, dissolution is not possible and the polymer remains unaffected by the solvent.[24]

The existence of this condition is the responsible factor that makes PU1 and PU2 insoluble in most solvents under study. Further, the solubility parameter between the polymer and solvents may be much different, which is another contributing factor for the lack of dissolution of the polymer with different solvents under study.

5.3.2.6 *SWELLING STUDIES*

The polymer imbibes the solvent and undergoes swelling to an extent determined by the nature of the polymer and the solvent when a cross-linked polymer is immersed in an appropriate solvent. Results obtained on swelling studies are presented in Figures 5.8(a,b) which show that equilibrium is obtained in 25 days.

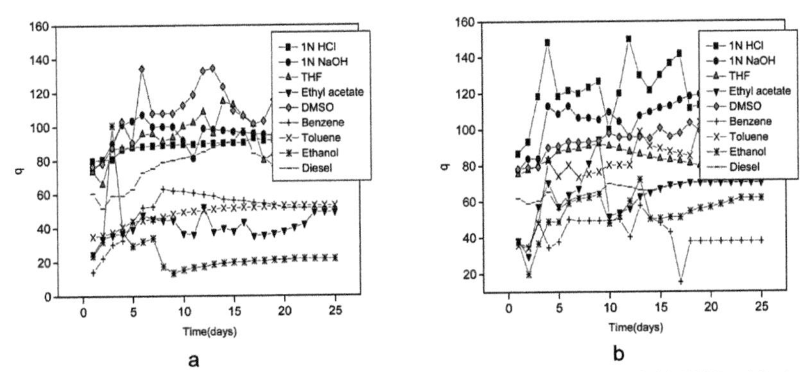

FIGURE 5.8 Plot of variation of percentage swelling of (a) PU1 and (b) PU2 with time.

The random variation in q values before attaining equilibrium is due to the extraction of the sol fraction. PU1 has less swelling property due to its high cross-linking.[25] Figure 5.3 also proves that the polymer samples show greater extent of swelling in nonpolar solvents compared to polar solvents.

5.3.2.7 DENSITY AND SPECIFIC GRAVITY

The production cost and profitability of the manufacturing process is affected by the significant factors such as density and specific gravity.

A reduction in the density reduces costs of raw materials and correspondingly, cost of manufacturing.[26] Table 5.2 presents the obtained density and specific gravity data which shows that PU1 is less dense than PU2. The difference in density was due to the fact that the cross-linking density in PU1 is greater than in PU2.

TABLE 5.2 Density and Specific Gravity of PU1 and PU2.

Sample code	Density (g/cm³)	Specific gravity
PU1	0.6970	0.6997
PU2	0.8699	0.8734

5.4 CONCLUSION

It is revealed by this study that cashew nut husk can be successfully used as a source for polyol in the PU industry and can be a potential substitute for their commercial petrochemical counterpart. India processes about 1,138,000 t of raw cashew nut seeds, out of which 3% of its weight is husk. Thus, the usage of renewable industrial waste in the production of PU will prove to be more economical. The bio-based PU obtained from cashew nut husk has low density, excellent chemical resistance and is insoluble in most solvents, showing its promising applications for exteriors in construction, automobiles, encapsulates and in packaging. But the major drawback being, its sensitivity to moisture. In future, work can be carried out to improve their water absorption properties so that they will provide excellent economic performance for acceptance in large quantity market and become potential counterparts to the commercially existing PUs.

5.5 ACKNOWLEDGEMENT

I thank the authorities of SCTIMST Thiruvananthapuram for providing me laboratory facilities for sample testing.

REFERENCES

1. Mohod, A.; Jain, S.; Powar, A. G. Cashew Nut Shell Waste: Availability in Small-Scale Cashew Processing Industries and Its Fuel Properties for Gasification. **2011**, ISRN Renewable Energy, 4.
2. Lokeswari, N.; Raju, K. J.; Pola, S.; Bobbarala, V. Extraction of Tannins from *Anarcardium occidentale* and Effect of Physical Parameters. *J. Pharm. Res.* **2010**, *3*(4), 90–908.
3. Plastics.com, **2011**. http://www.sdplastics.com/ polyuret.html [accessed July 13, 2011].
4. Lincoln, J. D.; Shapiro, A. A.; Earthman, J. C; Saphores, J. D. M.; Ogunseitan, O. A. Design and Evaluation of Bioepoxy-Flax Composites for Printed Circuit Boards. *IEEE Trans. Electron. Packag. Manuf.* **2008**, *31*(3), 211–220.
5. Lligadas, G.; Ronda, J. C.; Galia, M.; Cádiz, V. Plant Oils as Platform Chemicals for Polyurethane Synthesis: Current State-of-the- Art. *Biomacromolecules* **2010**, *11*(11), 2825–2835.
6. Wei, Y. P.; Cheng, F.; Ping, L. I. H.; Gao, Y. U. J. Synthesis and Properties of Polyurethane Resins from Liquefied Benzylated Wood. *Chin. Chem. Lett.* **2005**, *16*(3), 401–404.
7. Cateto, C.; Barreiro, F.; Rodrigues, A.; Belgacem, N. CP1042, International Conference on Times of Polymers (TOP) and Composites,© 2008 American Institute of Physics.

8. Raquez, J. M.; Deleglise, M.; Lacrampe, F.; Krawczak, P. Thermosetting (bio) Materials Derived from Renewable Resources: A Critical Review. *Prog. Polym. Sci.* **2010,** *35*, 487–509.
9. Pizzi, A. Condensed Tannins for Adhesives. *Ind. Eng. Chem. Prod. Res. Dev.* **1982,** *21*, 359–369.
10. Jin-Jie, G. E. Synthesis of Biodegradable Polyurethane Foams from Condensed Tannin and Bark of *Acacia mearnsii. Bull. Kyushu Univ.* **1998,** *79*, 21–85.
11. Entelis, S. G.; Evreinov, V. V.; Kuzaev, A. I. *Reactive Oligomers*. Brill Publishers: Moscow, Russia, 1988; pp 100, 129.
12. Manjula, K. S.; SatheeshKumar, M. N.; Soare, B. G.; Picciani, P.; Siddaramaiah, P. Biobased Chain Extended Polyurethane and Its Composites With Silk Fiber. *Polym. Eng. Sci.* **2010,** *50*(4)851–856.
13. Shah, V. *Hand book of Polymer Technology*. Inter Science Wiley publications: New Jersy, 1954.
14. Džunuzov, J. V.; Pergal, M. V.; Jovanovi, S.; Vodnik, V. V. Synthesis and Swelling Behavior of Polyurethane Networks Based On Hyperbranched Polymer. *HemijskaIndustrija* **2011,** *65*, 637–644.
15. Grace, W. R. Chromatogram Database, Catechins and Gallic Acid in Green Tea: Grace Davison Discovery science, 2010. www.Discoverysciences.com (accessed: 11 January 2011).
16. Kumar, K. P. V.; Sethuraman, M. G. Studies on Oleoresinous Arnishes and Their Natural Precursors. *Prog. Org. Coat.* **2004,** *49*, 244–251.
17. Kalsi, P. S. Spectroscopy of Organic Compounds. 6th Edition.; New age International publishers: New Delhi, **2010;** pp 65–410.
18. Mythili, C. V.; Retna, A. M.; Gopalakrishnan, S. Synthesis, Mechanical, Thermal and Chemical Properties of Polyurethanes Based on Cardanol. *Bull. Mater. Sci.* **2004,** *27*(3)235–241.
19. Trox, J.; Vadivel, V.; Vetter, W.; Stuetz, W.; Kammerer, R. D.; Carle, R.; Scherbaum, V.; Golaa, U.; Nohr, D.; Biesalski, H. K. Catechin and Epicatechin in Testa and Their Association with Bioactive Compounds in Kernels of Cashew Nut (*Anacardium occidentale* L.). *Food Chem.* **2011,** *128*, 1094–1099.
20. Pizzi, A. Tannin Based Polyurethane Adhesive. *J. Appl. Polym. Sci.* **1979,** *23*, 1889–1891.
21. Sunija, A. J.; Siva Ilango, S.; Vinod Kumar, K. P. Synthesis and Characterization Of Bio-Based Polyurethane From Benzoylated Cashew Nut Husk Tannins. *Bull. Mater. Sci.* **2014,** *37*(3)735–741.
22. Pistor, V.; Conto, D. D.; Ornaghi, F. G.; Zattera, A. J. Microstructure and Crystallization Kinetics of Polyurethane Thermopastics Containing Trisilanol Isobutyl POSS. *J. Nanomater.* **2012,** 1–8.
23. Tan, S. G.; Chow, W. S. Thermal Properties, Curing Characteristics and Water Absorption of Soybean Oil-Based Thermoset. *Express Polym. Lett.* **2011,** *5*(6)480–492.
24. Saijun, D.; Nakason, C.; Kaesaman, A.; Klinpituksa, P. Water Absorption and Mechanical Properties of Water-Swellable Natural Rubber. *Songklanakarin J. Sci. Technol.* **2009,** *31*, 561–565.

25. Gowarikar, V. R.; Vishwanathan, N. V.; Jayadev Sreedhar. *Polymer Chemistry*; New Age International Publishers Limited: New Delhi, 2005; pp 332–336.
26. Barikani, M.; Hepburn, C. V. Determination of Cross Linked Density of Swelling in Castable Polyurethane Elastomer Based on ¼- Cyclohexane Diisocyanate and Para-Henylenediisocyanate. *Iran. J. Polym. Sci. Technol.* **1992,** *1*(1)1–5.
27. Kažys, R.; Rekuvien, R. Viscosity and Density Measurement Methods for Polymer Melts. *Ultragarsas (Ultrasound)* **2011,** *66*(4)20–25.

CHAPTER 6

RECYCLING OF TEXTILE MILL (CELLULOSIC) WASTE INTO CARBOXYMETHYL CELLULOSE FOR TEXTILE PRINTING

JAVED SHEIKH[1,*] and V. D. GOTMARE[1]

[1]*Department of Textile Technology, Indian Institute of Technology (IIT), Delhi, India,* *E-mail: javedtextech@gmail.com*

[2]*Department of Textile Manufactures, V.J.T.I., Mumbai, India*

CONTENTS

ABSTRACT

In the current investigation, the waste of textile mill (cotton fibre waste) was converted into carboxymethyl cellulose (CMC) and then was utilized as a thickener in printing of cotton using vat dyes. The rheology of the thickener, which plays an important role in printing, was also studied in comparison with the commercial CMC that is available in market. The prints were analysed by measuring colour value, bending length and fastness to washing, crocking and light. Results suggest that CMC obtained by recycling of mill waste can substitute the sound CMC as a thickener in printing.

6.1 INTRODUCTION

Large quantity of cotton-fibre waste is generated on a daily basis in cotton-spinning mills. This waste can be easily converted into carboxymethyl cellulose (CMC) which in turn finds various applications in textiles. Carboxymethyl cellulose is widely used in sizing and printing of textiles. The printing of textile is a process of colouring textiles, and it is used to produce coloured patterns with sharp boundaries on textile materials. As thickeners provide viscosity to the paste, they play an important role in printing. Moreover, the quality of final prints is influenced by the rheology of thickener paste as well as availability of various desirable properties in thickeners. Many researchers studied the important aspects of rheology of thickeners.[1–4] Conversion of cellulosic waste into CMC was reported by some of the researchers.[5–7] Conversion of terry towel waste into CMC and its application in textile printing was reported from our laboratory.[8] Rheology of thickeners plays a very important role in textile printing as it decides the suitability of thickener, amount of paste applied on textile material and quality of prints. Rheology study of thickeners, extracted from germinated grains, was also reported from our laboratory.[3,9] In the current work, cotton waste was converted into CMC, and the efficacy of the recycled product was tested as a thickener in textile printing.

6.2 MATERIALS AND METHODS

6.2.1 MATERIALS

The waste of cotton fibre was obtained from spinning department of our institute, and 100% cotton fabric was used for printing (gsm-128). All chemicals used were of laboratory grade. Vat dyes used were supplied by Atul (P) Ltd.

6.2.2 *METHODS*

6.2.2.1 *SYNTHESIS OF CMC AND CHARACTERIZATION*

Cotton-fibre waste was finely cut to get fibrous mass. Carboxymethyl cellulose was synthesized following the procedure described in literature.[10,11] The CMC synthesized by the laboratory was analysed for the CMC content[12] and degree of substitution.[13] The Fourier transform infrared (FTIR) spectra of laboratory-synthesized and commercial CMC samples were recorded using an FTIR spectrophotometer (Shimadzu 8400s, Japan).

6.2.2.2 *RHEOLOGY STUDY OF CMC IN COMPARISON WITH COMMERCIAL CMC*

A standard CMC thickener (5%) was prepared by combining five parts of CMC and 95 parts of water. In the case of laboratory-synthesized CMC, paste of 4.5, 5 and 5.5% were made for comparison. Different factors affecting the viscosity of these thickener pastes, such as shear rate, solid content of thickener and time of shearing (shear sensitivity), were studied using a 'Brookfield Synchrolectric Viscometer (Model RV)'.

6.2.2.3 *PRINTING OF COTTON USING VAT DYES*

Printing of vat dyes was done by pre-reduction method—(Potash Rongalite Method).

6.2.2.4 *EVALUATION OF PRINTED FABRICS*

The printed samples were evaluated by reflectance method for the depth of colour by using a 10-degree observer. The absorbance of the dyed samples was measured on Spectraflash SF 300 (Datacolor International, U.S.A.). Bending length of printed samples, which is inversely proportional to the softness of the prints, was determined using a Shirley stiffness tester.[14] The test for colour fastness to washing was carried out using ISO IV methods.[15] The printed samples were tested for dry and wet crocking (rubbing) using a 'crock-meter' with 50 strokes of crocking. This test was conducted for colour fastness to light according to ISO 105/B02.[16]

6.3 RESULTS AND DISCUSSION

The laboratory-synthesized CMC (CMC content—85.10%, DS—2.81) was characterized to validate its preparation. The FTIR spectrum of the CMC (Fig. 6.1) when compared with the waste of the cotton fibre clearly indicates the introduction of peaks at 1419 and 1591 cm^{-1} which is due to carboxymethylation of cellulose backbone.

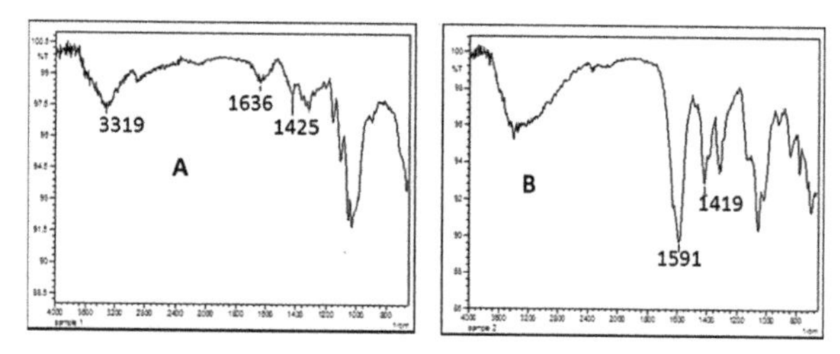

FIGURE 6.1 Fourier transform infrared spectroscopy spectra of cotton fibre waste (a) and recycled carboxymethyl cellulose (b).

6.3.1 *RHEOLOGY OF CMC AS A THICKENER AND PRINTING*

Rheology, the study of flow, of thickener paste was studied by measuring the effect of solid content of thickener, shear rate, shearing time on the viscosity of paste. The initial attempt was to vary solid content to get printable viscosity of the thickener paste. Commercial CMC showed printable viscosity at 5% solid content (5 g powder dissolved in 95 g of water). The solid contents of lab-made CMC usually varied from 4.5 to 5.5%, and the viscosity was measured at a wide range of shear rates (Table 6.1).

TABLE 6.1 Effect of Solid Content on Viscosity of Commercial and Lab-Made Carboxymethyl Cellulose (CMC) at Different Shear Rates.

Shear rate (RPM)	Viscosities of CMC paste			
	Commercial (C) (5%)	Laboratory synthesized (L)		
		4.5%	5%	5.5%
5	650	520	650	665
10	375	295	375	380
20	217.5	160	212.5	220
50	112	78	108	116
100	68	54	62	72

At all shear rates that were studied, a general increase in viscosity of paste occurred with the increase in solid content. This is obvious as the thickener being viscosity building material, higher resistance is offered to spindle movement at higher solid content. Results obtained in Table 6.1 showed equivalent viscosities of commercial and lab-made CMC at 5% solid content. The results also indicate a decrease in viscosity of paste (both commercial and lab-made) with an increase in shear rate. Such flow behaviour of thickener paste is known as non-Newtonian, pseudoplastic or shear thinning. For the transfer of paste through screen to the surface of fabric under the action of squeegee (pressure), this is very essential. However, the flow should be restricted in absence of pressure (during stoppage of machine) to prevent staining on fabric.

To compare the flow behaviour of commercial and lab-made CMC at equivalent solid content (5%), the log-viscosity was plotted versus log shear rate. The shear thinning index (STI) values which are calculated from graph and the K-values that is viscosity at 1 rpm give an indication of flow property. This is clearly demonstrated in Table 6.2. The STI values of commercial and lab-made CMC were found to be 0.75 and 0.79, respectively. In general, higher STI values indicate higher shear thinning behaviour which in turn is expected to give more transfer of paste to surface of fabric and hence higher colour values. K-values of commercial and lab-made CMC were similar.

TABLE 6.2 Shear-Thinning Index (STI) and K-Values for Laboratory Synthesized and Commercial CMC.

	STI and K-Values of CMC paste			
	Commercial (C) (5%)	**Laboratory synthesized (L)**		
		4.5%	**5%**	**5.5%**
STI	0.75	0.74	0.79	0.73
K[a]	2183.33	1950	2200	2316.67

[a]average value of three readings

The effect of shearing time on viscosity of CMC pastes (commercial and lab made 5%) was studied, and the entire effect according to the timings has been summarized in Tables 6.3 and 6.4.

TABLE 6.3 Effect of Shearing Time on Viscosity of Commercial CMC at Different Shear Rates for 5% Solid Content.

Shear rate (RPM)	Viscosity at different shearing time (poise)						
	1 min	2 min	3 min	4 min	5 min	8 min	10 min
5.0	660	650	650	650	650	650	650
10.0	375	375	375	375	375	375	370
20.0	217.5	217.5	217.5	217.5	217.5	217.5	215
50.0	112	112	112	112	112	110	110
100.0	68	66	66	66	66	66	66

TABLE 6.4 Effect of Shearing Time on Viscosity of Lab-Made CMC at Different Shear Rates for 5% Solid Content.

Shear rate (RPM)	Viscosity at different shearing time (poise)						
	1 min	2 min	3 min	4 min	5 min	8 min	10 min
5.0	650	645	645	645	645	645	645
10.0	375	375	375	370	365	365	365
20.0	215	212.5	212.5	212.5	212.5	212.5	212.5
50.0	108	106	106	106	106	106	106
100.0	62	62	62	62	62	62	62

Once the paste is sheared at a constant shear rate, an initial decrease in viscosity occurred followed by levelling of viscosity. This might be attributed to the initial displacement of polymer chain molecules from one plane to the other, breaking the intermolecular H-bonds which in turn bring down the viscosity. However, when such an operation is continued, further effect of the time of shearing gets nullified, and the viscosity does not show any change. The trends were found to be similar for both commercial and lab-made CMC.

6.3.2 ANALYSIS OF FABRICS PRINTED WITH VAT DYES USING CMC AS A THICKENER

To correlate rheology behaviour with actual effect in printing, cotton fabric was printed using vat dyes by using both commercial and lab-made CMC. The result analyses of printed fabrics are summarized in Table 6.5.

TABLE 6.5 Analysis of Samples Printed Using CMC Laboratory Sample (L) and Commercial Sample (C) as a Thickener (Dye Used—Novatic Rubine).

Std./batch	K/S[a]	Fastness properties				Bending length[b]
		Washing		Rubbing	Light	
		Change	Staining			
100% C (0.75)	1.00	5	5	4-5	7	1.03
30L: 70C	1.15	5	5	4-5	8	1.03
50L: 50C	1.23	5	5	4-5	8	1.00
70L: 30C	1.30	5	5	5	8	0.98
100% L (0.79)	1.40	5	5	5	8	0.98

[a]average value of three readings

[b]average value of four readings

Colour values (K/S) for fabric printed using lab-made CMC, as a thickener, were slightly higher than those for the fabrics printed using commercial CMC. These two thickener pastes when blended imparted K/S values, which increased marginally with the increase in the proportion of lab-made CMC paste in blend. This validates the correlation between STI of thickener and colour of the final printed fabric. In general, higher STI results in transfer of more amount of paste onto the fabric surface resulting in fixation of more colours in the design portion that tends to reflect higher colour values.

It is found to be increasing in the case of blends, even though in small magnitude, with the proportion of laboratory-made CMC paste in the blend increased. This shows a great promise for the utilization of spinning mill (cotton fibre) waste for manufacturing of CMC.

Stiffness of the printed fabric, affected by residual thickeners after printing, is evaluated, and the results are presented in Table 6.5. Bending length, which is inversely proportional to the stiffness of fabrics, was comparable, and no significant effect of variety of thickener on stiffness was observed.

Fastness properties of printed fabric, which indicate resistance of colour on the fabric towards removal, were evaluated, and these showed similar washing fastness ratings (IS0-III) for both kinds of CMC and even for their binary blends (Table 6.5). The all-round fastness properties of vat dyes make it extremely useful and popular. The washing fastness of prints, in both cases of commercial and lab-made CMC, was similar and was excellent in both the cases. In the case of rubbing and light fastness, similar kinds of comparative results were obtained.

6.4 CONCLUSION

The waste of spinning cotton fibre was successfully converted into CMC and utilized as a thickener in printing of textiles. The samples that were laboratory-synthesized showed equivalent qualities in terms of rheology of thickener pastes, colour strength of the printed samples, stiffness and fastness properties. This explores the new arena of recycling cellulosic waste which is available abundantly for manufacturing of useful chemicals.

KEYWORDS

- **cotton fibre waste**
- **carboxymethylation**
- **printing**
- **vat dyes**

REFERENCES

1. Prasil, M.; Dang, T. L. Rheological Behaviour of Disperse Dye Printing Pastes. *Vladkha a Textl.* **1996,** *3*(3), 89–91.
2. Sostar, S. Rheological Properties of Printing Pastes-Bases of Rheology-I. *Texstilec.* **1997,** *40*(3–4), 59–65.
3. Teli, M. D.; Sheikh, J. Rheological Study of Germinated and Non-Germinated Maize Starch vis a vis Printing. *Carbohydr. Polym.* **2011,** *86*(2), 897–902.
4. Teli, M. D.; Shanbag, V.; Dhande, S. S.; Singhal, R. S. Rheological Properties of *Amaranthus Paniculates* (Rajgeera) Starch vis-à-vis Maize Starch. *Carbohydr. Polym.* **2007,** *69*(1), 116–122.
5. Fakrul Alam, A. B. M.; Mondal, M. I. H. Utilization of Cellulosic Wastes in Textile and Garment Industries. I. Synthesis and Grafting Characterization of Carboxymethyl Cellulose from Knitted Rag. *J. Appl. Polym. Sci.* **2013,** *128*(2), 1206–1212.
6. Mondal, M. I. H.; Yeasmin, M. S.; Rahman, M. S. Preparation of Food Grade Carboxymethyl Cellulose from Corn Husk Agrowaste. *Int. J. Biol. Macromol.* **2015,** *79,* 144–150.
7. Joshi, G.; Naithani, S.; Varshney, V. K.; Bisht, S. S.; Rana, V.; Gupta, P. K. Synthesis and Characterization of Carboxymethyl Cellulose from Office Waste Paper: A Greener Approach Towards Waste Management. *Waste Manage.* **2015,** *38*(1), 33–40.

8. Sheikh, J.; Brahmecha, I.; Teli, M. D. Recycling of Terry Towel (Cellulosic) Waste into Carboxymethyl Cellulose(CMC) for Textile Printing. *Fibers Polym.* **2015**, *16*(5), 1113–1118.
9. Teli, M. D.; Sheikh, J.; Shah, R., Rheology Study of Starch Extracted from Germinated Ragi and its Application in Textile Printing, *Int. J. Fashion Technol. Text. Eng.* **2014,** *2*(1), *DOI: 10.4172/2329-9568.1000105.*
10. Barba, C.; Montané, D.; Rinaudo, M.; Farriol, X. Synthesis and Characterization of Carboxymethylcelluloses (CMC) from Non-Wood Fibers I. Accessibility of Cellulose Fibers and CMC Synthesis. *Cellulose* **2002,** *9,* 319–326.
11. Browning, B. L. *Methods of Wood Chemistry;* Interscience: New York/London, 1967; *Vol. II*, pp 490–493.
12. Toğrul, H.; Arslan, N. Production of Carboxymethyl Cellulose from Sugar Beet Pulp Cellulose and Rheological Behaviour of Carboxymethyl Cellulose. *Carbohydr. Polym.* **2003,** *54,* 73–82.
13. Barai, B. K.; Singhal, R. S.; Kulkarni, P. R. Optimization of a Process for Preparing Carboxymethyl Cellulose from Water Hyacinth (*Eichornia Crassipes*). *Carbohydr. Polym.* **1997**, *32,* 229–231.
14. Booth, J. E. *Principles of Textile Testing;* Butterworths: London, 1983.
15. Trotmann, E. R. *Dyeing and Chemical Technology of Textile Fibres;* Charles Griffin and Company ltd: England, 1984.
16. ISO Technical Manual, Geneva, Switzerland (2006).

CHAPTER 7

SUSTAINABLE NOISE CONTROL MATERIALS BASED ON BAMBOO/ UNSATURATED POLYESTER COMPOSITES: ANALYSIS OF FACTORS AFFECTING THE SOUND ABSORPTION COEFFICIENTS

ELAMMARAN JAYAMANI[1], SININ HAMDAN[2], and MUHAMMAD KHUSAIRY BIN BAKRI[1]

[1]*Faculty of Engineering, Computing and Science, Swinburne University of Technology Sarawak Campus, 93350 Kuching, Sarawak, Malaysia*

[2]*Department of Mechanical and Manufacturing Engineering, Universiti Malaysia Sarawak, 94300 Kota Samarahan, Sarawak, Malaysia*

CONTENTS

ABSTRACT

The aim of the current chapter is to develop sustainable composites made of bamboo fibres as foundation materials for noise control applications. Fibres extracted from bamboo plants were obtained and sodium hydroxide (NaOH) was used to treat them. These fibres were used as reinforcement for unsaturated polyester, with 5/95, 10/90, 15/85 and 20/80 blend ratio by weight. The composites were fabricated by using compression moulding technique. According to the American Society for Testing and Materials, ASTM E1050-12 standard, the prepared composites were subjected to acoustical tests in a two-microphone transfer function impedance tube device. Sound absorption coefficients and the factors affecting sound absorption coefficients of composites were investigated. It was indicated by the obtained results that the sound absorption depends on the fibre content and surface modifications. The untreated and treated bamboo reinforced composites demonstrate a maximum sound absorption coefficient of 0.095 and 0.120, respectively, at the frequency of 5500 Hz. The Fourier transform infrared spectroscopy analysis showed that the removal of lignocellulosic components from the fibre surface influenced the increase of the sound absorption. The hollow lumen structures of bamboo fibres and fibre distributions in the polyesters are revealed by morphological studies.

7.1 INTRODUCTION

Noise pollution is one of the key problems of environmental pollution in the world. Due to the advanced development of manufacturing, automobile and electro-chemical industries, noise pollution has become an alarming issue that needs to be addressed immediately to maintain a good acoustic environment. Developing efficient sound absorption materials may be one important factor for solving this problem.[28] Therefore, the need arises for a thorough research for developing efficient environment-friendly sound absorbing materials and structures. Porous materials are widely used as sound absorbing substances in noise control engineering. In a network of interconnected pores, these types of materials propagate sound in such a way that viscous and thermal effects cause dissipation of acoustic energy.[9]

It may consist of several synthetic materials from glass wool to minerals and agricultural-based foams and fibres.[11] Typical types of porous absorber materials are carpets, curtains, acoustic (open cell) foams, acoustic tiles, mineral and cotton wools such as fibreglass. However, serious health issues such as eyes, skin, lungs and other problems may be caused by these types of materials. Recently, many researchers came up with alternative materials that are friendlier to human health than the synthetic fibres.[1] To combine high thermal and acoustic performances with a low impact on the environment and human health, an increasing attention has been focused and turned towards natural fibres as alternatives than the synthetic fibres. This has resulted in building further awareness about the use of natural fibre based materials, mainly composites.

The production of 100% natural fibre based materials as a substitute for petroleum-based products is not considered an economical solution. Combining petroleum and bio-based resources to develop a cost-effective product with diverse applications is a feasible solution.[16] A lot of researchers reported that the polymers are used as sound absorbing materials due to their wide versatility and relatively easy process.[25] Higher sound absorption coefficient was exhibited by composite boards made of wood particles and random-cut rice straws than the plywood, fibreboard and particleboard for the frequency range of 500–8000 Hz.[26] Hong et al.[12] described that the composite structure with a combination of rubber particle, perforated panel, polyurethane foam, glass wool and porous materials were found to exhibit a significant sound reduction.

The sound absorption of an industrial waste, made and developed from the processed tea leaves, was investigated by Ersoy and Küçük.[10] It was explained by them that a 1 cm thick tea leaf fibre waste material with backing provides sound absorption that is almost equivalent to the six layers of woven textile cloth.[10] Furthermore, it was found that coconut coir fibres that were compressed into bales and mattress sheet demonstrate good sound absorption coefficient.[21] From an environmental-protection viewpoint, the natural bamboo fibres were typically best to be used for sound absorption materials. Bamboo fibres formed into a fibreboard yield superior sound absorption properties when compared with plywood materials with similar density.[18] Yang et al.[27] investigated the possibility of using waste tyre composites reinforced with rice straw as construction materials. They came up with the conclusion that the composite boards demonstrate good electrical insulation, anti-caustic, acoustical insulation and anti-rot properties.[27]

In another study, Nor et al.[22] also investigated the effect of different factors on the acoustic absorption of coir fibre. It is observed in the results that the layer thickness and fibre diameter have a significant effect on the sound absorption. Furthermore, it also indicated that the bulk density does not have a major effect on the sound absorption. The sound absorption coefficients of kapok fibrous assemblies with different thickness, bulk density, orientation and fibre length were investigated by Xiang et al.[24] They established that kapok fibre has an excellent acoustic damping because of its natural hollow structure. They also found that the sound absorption coefficients of kapok fibrous assemblies are significantly affected by the thickness, bulk density and arrangement of kapok fibres. However, it is less dependent on the fibre length. *Arenga pinnata* fibre is considered a promising raw material for sound absorbing materials with low cost, biodegradability and light weight.[14] However, it has been found that the scientific data on the sound absorption coefficients of bamboo fibre composites is very limited. Moreover, limited research is available on the effect of the sound absorption coefficient due to the alkaline treatment. Thus, there is a need of implementing more research on composite materials and natural fibres for better understanding the effect of alkaline treatment on the sound absorption coefficients. Hence, the present study has been carried out to develop and investigate the sound absorption characteristics of bamboo/unsaturated polyester composites.

7.2 EXPERIMENTAL AND FABRICATING METHOD

7.2.1 MATERIALS

The bamboo fibres used in this study were obtained from the Suzhou Shenboo Textile Co. Ltd., Suzhou, China. Table 7.1 shows the physical, chemical and mechanical properties of bamboo fibre. Commercially available unsaturated polyester resin with the trade name of 'Reversol P9509' supplied by Revertex Malaysia Sdn. Bhd., Johor, Malaysia was used as matrix materials. This type of resin has high rigidity, low reactivity and is thixotropic for general purpose on orthophthalic characteristics. For curing purposes, the matrix needs to be mixed with a curing catalyst, namely methyl ethyl ketone peroxide (MEKP) at a concentration of 1wt% by weight ratio of the matrix. The MEKP was supplied by Revertex Malaysia Sdn. Bhd., Johor, Malaysia.

TABLE 7.1 The Physical, Chemical and Mechanical Properties of Bamboo Fibres.

Natural fibre	Physical, chemical and mechanical properties of natural fibres				References
Bamboo	Cellulose (%)	Lignin (%)	Microfibriller Angle (°)	Diameter (μm)	[7]
	60	32	2–10	88–125	
	Density (kg/m^3)	Moisture (%)	Tensile strength (MPa)	Young's modulus (GPa)	
	800	9	441	359	

7.2.2 METHODS

7.2.2.1 ALKALINE TREATMENT

For natural fibres, the alkaline treatment is a low-cost surface treatment. As far as the adhesive characteristics of the fibres are concerned, the concentration of sodium hydroxide solution, temperature and treatment times play a vital role in achieving the optimum efficiency of the fibre. This treatment helps in removing a certain amount of wax, lignin, hemicellulose, cellulose, pectin, oils that cover the external surface of the fibre cell wall and ruptures the cell wall.[20] Furthermore, it also depolymerizes the native cellulose structure and exposes short length crystallites.[20] The caustic soda type of sodium hydroxide (NaOH) with product code 'S/4920/AP1' was supplied by Fisher Scientific, UK and was used to treat the fibres. The fibres were immersed in a 5 wt% of sodium hydroxide solution at room temperature for 48 h. The sodium hydroxide solution was obtained by mixing this 5 wt% of sodium hydroxide in distilled water. The immersed fibres were cleaned with distilled water and dried in an oven at 60°C for 48 h.

7.2.2.2 SPECIMEN PREPARATION

By using the cold press moulding technique to test the influence of fibre loading on sound absorption, specimens were prepared by the following fibres to matrix weight percentage ratios of 5:95, 10:90, 15:85 and 20:80. To fabricate the specimens, a circular mould with a diameter of 25 mm, 80 mm and thickness of 5 mm was used. The inner surfaces of the mould were waxed with a thin layer of carnauba wax as a release agent. The known weight of unsaturated polyester resin is mixed with 1 wt% of MEKP catalyst. It helps to initiate polymerization of polyester resin and provides smoothness in the solution mixture that promotes better stirring. The known weights of the bamboo fibres were then mixed with a solution of catalyst and polyester resin. The mixture was then placed in the mould that had already been waxed.

The filling process of the mixture to mould must be done slowly to make sure that the mould cavity is properly filled. The mould cavities normally boost the building of the bubble or the void structure when compressing the mixture. Besides, a steel roller was also used to distribute the fibres uniformly and to release the air bubbles from the composites. Until a required thickness was achieved, this procedure was repeated continuously. A pressure of about 7 MPa was applied to the mould to ensure that the bubbles were forced out. Thus, the efficient way to remove the bubbles is by placing the mould in between the two compression plate and compressing it by using a compression machine. The compression machine with product code 'LS-22071' was supplied by Lotus Scientific Sdn. Bhd., Selangor, Malaysia. The composite was then left to cure for about 24 h at room temperature. Lastly, the mould was opened to remove the composites.

7.2.2.3 TESTING

According to the American Society for Testing and Materials,[4] standard, the sound absorption coefficients of the composites were measured by using two-microphone transfer function impedance tube method. The small and large impedance tube devices have a diameter of 25 and 80 mm. The large tube can operate in the frequency range of 500–1800 Hz, and the small tube can operate in the frequency range from 500–6000 Hz. Later, the small and large tube measurements were combined to determine the sound absorption rate for the frequency range of 500–6000 Hz.

By using a scanning electron microscope with a field emission gun and an accelerating voltage of 15 kV, the morphological studies of the untreated and alkaline-treated composites were observed. The scanning electron microscope with product code 'TM3030 Tabletop Scanning Electron Microscope' was supplied by Hitachi Ltd., Tokyo, Japan. It was done according to ASTM E2015-04[6] standard. A Fourier transform infrared spectroscopy was used to collect the Fourier transform infrared spectrum and to understand the functional groups of the untreated and alkaline-treated fibre composites. The Fourier transforms infrared spectroscopy with product code 'FTIR-8101 Shimadzu Spectrometer' was supplied by Shimadzu Corporation, Kyoto, Japan. The test was done according to ASTM E168-06[3] and ASTM E1252-98[5] standard.

7.3 RESULTS AND DISCUSSIONS

7.3.1 EFFECT OF FIBRE LOADING ON SOUND ABSORPTION

Figure 7.1 shows the results of the sound absorption coefficients for composites with varied fibre loading. It is vividly observed that the patterns of most curves were almost identical. Furthermore, it also evidently shows that fibre loading influenced the sound absorption. There is a considerable portion of porous structure that tends to absorb certain sound waves due to the random distribution of bamboo fibres in the matrix. The highest sound absorption coefficients obtained is for 20 wt%, and the lowest sound absorption coefficients obtained is for 5 wt%. The addition of various percentages of bamboo fibres had different responses to applied sound energy. The density and porosity of the produced composite boards were believed to be the key factors that determine the sound absorption coefficients of the screened samples.[17]

In the graph that was obtained in Figure 7.1, it showed that the sample tends to absorb more sound at higher frequency. Koizumi et al.[18] also showed the increase of sound absorption in the middle and higher frequency as the density of the sample increased. It is known that the number of fibres increases as per unit area when the apparent density is large. Furthermore, energy loss started to increase as the surface friction increased, thus the sound absorption increased. Jiang et al.[15] also found out that as increasing in seven holes hollow polyester fibre composite material content, the sound absorption coefficients of composite materials increased. The air molecules at the surface of the material and within the pores of the material are forced to vibrate when a composite is exposed to incident sound waves and, due to this, some of the energy was lost. This is because a part of the energy of the air is converted into heat due to viscosity and thermal losses at the walls of the interior pores and tunnels within the materials.[2]

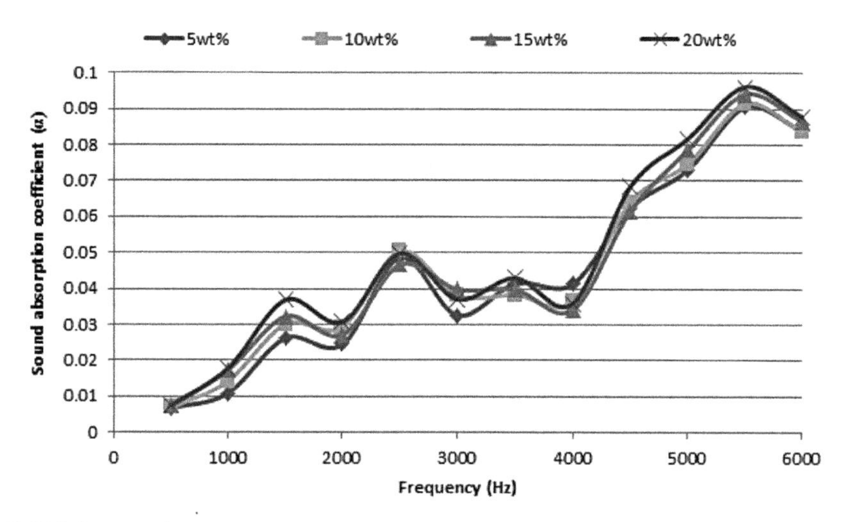

FIGURE 7.1 The comparison of sound absorption coefficients of composite materials for different fibre loading with untreated fibres.

It also can be seen that as the frequency increases, there is an increase in the sound absorption coefficients of the composite materials. However, at certain frequency range, it started to decrease and then increased again (fluctuating). This was because of the specific characteristics of bamboo fibres that tend to reflect sound at some certain frequencies and also to absorb sound at some certain frequency, especially in the middle and higher frequencies. Different fibre diameter size of bamboos are another cause of this fluctuating result.

7.3.2 EFFECT OF ALKALINE TREATMENT ON SOUND ABSORPTION

The results of sound absorption coefficients for alkaline-treated fibre composites with varying fibre loading are shown in Figure 7.2. The effect of alkaline treatment can be seen to increase the sound-absorption coefficients. Among all the composite samples, 20 wt% showed the highest sound absorption performance with a sound absorption coefficient of 0.12 at the frequency of 5500 Hz, which was higher than that of the untreated fibre composites with the coefficient of 0.093 Hz at 5500 Hz.

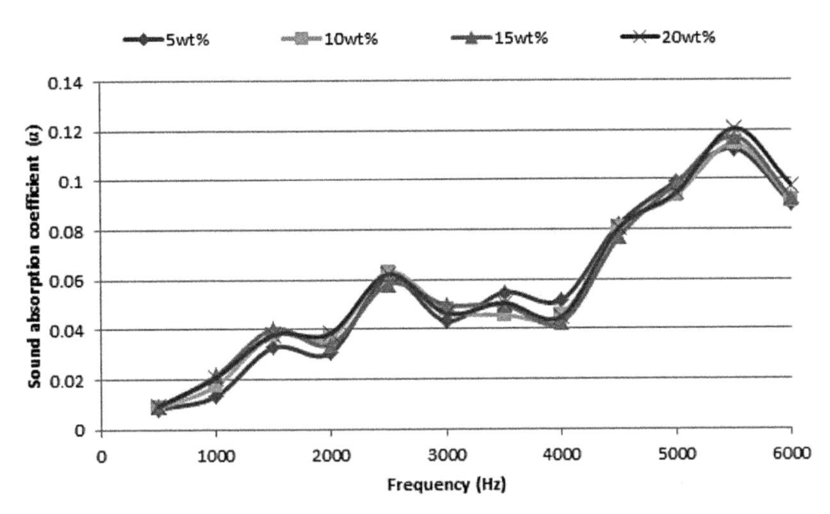

FIGURE 7.2 The comparison of sound absorption coefficients of composite materials for different fibre loading with alkaline treated fibres.

Diameter size is an important microscopic parameter of a fibre. The fibre diameter is directly related to the sound absorbing characteristics of the material.[2] Alkaline treatment alters the diameter of the fibres, causes the changes in sound absorption coefficients of the composites. Other than that, the alkaline treatment also caused swelling on the fibre structure that caused a variety of diameter size along the fibre structure. In the untreated fibres, the lignocellulosic contents such as pectin, lignin and hemicellulose and other lower molecular weight materials are present that can form a dense layer on the surface of bamboo fibres that cause higher reflection.[8] Thus, the content of lignin, hemicellulose and pectin from the outer and indirectly inner layer of the fibre is removed by alkaline treatment. This forms a porous structure on the surface and indirectly inner structure of the fibres so that the reflection is lower and the sound absorption is higher. This could be attributed to the different level of voids inside the composite materials; the inter-fibre micro-voids in the bamboo fibre could help in efficient absorption of the sound energy, whereas the inter-air pockets in the composites will allow the sound wave to pass through easily.[13]

7.3.3 MORPHOLOGICAL AND SPECTRAL ANALYSIS

Figures 7.3(a) and (b) show the hollow lumen structures and random distribution of bamboo in the unsaturated polyester/bamboo composites, respectively. It is found out that the bamboo fibres are made up of a bundle of hollow sub-fibres that have lumen inside. Bamboo fibres contain numerous connected open air cavities, and those air cavities might be the major contributors of sound absorption. It is obvious from Figure 7.3(b) that fibres were entangled in the composite. This implies that there are a considerable number of fibre network structures in the composites.[15] When the sound energy is incident on the surface of porous structure of composites, the air motion and compression in micropores caused by sound wave vibration may cause the friction with the micropores wall. The air close to the micropore wall cannot move easily due to this. On account of the friction and viscous forces, part of the sound energy can be converted to heat and cause sound energy reduction or dissipation occurs.[8] The voids between the fibres and matrix would also facilitate sound absorption.

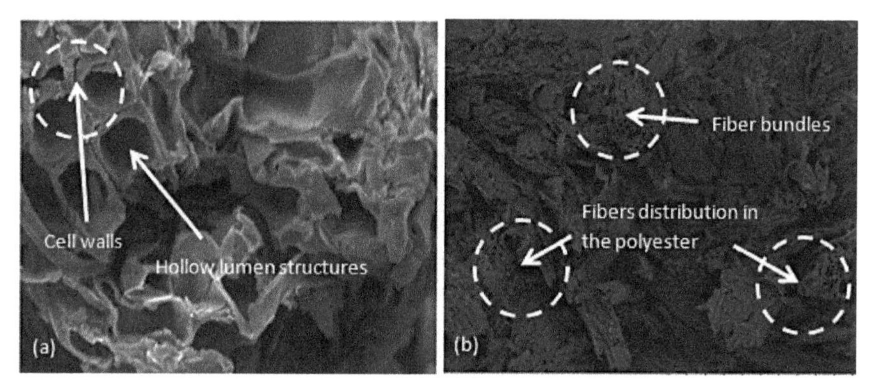

FIGURE 7.3 The scanning electron microscope micrographs of bamboo fibres (a) hollow lumen structures of bamboo fibres and; (b) fibre bundles in polymer matrix.

Infrared spectroscopy is considered a direct, simple and sensitive characterization technique. Therefore, it is widely used to elucidate the structure and interactions in different materials that may include the composites as well.[23] A lignocellulosic fibre consists of bundles of hollow cellulose fibrils. Their cell walls are reinforced with spirally oriented cellulose in a hemicellulose and lignin matrix.[19] Figures 7.4(a) and (b) show the Fourier transform infrared spectra of the untreated and alkaline-treated fibres. The C=O (carbonyl) peak at 1788.01 cm^{-1} in the untreated fibres disappeared at alkaline-treated fibres due to the removal of the reducible hemicellulose from the fibre surfaces. In addition, the peak at 1323.17 cm^{-1} reduced during the alkaline treatment. This peak is a syringe ring breathing with C–O stretching (lignin) and CH_2 wagging in cellulose. The removal of this lignocellulosic material makes composites less dense, so the sound wave reflection is lower; the sound absorption coefficient is higher.

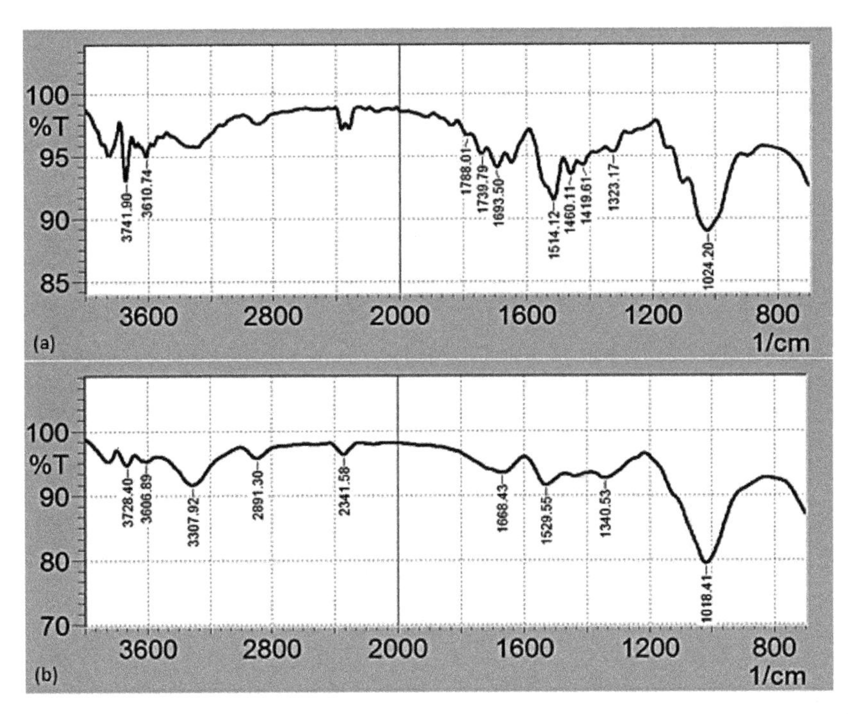

FIGURE 7.4 The infrared spectra of (a) untreated and; (b) alkaline-treated polyester/bamboo composites in spectral regions of 4000–800 cm^{-1}.

7.4 CONCLUSION

In a nutshell, the factors affecting the sound absorption coefficients of bamboo fibre/unsaturated polyester composites were studied thoroughly. The results of the sound absorption coefficient measurement showed that the composites with alkaline-treated fibres have better sound absorption than the untreated ones. On the sound absorption coefficients of composite materials, a higher fibre loading has a positive effect. The spectral studies revealed the removal of hemicellulose and lignin from the fibre due to alkaline treatment. Thus, this caused lesser dense layer structure in the composites and increased sound absorption. In addition, morphology studied by the scanning electron microscope revealed the hollow lumen structures of bamboo fibres and the distribution of bamboo fibres in the composites. The main reason behind better sound absorption is these special structures and the distribution. The results of this research distinctly show the link between alkaline treatment effects on the infrared spectral analysis, morphological analysis and sound absorption coefficients of natural fibres reinforced polymer matrix composites.

KEYWORDS

- **bamboo**
- **composites**
- **sound absorption coefficients**
- **alkaline treatment**

REFERENCES

1. Al-Rahman, R.; Raja, I. R.; Abdul Rahman, R.; Ibrahim, Z. Acoustic Properties of Innovative Material from Date Palm Fibre. *Am. J. Appl. Sci.* **2012**, *9,* 1390–1395.
2. Arenas, J. P.; Crocker, M. J. Recent Trends in Porous Sound-Absorbing Materials. *Sound Vib.* **2010,** *44,* 12–18.
3. ASTM E168-06. *Standard Practices for General Techniques of Infrared Quantitative Analysis*; ASTM International: West Conshohocken, PA, 2006.
4. ASTM E1050-12. *Standard Test Method for Impedance and Absorption of Acoustical Materials Using a Tube, Two Microphones and a Digital Frequency Analysis System;* ASTM International: West Conshohocken, PA, 2012.
5. ASTM E1252-98(2013)e1. *Standard Practice for General Techniques for Obtaining Infrared Spectra for Qualitative Analysis;* ASTM International: West Conshohocken, PA, 2013.
6. ASTM E2015-04(2014). *Standard Guide for Preparation of Plastics and Polymeric Specimens for Microstructural Examination;* ASTM International: West Conshohocken, PA, 2014.

7. Chattopadhyay, S. K.; Khandal, R. K.; Uppaluri, R.; Ghoshal, A. K. Bamboo Fiber Reinfored Polypropylene Composites and Their Mechanical, Thermal and Morphological Properties. *J. Appl. Polymer Sci.* **2011,** *119,* 1619–1626.
8. Chen, D.; Li, J.; Ren, J. Study on Sound Absorption Property of Ramie Fiber Reinforced Poly(l-lactic acid) Composites: Morphology and Properties. *Composites Part A* **2010**, *41,* 1012–1018.
9. Cox, T. J.; D'antonio, P. *Acoustic Absorbers and Diffusers: Theory, Design and Application*; CRC Press: New York, 2009.
10. Ersoy, S.; Küçük, H. Investigation of Industrial Tea-Leaf-Fibre Waste Material for its Sound Absorption Properties. *Appl. Acoust.* **2009**, *70,* 215–220.
11. Fouladi, M. H.; Ayub, M.; Mohd Nor, M. J. Analysis of Coir Fiber Acoustical Characteristics. *Appl. Acoust.* **2011**, *72,* 35–42.
12. Hong, Z.; Bo, L.; Guangsu, H.; Jia, H. A Novel Composite Sound Absorber with Recycled Rubber Particles. *J. Sound Vib.* **2007**, *304,* 400–406.
13. Huda, S.; Yang, Y. A Novel Approach of Manufacturing Light-Weight Composites with Polypropylene Web and Mechanically Split Cornhusk. *Ind. Crops Prod.* **2009,** *30,* 17–23.
14. Ismail, L.; Ghazali, M. I.; Mahzan, S.; Zaidi, A. M. A. Sound Absorption of Arenga Pinnata Natural Fiber. *World Acad. Sci., Eng. Technol.* **2010**, *67*, 804–806.
15. Jiang, S.; Xu, Y.; Zhang, H.; White, C. B.; Yan, X. Seven-Hole Hollow Polyester Fibers as Reinforcement in Sound Absorption Chlorinated Polyethylene Composites. *Appl. Acoust.* **2012**, *73,* 243–247.
16. John, M. J.; Anandjiwala, R. D. Recent Developments in Chemical Modification and Characterization of Natural Fiber-Reinforced Composites. *Polym. Compos.* **2008**, *29,* 187–207.
17. Karademir, A.; Yenidoğan, S.; Aydemir, C.; Kucuk, H. Evaluation of Sound Absorption, Printability, and Some Mechanical Properties of Thin Recycled Cellulosic Sheets Containing Wool, Ceramic Fiber, and Cotton Dust. *Int. J. Polym. Mater.* **2012,** *61,* 357–370.
18. Koizumi, T.; Tsujiuchi, N.; Adachi, A. The Development of Sound Absorbing Materials Using Natural Bamboo Fibers. *High Perform. Struct. Compos.* **2002**, *4,* 157–166.
19. Mosiewicki, M. A.; Marcovich N. E.; Aranguren. M. I. Characterization of Fiber Surface Treatments in Natural Fiber Composites by Infrared and Raman Spectroscopy. *Interface Eng. Nat. Fibre Compos. Max. Perform.* **2011**, *1,* 117–145.
20. Mohanty, A. K.; Misra, M.; Drzal, L. T. Surface Modifications of Natural Fibers and Performance of the Resulting Biocomposites: An Overview. *Compos. Interfaces* **2001**, *8,* 313–343.
21. Nor, M. J. M.; Jamaludin, N.; Tamiri, F. M. A Preliminary Study of Sound Absorption Using Multi-Layer Coconut Coir Fibers. *Electron. J. Tech. Acoust.* **2004,** *3,* 1–8.
22. Nor, M. J. M.; Ayub, M.; Zulkifli, R.; Amin, N.; Fouladi M. H. Effect of Different Factors on the Acoustic Absorption of Coir Fiber. *J. Appl. Sci.* **2010**, *10,* 2887–2892.
23. Părpăriţă, E.; Darie, R. N.; Popescu, C.-M.; Uddin, M. A.; Vasile, C. Structure–Morphology–Mechanical Properties Relationship of Some Polypropylene/Lignocellulosic Composites. *Mater. Des.* **2014**, *56,* 763–772.
24. Xiang, H.-F.; Wang, D.; Liu, H-C.; Zhao, N.; Xu, J. Investigation on Sound Absorption Properties of Kapok Fibers. *Chin. J. Polym. Sci.* **2013**, *31,* 521–529.

25. Xu, Z.; Wang, B.; Zhang, S.; Chen, R. Design and Acoustical Performance Investigation of Sound Absorption Structure Based on Plastic Micro-Capillary Films. *Appl. Acoust.* **2015**, *89,* 152–158.
26. Yang, H.-S.; Kim, D.-J.; Kim, H.-J. Rice Straw–Wood Particle Composite for Sound Absorbing Wooden Construction Materials. *Bioresour. Technol.* **2003**, *86,* 117–121.
27. Yang, H.-S.; Kim, D.-J.; Lee, Y.-K.; Kim, H.-J.; Jeon, J.-Y.; Kang, C.-W. Possibility of Using Waste Tire Composites Reinforced with Rice Straw as Construction Materials. *Bioresour. Technol.* **2004**, *95,* 61–65.
28. Zhang, C. H.; Hu, Z.; Gao, G.; Zhao, S.; Huang, Y. D. Damping Behavior and Acoustic Performance of Polyurethane/Lead Zirconate Titanate Ceramic Composites. *Mater. Des.* **2013,** *46,* 503–510.

CHAPTER 8

MORPHOLOGICAL AND DYNAMIC MECHANICAL PROPERTIES OF CHEMICALLY TREATED BAUHINIA RACEMOSA/GLASS FIBRE POLYMER COMPOSITES

J. RONALD ASEER[1,*], K. SANKARANARAYASAMY[2], P. JEYABALAN[3], and PRIYA K. DASAN[4]

[1]*School of Mechanical Engineering, Galgotias University, Greater Noida, India, *E-mail: j.ronaldaseer@galgotiasuniversity.edu.in, Tel: +91-9818510793*

[2]*Department of Mechanical Engineering, NIT, Tiruchirappalli, India*

[3]*Department of Civil Engineering, NIT, Tiruchirappalli, India*

[4]*Material Chemistry Division, SAS, VIT University, Vellore, India*

CONTENTS

ABSTRACT

In the present study, researchers considered natural fibres, extracted from the bark of Burmese silk orchid [*Bauhinia racemosa* (BR) Lam.], through ancient Indian methodology. The chemical treatment of BR fibre was done by using sodium hypochlorite NaClO/H_2O (1:1) at 60°C. The effect of chemical treatment on morphological and water absorption properties of BR was studied. The study is further extended to analyse the influence of the BR content on the mechanical and viscoelastic properties of short randomly oriented BR/glass fibre vinyl ester composite. The treated fibre shows improved morphological and reduced water-uptake properties. Mechanical properties of composite show higher impact strength and hardness value of 158±19.8 kJ/m^2 and 53±4, respectively, with addition of glass fibre. The dynamic mechanical analysis results show increased storage and loss modulus in which tanδ is decreased. Hybrid composites with 30% of BR replacing the glass fibre exhibited better properties than the other composites.

8.1 INTRODUCTION

Natural fibres are renewable source which also reinforce polymer-based composite materials. In various engineering applications, such as automobile, aircraft, building and so on, these eco-friendly natural fibres are used as replacement for glass fibres. Their significant advantages such as low density, low cost, recyclability and biodegradability, when compared with those of conventional inorganic man-made fillers are found to increase their commercial and research prospective.[1,2] Many researchers have been developing various chemical treatments such as alkalization, acetylation, cyanoethylation, silane coupling and heating of fibre to modify the surface characteristics of the fibres.[3,4] Bauhinia trees are available in Tamil Nadu, Karnataka, Andhra Pradesh and so on. The lobed leaves usually measure 10–15 cm across. The *Bauhinia racemosa* (BR) Lam. fibres are extracted from the bark of the Bauhinia trees by retting. They are then dried in the sunlight for 24 h. Unlike other lignocellulosic fibres, BR Lam. still lacks further studies. But, the leaves of Bauhinia trees are used in the production of beedi, a thin Indian cigarette.[5] The density of BR Lam. fibre is 1.3 g/cm^3 which is suitable for lightweight application such as thermal-insulation panels. The tensile strength, Young's modulus and elongation at break are 430–640 MPa, 23 GPa and 1.8–2.7%, respectively. At present, BR Lam. fibre is considered as waste, and its industrial potential is untapped. Hence, it can be a good choice of reinforcement for polymer composites. Hybridization of composite provides flexibility to the design engineer in order to modify the material properties based on the requirements.[6] Mishra et al.[7] analyse the mechanical properties of pineapple leaf and sisal fibre-reinforced polyester composites which contain low-glass fibre volume content. In addition, the comparison of the mechanical properties of phenol formaldehyde composites reinforced with banana and glass fibres are studied by Joseph et al. Mechanical property evaluation of sisal–jute–glass fibre-reinforced polyester composites are analysed by Ramesh et al.[9] The analysis of mechanical properties of hybrid Burmese silk orchid and glass fibres composite material is done by Lakshmanan et al. It has shown that the value of mechanical properties will increase with increase in length of fibre reinforcement.[10]

Dynamic mechanical analysis (DMA) predicts the long-term mechanical behaviour in a wide range of temperature/time, also allowing correlation between microscopic and macroscopic properties.[11] The dynamic and static mechanical properties of randomly oriented, intimately mixed, short banana/sisal hybrid fibre-reinforced polyester composites are determined by Idicula et al.[12] The results show that storage modulus was found to increase with fibre volume fraction above glass transition temperature (T_g) of the matrix, and maximum value is obtained at a volume fraction (V_f) of 0.40. Valeria and Marie[13] investigated the DMA of unsaturated polyester resin, and the result shows that 1,3-amino propyl triethoxy silane incorporated to the resin by reaction of its amino group with glycidyl methacrylate has its flexibility improved by the addition of a chain extender. Saha et al.[14] studied the dynamic mechanical properties of the jute/polyester composites. The storage modulus and the thermal transition temperatures of the composites shifted to higher values for cyanoethylated jute fibre. Further, the results indicated that the incorporation of sisal–jute fibre with glass fibre-reinforced polymer (GFRP) can improve the properties and used as an alternate material for GFRP composites. The objective of this work is to analyse the mechanical and viscoelastic properties of chemically treated BR/glass composites by keeping the overall fibre volume fraction ($\%V_f$) constant and varying the BR/glass ratio.

8.2 MATERIALS AND METHODS

8.2.1 RAW MATERIALS

The BR fibres were collected from Hogenakkal, Krishnagiri district, Tamil Nadu, India (Fig. 8.1). The chopped BR fibres were dipped in NaClO/H_2O solution of 1:1 proportion under heating (60°C) for 4 h. Followed by this, the fibres were cleaned with distilled water and dried in a hot air oven (60°C) for 24 h.[15] Isophthalic vinylester resin and the catalyst methyl ethyl ketone peroxide were obtained from Mayore polymers, Bangalore. The accelerator of cobalt napthanate was added as 1% with the resin and the catalyst. The glass fibre of unidirectional mat having 300 GSM was used.

8.2.2 FABRICATION OF COMPOSITES

The composite materials were fabricated by hand lay-up process. Chopped BR fibre of 30 mm length was used to prepare the specimen. The composite consisted of total three layers in which glass fibre layers were fixed on top and under bottom of the specimen. The middle layer was filled by BR fibre. The processed composite was pressed by hydraulic press and dried for 24 h. The composites were prepared by keeping the overall fibre volume fraction (20, 30 and 40 vol.%) constant and the volumetric ratio between BR/glass fibre varied by 100:0, 75:25, 50:50, 25:75 and 0:100 (Table 8.1).

FIGURE 8.1 Raw *Bauhinia racemosa* (BR) fibre.

8.2.3 MECHANICAL AND DMA

Barcol hardness was measured using a HPE Bareiss Durometer by using ASTM D2583 and an average of 10 measurements. Izod impact testing was carried out on unnotched specimens (60 × 12.7 × 3 mm) using a Digital impact tester according to ASTM D256. Thermal analyser DMA Q800 V20.6 Build 24 was used to study the DMA. Specimen dimensions used were 35 × 14.27 × 3.72 mm. In a nitrogen atmosphere with frequency of 1 Hz (oscillating amplitude 0.1 mm) at a heating rate of 50°C/min and temperature range from 25 to 300°C, the specimens were tested. Five samples were used for this work.

8.3 RESULTS AND DISCUSSION

8.3.1 MORPHOLOGICAL PROPERTIES ANALYSIS

The surface morphology and cross section of BR fibres were analysed through scanning electron micrographs. The surface of the treated BR fibre (Fig. 8.2(b)) was rougher than that of the raw fibre (Fig. 8.2(a)) and also showed fibrillation due to the removal of hemicellulose, lignin and other constituents.[16] The water absorption values of treated fibres were less than that of raw fibres, which agreed with the previous report.[17] A 60% weight increase was observed in raw fibre, whereas only 44% increase of weight was observed in NaClO-treated BR fibre. Due to the removal of lignin and hemicellulose, treated fibre showed less water absorption.

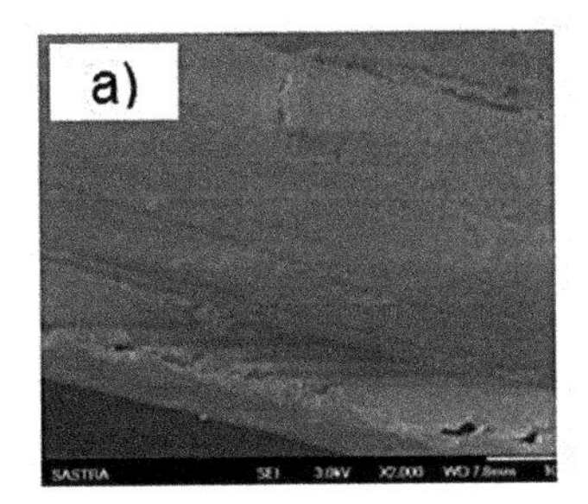

FIGURE 8.2 BR fibre (a) raw, (b) treated.

8.3.2 EFFECT OF FIBRE CONTENT ON MECHANICAL PROPERTIES

The Barcol hardness and impact strength of the composites are given in Table 8.1.

TABLE 8.1 Barcol Hardness and Impact Strength of the Composites.

Sample code	Fibre volume fraction (% in volume)	Bauhinia racemosa fibre content (% in volume)	Glass fibre content (% in volume)	Hardness	Impact strength (kJ/m²)
20/100/0	20	100	0	15±3	30±6.5
20/75/25	20	75	25	28±3	70±7.5
20/50/50	20	50	50	32±3	87±16.7
20/25/75	20	25	75	36±2	88±5.8
20/0/100	20	0	100	48±4	101±16.5
30/100/0	30	100	0	16±3	31±11.1
30/75/25	30	75	25	25±3	80±9.3
30/50/50	30	50	50	35±4	120±14.1
30/25/75	30	25	75	40±4	147±17.1
30/0/100	30	0	100	53±4	158±19.4
40/100/0	40	100	0	10±3	26±5.3
40/75/25	40	75	25	26±3	75±4.5
40/50/50	40	50	50	34±3	105±7.8
40/25/75	40	25	75	32±3	110±12.1
40/0/100	40	0	100	50±3	122±15.7

Hardness values were influenced by the BR fibres and vinyl ester resin that were near the surface in which the test was conducted.[18] It increased higher content of glass fibre and decreased higher content of BR fibres. This is due to the rigidity of glass fibre incorporated in the composites. Impact strength increases with glass fibre incorporation, which shows better adhesion with resin compared with BR fibre. It also increases overall fibre volume content (30% V_f) as there is relatively more energy dissipation at the fibre/resin interface by fibre pull-out/debonding subsequent impact event.[19] The poor adhesion of BR fibres is due to its polar character. Premature failure of composites may cause this weakening of the fibre/matrix interface.[20]

8.3.3 EFFECT OF FIBRE CONTENT ON STORAGE MODULUS (E')

The variation of the storage modulus as a function of temperature is shown in Figure 8.3. It is observed that storage modulus of the composites in the glassy state increases with glass fibre incorporation. This is due to the rigidity and the glass fibre/matrix adhesion that adds for higher modulus value in this region. The main transition for amorphous polymers from the glassy state to the elastomeric consists of different types of molecular motion, such as cooperative short-range diffusional motion of the chain segments,[19] sub-Rouse and Rouse models.[21]

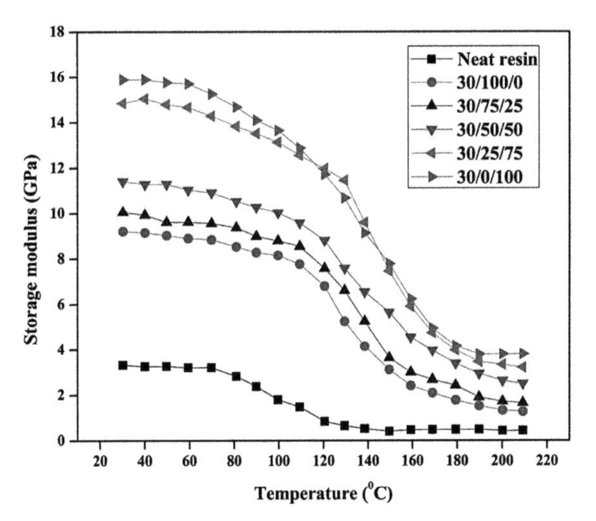

FIGURE 8.3 Variation of the storage modulus as a function of temperature.

The reinforcement effect is more effective in the glassy region than in the elastomeric one. For all composites, the modulus values in the glassy region are higher upon glass incorporation. But values are varied in the rubbery state region due to the heterogeneity of natural fibre and non-uniformity in fibre distribution. In the elastomeric region, modulus is also higher for the samples containing higher glass fibre content.[22] The E′ values are increased with the increase of the BR fibre content in the composites. This is due to the higher stress transfer at the interface imparted by the BR fibres. It is shown that the changes in the E′ values considerably dropped at temperatures between 100 and 180°C for the composites.

8.3.4 EFFECT OF FIBRE CONTENT ON LOSS MODULUS

Figure 8.4 shows the variation of the loss modulus E″ of the neat resin and the composites with the temperature. The vinyl ester resin has the peak loss modulus at 100°C, which is attributed to the mobility of the resin molecules.[23] This peak is also considered as the glass transition temperature (T_g) of the resin.[14,24,25] The incorporation of glass fibre causes broadening of the loss modulus peak (Fig. 8.4). This can be related to the inhibition of relaxation processes (more volume in interface) in the composites which decrease mobility at the surface of the fibre and a higher number of chain segments upon fibre addition. The observed broadening may be explained by the change in the physical state of the fibre, matrix and immobilized polymer layer matrix upon glass incorporation.[19] Table 8.2 shows the E″ max values and the E″ max peak temperatures of the vinyl ester resin and the BR fibre-reinforced composites. The E″ value corresponding to the T_g is lowest for the resin (142 MPa), but increased significantly in the composites. A similar increase in the T_g and the loss modulus due to the incorporation of the reinforcing fibres was also reported earlier.[14]

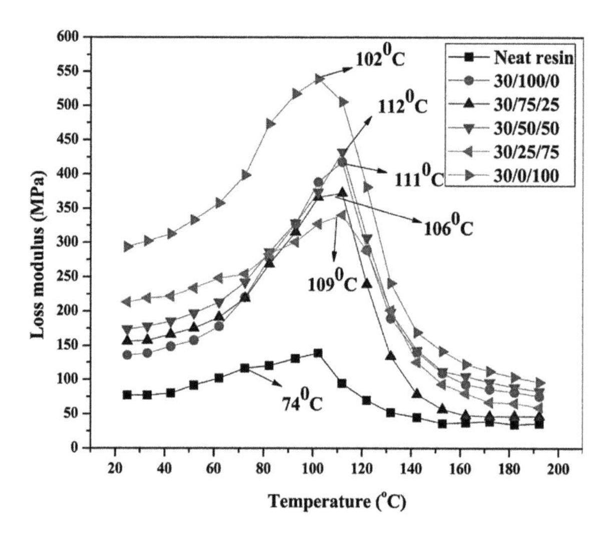

FIGURE 8.4 Variation of the loss modulus E″ of the neat resin and the composites with the temperature.

8.3.5 EFFECT OF FIBRE CONTENT ON DAMPING (TAN DELTA)

The variation of the tan delta of the vinyl ester resin and the composites as a function of temperature is shown in Figure 8.5. The tan delta is high for the neat resin due to the high reduction of the storage modulus values on increasing the temperature. Fibre incorporation has restricted the mobility of the polymer molecules, which resulted in the rise of storage modulus values and reduction of the viscoelastic lag between the stress and the strain. Hence, the tan delta values are decreased in the composites.[14,26] In addition, tan delta values are reduced in the composites compared with the neat resin as there is less matrix by volume to dissipate the vibrational energy.[19] Moreover, it can be observed that damping is lower for the composites than for the neat polyester matrix (Table 8.2). This can happen if the fibres are carrying a greater amount of the load, dissipating a small amount of it at the fibre/matrix interface. However, no particular trend is followed by the glass transition temperature. It may be due to the high void content of the composite or due to lower fibre content present in that composite.[27]

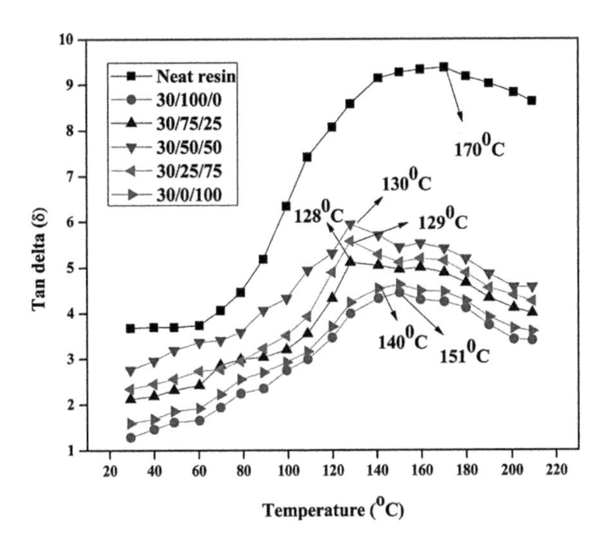

FIGURE 8.5 Variation of the tan delta of the resin and the composites as a function of temperature.

TABLE 8.2 Dynamic Mechanical Analysis of Neat Resin and Composites.

Sample code	Loss modulus, E″max (Mpa)	Damping tan deltamax	Temperature (°C) E″ max	Glass transition temperature (T_g) (°C) tan delta max
Neat resin	142	9.4	102	170
30/100/0	416	4.4	111	151
30/75/25	370	5.1	106	128
30/50/50	436	5.9	112	130
30/25/75	338	5.6	109	129
30/0/100	540	4.6	103	140

8.4 CONCLUSION

In the current study, mechanical and viscoelastic properties of chemically treated BR/glass fibre composites are investigated. Improved morphological and less water absorption properties are shown by the chemical treatment. Hardness and impact strength properties show the same trend due to the intrinsic characteristics of the glass fibre such as stronger adhesion to the resin and higher energy dissipation at the fibre/resin interface compared with BR fibre. From DMA, it is observed that the storage and loss modulus increased for higher glass fibre content. This is generally due to a greater degree of restriction imposed by the glass fibre to the resin, which allows a greater stress transfer through the vinyl ester matrix/fibre reinforcement interface. In addition, the storage modulus decreased with the increase in the temperature. It has a sharp drop between the temperature 100 and 178°C. The mechanical and viscoelastic properties show that BR fibre can be a good substitute for making biocomposites which is used in automobile, aerospace and wall-panelling applications.

KEYWORDS

- ***Bauhinia racemosa***
- **composites**
- **viscoelastic properties**
- **glass fibre**

REFERENCES

1. Ho, M. P.; Wang, H.; Lee, J. H.; Ho, C. K.; Lau, K. T.; Leng, J.; Hui, D. Critical Factors on Manufacturing Processes of Natural Fibre Composites. *Compos. Part B.* **2012,** *43*, 3549–3562.
2. Bledzki,; Gassan. Composites Reiforced with Cellulose Based Fibers. *Prog. Polym. Sci.* **1999,** *24*, 221–274.
3. Bledzki, A. K.; Reihmane, S.; Gassan, J. Properties and Modification Methods for Vegetable Fibers for Natural Fiber Composites. *J. Appl. Poly. Sci.* **1996,** *59*, 1329–1336.
4. Mohanty, A. K.; Singh, B. C. Redox-initiated Graft Copolymerization onto Modified Jute Fibers. *J. Appl. Poly. Sci.* **1987,** *34*, 1325–1327.
5. http://www.efloraofgandhinagar.in/tree/bauhinia-racemosa (accessed on September 1, 2016).
6. Velmurugan, R.; Manikandan, V. Mechanical Properties of Palmyra/Glass Fiber Hybrid Composites. *Compos. Part A – Appl. Sci.* **2007,** *38*, 2216–26.
7. Mishra, S.; Mohanty, A. K.; Drzal, L. T.; Misra, M.; Parija, S.; Nayak, S.K.; et al. Studies on Mechanical Performance of Biofibre/Glass Reinforced Polyester Hybrid Composites. *Compos. Sci. Technol.* **2003,** *63*, 1377–85.
8. Joseph, S.; Sreekala, M. S.; Oommen, Z.; Koshy, P.; Thomas, S. A Comparison of The Mechanical Properties of Phenol Formaldehyde Composites Reinforced with Banana Fibres and Glass Fibres. *Compos. Sci. Technol.* **2002,** *62*, 1857–1868.
9. Ramesh, M.; Palanikuma, K.; Reddy, H. K. Mechanical Property Evaluation of Sisal–Jute–Glass Fiber Reinforced Polyester Composites. *Compos. Part B.* **2013,** *48*, 1–9.
10. Lakshmanan, V.; Ramesh Kumar, C.; Gopal P. Analysis of Mechanical Properties of Hybrid Burmese Silk Orchid and Glass Fibers Composite Material. *Sci. Technol. Arts Res. J.* **2015,** *4*, 241–246.
11. Cassu, S. N.; Felisberti M. I. Comportamento Dinamico-Mecanico e Relaxacoes em Polimeros e Blendas Polimericas. *Quim. Nova.* **2005,** *28*, 255–263.
12. Idicula, M.; Malhotra, S. K.; Joseph, K., Thomas, S. Dynamic Mechanical Analysis of Randomly Oriented Intimately Mixed Short Banana/Sisal Hybrid Fibre Reinforced Polyester Composites. *Compos. Sci. Technol.* **2005,** *65*, 1077–1087.
13. Valeria, M. R.; Maria, I. F. *J. Appl. Poly. Sci.* **2001,** *81*, 3272.
14. Saha, A. K.; Das, S.; Bhatta, D.; Mitra, B. C. Study of Jute Fibre Reinforced Polyester Composites by Dynamic Mechanical Analysis. *J. Appl. Poly. Sci.* **1999,** *71*, 1505–1513.
15. Khan, F. Ahmad, S. R. Chemical Modification and Spectroscopic Analysis of Jute Fibre. *Polym. Degrada. Stabil.* **1996,** *52*, 335–340.
16. Gassan, J.; Bledzki, A. K. Possibilities for Improving the Mechanical Properties of Jute/Epoxy Composites by Alkali Treatment. *Compos. Sci. Technol.* **1999,** *59*, 1303–1309.
17. Jimenez, A.; Bismarck, A. Wetting Behavior, Moisture Uptake and Electro kinetic Properties of Lignocellulosic Fibers. *Cellulose* **2007,** *14*, 115–127.
18. Reddy, G. V.; Naidu, S. V. A Study on Hardness and Flexural Properties of Kapok/Sisal Composites. *J. Reinf. Plast. Compos.* **2009,** *28*, 2035–2044.

19. Ornaghi Jr, H. L.; Silva, H. S. P.; Zattera, A. J.; Amico, S. C. Hybridization Effect on The Mechanical and Dynamic Mechanical Properties of Curaua Composites. *Mat. Sci. Eng. A*. **2011,** *528*, 7285–7289.
20. Gregorova, A; Hrabalova, M.; Kovalcik, R.; Wimmer, R. Surface Modification of Spruce Wood Flour and Effects on The Dynamic Fragility of PLA/Wood Composites. *Polym. Eng. Sci.* **2011,** *51*(1), 43–50.
21. Jinrong, W.; Huang, G.; Liangliang, Q.; Zheng, J. Correlations between Dynamic Fragility and Dynamic Mechanical Properties of Several Amorphous Polymers. *J. Non-Cryst. Solids*. **2009,** *355*, 1755–1759.
22. Junior, J. H. S. A.; Junior, H. L. O.; Amico, S. C.; Amado, F. D. R. Study of Hybrid Intralaminate Curaua/Glass Composites. *Mater. Des.* **2012,** *42*, 111–117.
23. Hon David, N. S.; Shiraishi N. *Wood and Cellulose Chemistry;* Mercel Dekker Incorporation: New York and Basel, 1991.
24. Rana, A. K.; Mitra, B. C.; Banerjee, A. N. Short Jute Fibre-Reinforced Polypropylene Composites: Dynamic Mechanical Study. *J. Appl. Poly. Sci.* **1999,** *71*, 531–539.
25. Saha, N.; Banerjee, A. N.; Mitra, B. C. Dynamic Mechanical Study on Unidirectional Polyethylene Fibre-Pmma and Glass Fibre-Pmma Composite Laminates. *J. Appl. Poly. Sci.* **1996,** *60*, 657–662.
26. Ghosh, P.; Bose, N. R.; Mitra, B. C.; Das, S. Dynamic Mechanical Analysis of FRP Composites Based on Different Fibre Reinforcemnts and Epoxy Resin as the Matrix Material. *J. Appl. Poly. Sci.* **1997,** *62*, 2467–2472.
27. Joseph, P. V.; Joseph, K.; Thomas, S. Effect of Processing Variables on The Mechanical Properties of Sisal-Fiber-Reinforced Polypropylene Composites. *Compos. Sci. Technol.* **1999,** *59*, 1625–1640.

CHAPTER 9

OPTIMIZATION OF PLASMA MODIFICATION FOR LOW-TEMPERATURE DYEING OF SILK FABRIC

M. D. TELI[1,*], PINTU PANDIT[1,2], and KARTICK K. SAMANTA[3,4]

[1]*Department of Fibres and Textile Processing Technology, Institute of Chemical Technology, Nathalal Parekh Marg, Matunga, Mumbai 400019, India, *E-mail: mdt9pub@gmail.com*

[2]*E-mail: pintupanditict@gmail.com*

[3]*ICAR-Central Institute for Research on Cotton Technology, Matunga, Mumbai 400019, India.*

[4]*National Institute of Research on Jute and Allied Fibre Technology, Kolkata 700040, India.*

CONTENTS

ABSTRACT

Silk fabric was plasma treated in an indigenously developed atmospheric pressure plasma reactor in presence of helium/nitrogen (He/N_2) mixture at a discharge voltage of 5 kV and frequency of 21–23 kHz. The samples were treated in plasma for different times of 1–7 min at varying nitrogen gas flow rates in the range of 0–100 ml/min keeping the constant helium (He) flow rate at 450 ml/min. The effect of plasma treatment time and N_2 gas flow rates on acid dyeing of silk at different temperatures of 40–60°C has been studied in terms of exhaustion %, K/S, fixation % and strength loss %, by using Box–Behnken design (BBD). Box–Behnken design was used for the optimization of plasma treatment process and to evaluate the effects and interactions of the process variables, that is treatment time, change in concentration of gases and dyeing temperature on the acid dyeing of silk fabrics. The optimum conditions for maximum plasma treatment time (4 min), N_2 gas flow rate (50 ml/min) and dyeing temperature (60°C) of silk fabrics were established. Physical and chemical properties of the samples were analysed using scanning electron microscopy (SEM), energy dispersive X-ray (EDX) spectroscopy and secondary ion mass spectroscopy (SIMS) analysis. Surface etching and the presence of micro pits and craters on the treated fabric was shown by SEM. The plasma-treated sample showed an increase in amine groups, which helps in faster exhaustion of acid dye molecules even at a lower temperature at 40°C temperature, which was conventionally dyed at 90°C. Due to enhanced exhaustion and fixation, the plasma-treated sample showed deeper shade. From EDX and SIMS analysis, it seems that formation of more amino groups in silk protein after plasma treatment helped in better dyeing properties.

9.1 INTRODUCTION

In accordance with the sustainability parameters, natural products are increasingly being in demand and thus our researcher group (Teli and Pandit) has been working on many natural fibres including silk, wool, jute and linen.[1,8,9,13–16,18] Over the years, silk fibre is called the queen of fibres. It is the only fibre that is spun among all the natural fibres. It has been used in textile industry for more than 5000 years. The reasons behind it are that silk has unique lustre, handle, durability and dyeability. The silk dyeing was mostly done with synthetic acid dye, which needs high temperature, time and cost. Keeping the sustainability in mind, surface modification of the silk material is done by the plasma treatment without altering the bulk properties to increase the dye uptake and to impart functional finishes. Plasma reaction is a complex process involving both constructive and destructive type of uncontrollable reactions. Non-polymerizing gases such as O_2, N_2, H_2, He and Ar have mainly been used for improving hydrophilic property and colouration of natural and synthetic fibres.[2,3,5,7,17,18] After plasma treatment of silk fabric, slight flutes appeared and also fibrillar unit was more evident in treated silk.[6] Significant changes in the morphological and chemical properties of the plasma-treated textile surfaces have been reported.[11,21] The fibre nature changed from slight hydrophobic to hydrophilic, which is the key point for adsorption of aqueous dye solutions to achieve excellent dye uptake, good uniformity and good fastness levels. It is possible to obtain deeper shades using less concentration of dyestuffs at a lower temperature and shorter dyeing times if the plasma parameters during the treatment of textiles substrates are properly maintained. In this regard, we have indigenously developed atmospheric pressure cold plasma reactor in presence of helium/nitrogen (He/N_2) mixture at a discharge voltage of 5 kV and frequency of 21–23 kHz to achieve low-temperature dyeing of mulberry silk fabric. In this regard, in order to obtain optimum parameters for low-temperature dyeing of silk and the relation between different variables, the Box–Behnken statistical method can be used with a reduced number of trials and incorporate most of the main effects, as well as first-order interaction and quadratic relationships.[4,20] Design of experiments (DoE) is the process of planning experiments so that the results collected can be analysed by statistical methods, producing valid and objective conclusions. The design helps to find the optimal experimental point by predicting response values for all possible factor combinations. Researchers have carried out not only the optimization of degumming of Mulberry, Muga, Tassar and Eri silk fabric[12] but also the optimization of protein extraction using silk cocoon based on this Box–Behnken model.[10]

In the present investigation, an attempt has been made to optimize the plasma parameters such as nitrogen gas flow rate, plasma treatment time and the dyeing temperature for plasma-treated fabric using the Box–Behnken method for acid dyeing of mulberry silk fabric. The comparative study of optimized process parameters was expected to throw light on the effect of dyeing of silk fabric at a low temperature in comparison with untreated silk fabric.

9.2 EXPERIMENTAL

9.2.1 MATERIAL AND METHODS

The 100% mulberry plain-woven bleached silk fabric having 110 ends/inch and 100 picks/inch with an areal density of 39 GSM was used for the plasma treatment. Acid dye (Telon Blue M-GLW-Acid Blue 221) used in this experiment was supplied by Dystar India Pvt. Ltd. The samples were plasma treated in an indigenously developed atmospheric pressure low temperature plasma reactor in presence of helium/nitrogen (He/N_2) mixture for 1–7 min. The helium (He) gas flow rate was kept constant at 450 ml/min (He) and nitrogen (N_2) gas flow rate was varied in the range of 0–100 ml/min using two mass flow controllers supplied by Alicat Scientific, United States of America, during the treatment of plasma.

9.2.2 DYEING OF SILK

The untreated and plasma-treated silk fabrics were dyed at 2% shade by using acid dye in acidic condition (pH 4.5) with a material to liquor ratio of 1:40. The dyeing was carried out at a range of 40–60°C for 60 min. Dye bath exhaustion was measured by sampling the dye bath solution before and after dyeing in Perkin-Elmer Lambda 25 Ultraviolet–visible spectrophotometer. The percentage of exhaustion (E) of dye bath was calculated as per the following formula:

$$\% E = \frac{A_0 - A_t}{A_0} \times 100$$

where, A_0 and A_t represent the absorbance of the dye solution before and after the dyeing, respectively. Colour depth of the samples was evaluated by measuring the K/S values using SpectraScan 5100 + spectrophotometer at λ_{max} of 650 nm, and the fixation % is determined by using the formula,

$$\text{Fixation } (\%) = \frac{\text{K/S of dyed sample before washing}}{\text{K/S of dyed sample after washing}} \times 100$$

9.2.3 SCANNING ELECTRON MICROSCOPE ANALYSIS

Surface features of the silk fabrics were analysed using scanning electron microscope (SEM), Model—Philips-XL30 at a magnification of 1200×.

9.2.4 ENERGY DISPERSIVE X-RAY SPECTROSCOPY ANALYSIS

The surface chemistry of untreated and plasma-treated samples in terms of element present were analysed using Field Emission Gun-SEM.

9.2.5 SECONDARY ION MASS SPECTROSCOPY ANALYSIS

Surface chemical composition in terms of molecules present and their distribution in the untreated and plasma-treated mulberry silk samples was analysed using time of flight secondary ion mass spectrometer (ToF-SIMS), Model PHI TRIFT V nano TOF using gallium as a primary ion source.

9.2.6 DETERMINATION OF TENSILE STRENGTH

Effect of plasma treatment time and N_2 gas flow rates on breaking load of the silk fabric was evaluated by measuring breaking load of the samples in H5KS Single Column Universal Tester (Tinius Olsen) as per the ASTM D 5035-1995 method.

9.2.7 ASSESSMENT OF FASTNESS PROPERTIES

The dyeing was subjected to washing fastness test using ISO 105-C10:2006 (no. B), test method (ISO method II) where in the composite sample was treated in soap solution of 5 g/l at liquor ratio 1:50 at 50±2°C for 45 min in a wash fastness tester followed by washing and drying. Similarly, light fastness and rubbing fastness of the acid dye treated silk fabric samples were also assessed according to ISO 105-B02:2013 and ISO 105-X 12: 2002 methods, respectively.

9.2.8 *OPTIMIZATION OF PLASMA-TREATED ACID DYEING OF SILK FABRIC USING BOX–BEHNKEN DESIGN*

The most important parameters that affect the wettability of the silk fabrics are treatment time and rate of nitrogen gas. To study the combined effect of these factors, experiments were performed at different combinations of the physical parameters by using statistically designed experiments. Based on Analysis of variance (ANOVA) response, the three-dimensional (3D) response surface plots were drawn using the Design Expert 6.0 software. The influence of the variables on the response Y_i for a quadratic model can be adjusted using the following polynomial function:

$$Y_i = b_0 + \Sigma b_i X_i + \Sigma b_{ii} X_i^2 + \Sigma b_{ij} X_i X_j; \; i \geq j; \; i, j = 1, 2, 3$$

where, Y_i is the predicted response, X_i and X_j are the coded independent variables, b_0 is an independent term according to the mean value of the experimental plan, b_i are regression coefficients of the variables in their linear form, b_{ij} are regression coefficients of the interaction terms between variables and b_{ii} are regression coefficients of the quadratic form of variables. The resultant experimental data of response properties were compared with that of the predicted values. Moreover, linear regression plots between the observed and the predicted values of the response variables were drawn. Levels of variables chosen for the experiment are shown in Table 9.1.

TABLE 9.1 Levels of Variables Chosen.

Variables	Units	Low (−1)	Middle (0)	High(+1)
Time (A)	min	1	4	7
N_2 gas rate (B)	ml/min	0	50	100
Dyeing temp. (C)	°C	40	50	60

9.3 RESULTS AND DISCUSSION

9.3.1 *EXPERIMENTS DESIGNED AND CONDUCTED*

The DoE of plasma-treated dyeing of silk fabric was conducted using BBD according to the design matrix, and the results are presented as shown in Table 9.2.

TABLE 9.2 Design Matrix for Plasma-Treated Acid Dyeing of Silk Fabric.

Run	Time (min)	N_2 Flow rate (ml/min)	Dyeing temp. (°C)	Exhaustion (%)	K/S (λ_{max} 650 nm)	Fixation (%)	Strength loss (%)
1	4	50	50	97.68	3.78	91.37	7.89
2	4	50	50	98.41	3.77	90.62	7.84
3	1	0	50	88.62	3.19	80.75	0.68
4	4	50	50	97.14	3.78	90.65	7.85
5	4	100	60	99.22	4.86	96.15	13.89
6	1	50	60	96.21	4.16	92.75	1.48
7	7	100	50	98.45	4.41	94.18	18.75
8	1	100	50	95.62	3.84	89.3	3.34
9	7	50	40	91.08	3.64	91.32	13.18
10	1	50	40	86.65	2.70	82.85	1.61
11	7	0	50	94.42	3.54	87.84	8.91
12	4	0	40	84.85	2.88	81.76	7.38
13	7	50	60	99.02	4.85	95.61	13.62
14	4	50	50	97.18	3.81	91.01	7.85
15	4	50	50	97.42	3.64	90.65	7.89
16	4	100	40	92.69	3.44	93.52	12.19
17	4	0	60	94.01	4.15	91.29	6.11

9.3.2 *REGRESSION ANALYSIS AND MODEL SIGNIFICANCE*

After conducting the experiments, statistical analysis was done. Statistical analysis was tested by the Fisher's *F*-test for ANOVA and the results are given in Table 9.3 for exhaustion %, K/S, fixation % and strength loss %, respectively. There is only a 0.01% chance that a Model *F* value this large could occur due to noise. Values of 'Prob > *F*' less than 0.05 indicate model terms are significant. Values greater than 0.10 indicate that the model terms are not significant. If there are many insignificant model terms (not counting those required to support hierarchy), the model reduction may improve the model.

The 'Lack of Fit *F*-value' for 1.44, 5.33, 2.93 and 1.97 implies the Lack of Fit is not significant relative to the pure error. There is a 35.67,

TABLE 9.3 Analysis of Variance Results of Responses for Acid Dyeing of Silk Fabric.

Source		Exhaustion (%)		K/S		Fixation (%)		Strength loss (%)	
		F value	**Prob>F**	**F value**	**Prob>F**	**F value**	**Prob>F**	**F value**	**Prob>F**
Model		108.63	<0.0001	47.76	<0.0001	183.23	< 0.0001	52606.15	<0.0001
	A	98.46	<0.0001	64.22	<0.0001	347.03	< 0.0001	3.42E+05	< 0.0001
	B	226.69	<0.0001	76.88	< 0.0001	634.68	< 0.0001	95878.08	< 0.0001
	C	430.65	<0.0001	283.75	< 0.0001	443.83	< 0.0001	83.4	< 0.0001
	A^2	24.73	0.0016	2.994E-03	0.9579	44.96	0.0003	7167.69	< 0.0001
	B^2	48.44	0.0002	0.021	0.8881	42.05	0.0003	7862.95	< 0.0001
	C^2	115.03	<0.0001	2.38	0.1672	31.92	0.0008	3205.89	< 0.0001
	AB	6.9	0.0341	0.96	0.3608	6.24	0.0411	15703.52	< 0.0001
	AC	2.05	0.1951	1.23	0.3032	40.24	0.0004	98.97	< 0.0001
	BC	5.41	0.0530	0.44	0.5263	60.87	0.0001	2686.96	< 0.0001
Lack of Fit		1.44	0.3567	5.33	0.0698	2.93	0.1632	1.97	0.2609

6.98, 16.32 and 26.09% chance that 'Lack of Fit F-value' this large could occur due to noise for exhaustion %, K/S, fixation % and strength loss %, respectively. Hence, this model can be used to navigate the design space. A simple regression equation analysis was undertaken to fit the response function and predict the outcome of individual responses for plasma-treated acid-dyed silk fabric. The responses, namely exhaustion %, K/S, fixation % and strength loss %, were expressed in form of regression Equations (9.1–9.4), respectively. Factors A, B and C stand for a time, nitrogen gas flow rate and dyeing temperature, respectively.

Exhaustion (%) =

$$+97.57 + 1.98\,A + 3.01B + 4.15C - 1.37A^2 - 1.92B^2 - 2.96C^2 - 0.74AB - 0.40AC - 0.66\,BC \quad (9.1)$$

K/S =

$$+3.76 + 0.32A + 0.35B + 0.67C - 0.003A2 - 0.008B^2 + 0.085C^2 + 0.055AB - 0.063AC + 0.038BC \quad (9.2)$$

Fixation (%) =

$$+90.86 + 2.91A + 3.94B + 3.29C - 1.45A^2 - 1.40B^2 + 1.22C^2 - 0.55AB - 1.40AC - 1.72BC \quad (9.3)$$

Strength loss (%) =

$$+7.86 + 5.92A + 3.14B + 0.09C - 1.18A^2 + 1.24B^2 + 0.79C^2 + 1.79AB + 0.14AC + 0.74BC \quad (9.4)$$

9.3.3 RESULTS OBTAINED FROM RESPONSE SURFACE GRAPH

The dyeing of silk fabric by plasma process is adequately explained by high values of coefficient of determination, but it would be inappropriate to determine the optimal value using regression analysis only. The effect of various factors influencing various responses, such as exhaustion %, K/S, fixation % and strength loss % for silk dyeing, has been represented by (3D) response surfaces plots.

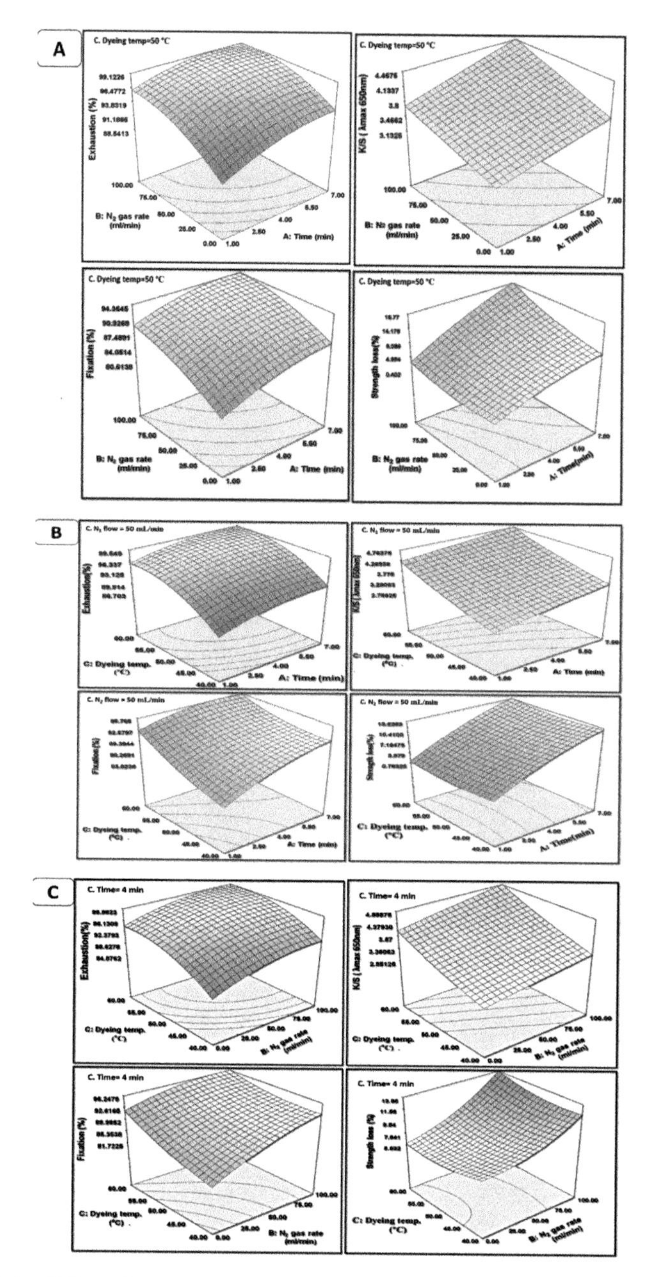

FIGURE 9.1 Effect of response for exhaustion %, K/S, fixation % and strength loss % on (A) dyeing temperature and nitrogen gas flow rate on actual factor for 4 min, (B) plasma treatment time and nitrogen gas flow rate on actual factor for 50°C and (C) effect of dyeing temperature and plasma treatment time on actual factor for 50 ml/min.

The plots provide a method to visualize the relationship between responses and experimental levels of each variable and the type of interactions between two test variables. Statistically significant model by Design-Expert programme generated the fitted response surface plot to understand the interaction of the parameters required for optimum exhaustion %, K/S, fixation % and strength loss %, respectively.

As the concentration of nitrogen flow rate, dyeing temperature and plasma treatment time increased, exhaustion %, K/S, fixation % and strength loss % also increased on the provided actual factor as shown in Figure 9.1, but the rate of increase in percentage for the obtained result depended on the input variables conditions. The absence of nitrogen gas flow rate for 0 ml/min, that is the presence of 450 ml/min flow rate of He gas only, showed the marginal effect on output variables. But in the presence of nitrogen gas, the effect goes on increasing with the nitrogen gas flow rate based on plasma treatment time. Dyeing temperature plays more important role in the case of an increase in exhaustion %. It can be seen that enlarging the nitrogen gas flow rate from 50 to 100 ml/min at 40, 50 and 60°C, respectively, resulted in a gradual improvement in the depth of colour on the fabric. It has been observed that increase in nitrogen gas flow rate helped to increase exhaustion %, which might be due to the formation of more amine and other polar groups. It might be correlated with the destruction of the outer layer due to the plasma etching and removal of a hydrophobic fatty layer from the surface of silk fibres as shown in SEM analysis, which in turn makes the fabric more accessible to water and dye molecules. Increase in the amount of dye fixation % was observed on increasing the nitrogen gas flow rate, which might be due to more amounts of dye sites generated on increase in nitrogen gas flow rate, which helps to retain more amount of dye after washing. Further, at constant nitrogen flow rate, increasing dyeing temperature also helped to retain more amounts of dye molecules inside the fabric. At high temperature, the kinetic energy of the dye molecules is higher. As a result, more amount of dye is moved to the interior of the fabric, increasing the dye fixation. A similar effect was also observed with respect to strength loss %. However, in the case of strength, our requirement was to achieve less strength loss % to maintain the inherent strength of the silk fabric as much as possible. It was observed that for N_2 gas flow rates of 100 ml/min, there was 13.8% loss in strength whereas, at 50 ml/min, there was only 7.84% loss in strength for 4-min plasma treatment time was observed. Higher strength loss might be due to the more etching of silk polymer with higher N_2 gas flow rate as shown in Figure 9.1(A). However, plasma treatment time was found to be a more dominant factor for loss in strength %. For 7-min plasma treatment time at 100 ml/min, N_2 gas flow rate strength loss was 18.75% due to more etching of the surface of the silk fabric in correlation with N_2 gas flow rate as shown in Figure 9.1(B). It can also be seen from the Figure 9.1(C) that prolonging the plasma treatment time at 60°C results in a gradual improvement in the depth of colour on fabric, which might be due to more penetration of the dye molecules into the fibre which shows increased swelling of fibre. However, beyond 4-min plasma treatment time, there was an only marginal increase in the dyeing of silk fabric. Samples showed higher exhaustion %, K/S value and fixation % as more amount of dye get adsorbed by the plasma-treated fabrics. Thus, at maximum flow rate of nitrogen gas and maximum treatment time, the highest percentage of exhaustion, fixation and K/S value were observed, which was good for our dyeing purpose. But at the same time, strength loss also increased. Hence, there was a need to optimize the input variables for better dyeing by maintaining the strength of the silk fabric.

9.3.4 VERIFICATION AND VALIDATION OF THE MODEL

In multi-response optimization, desirable weightage can be given to all responses, and for a combined influence of all responses, desirability can be determined by varying values of input parameters. The experiment with the highest desirability of the optimized sample based on the solutions was performed. Table 9.4 shows the predicted and experimental responses obtained on the basis of the highest desirability.

TABLE 9.4 Verification of Reynolds Stress Model for Acid Dyeing of Silk.

Process parameters, unit	Predicted	Experimental
Time (min)	4.00	4.00
N_2 gas rate (ml/min)	50.00	50.00
Dyeing temp. (°C)	60.00	60.00
Exhaustion (%)	98.76	99.49
K/S	4.51	4.46
Fixation (%)	95.37	94.95
Strength loss (%)	8.75	7.65

It can be observed that the predicted and experimental values of the responses of the optimized sample obtained by the solution with the highest desirability are almost similar as shown in Table 9.4. Hence, it can be inferred that the appropriate empirical model equations that are developed for acid dyeing of silk fabric process are valid. This verifies the suitability and significance of the response surface for BBD model to optimize the plasma-treatment process. The values predicted by the regression polynomial equation showed no significant difference from the experimental value which shows the validity of the predicted model.

9.3.5 COMPARISON OF UNTREATED SAMPLE WITH OPTIMIZED PLASMA-TREATED SAMPLE

At the same concentration and condition of acid dyeing of silk fabric, the comparison was done with the untreated sample and the optimized plasma-treated sample. Tables 9.5 and 9.6 show the comparison of the optimized sample with the untreated sample in terms of various performance parameters.

TABLE 9.5 Comparison of Untreated Sample with Optimized Plasma-treated Sample.

Sample	Exhaustion (%)	Fixation (%)	K/S	Strength loss (%)
Untreated	86.21	81.58	3.19	0.00
Optimized	99.49	94.95	4.46	7.65

It can be observed that the values of optimized plasma-treated silk fabric in terms of exhaustion %, K/S and fixation % were found to be higher in comparison with those obtained for untreated silk fabric after acid dyeing at same dye concentration of 2% shade, dyeing temperature at 60°C for time 60 min.

TABLE 9.6 Comparison of Fastness of Optimized Sample with Untreated Sample.

Sample	Fastness			
	Washing	Rubbing		Light
		Dry	Wet	
Untreated	4	4–5	4	5–6
Optimized	5	5	5	6–7

It can be observed that the washing, rubbing and light fastness for the untreated sample was found to be lower in comparison with the optimized plasma-treated sample.

9.3.6 PHYSICAL ANALYSIS OF UNTREATED AND OPTIMIZED PLASMA-TREATED SILK FABRICS USING SCANNING ELECTRON MICROSCOPY

Scanning electron microscopy micrograph of silk fabrics in Figure 9.2(a) shows the fibre surface of the untreated fabric. As shown in the figures, the untreated fabrics had a smooth surface, with no surface imperfections and no irregular surface compared with that of the treated fabric. The SEM image of the plasma-treated sample with optimized conditions such as 4-min treatment time and He/N_2 (450/50 ml/min) as shown in Figure 9.2(b) reveals that the surface of the material is rough and etched effectively due to the plasma treatment. On the surface of the fibre, the silk fabric treated with atmospheric plasma showed some cracks and particles. The etching leaves the surface of the fibres in the peeling off the pattern and non-uniform longitudinal flutes, small pits, micro-craters and striations are also visible on the surface. The increase in surface roughness might have helped in increasing surface area and capillary action resulting in better wicking of water and dye molecules which helps to increase in dye exhaustion percentage as compared with untreated silk fabric.

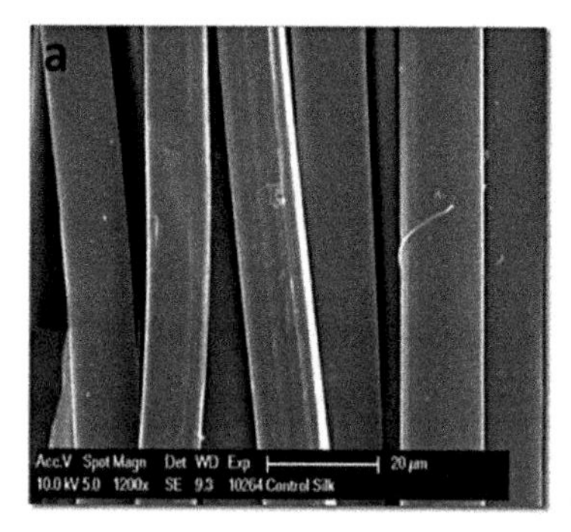

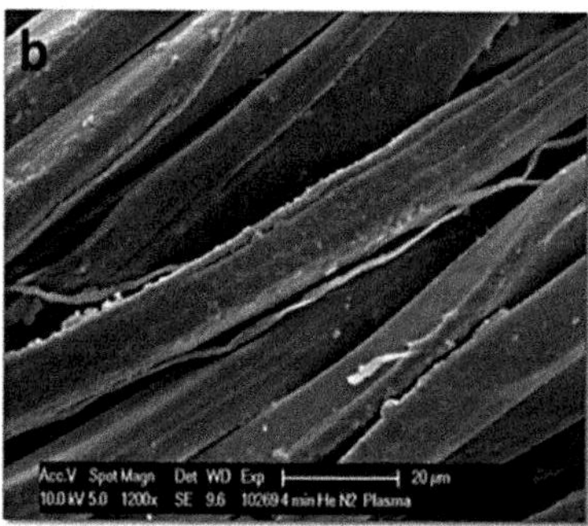

FIGURE 9.2 Scanning electron microscopy images (a) untreated (b) plasma-treated silk fabric.

9.3.7 CHEMICAL ANALYSIS OF UNTREATED AND PLASMA-TREATED SILK FABRIC USING ENERGY DISPERSIVE X-RAY SPECTROSCOPY

To study the change in atomic percentage, surface chemistry of the untreated and 4-min He (450 ml/min) and He/N_2 (450/50 ml/min) plasma-treated silk fabric was analysed under energy dispersive X-ray spectroscopy (EDX) spectroscopy. The elemental peaks are shown in Figure 9.3, and the atomic percentages are reported in Table 9.7. Energy dispersive X-ray shows the peaks for only C, N and O. In the untreated sample, the oxygen atomic percentage is 28.31% and it increased to 30.25% for He and 31.32% for He/N_2 plasma-treated fabric, respectively. Moreover, there was a decrease in carbon % for He/N_2 plasma to 47.50% from 50.66% for untreated silk fabric. The incorporation of more amount of oxygen and nitrogen in the plasma-treated sample might have helped in the formation of more amine and amide groups during the plasma treatment in the presence of helium and nitrogen gas. He/N_2 together help to give synergistic effect on the silk fabric. It might be due to the partial etching of the treated silk fabric in He/N_2 atmosphere and also for the generation of extra polar groups like COOH, NH_2 on the surface of the plasma-treated silk fabric. When plasma treatment was carried out an increase of O and N atoms and consequent decrease of C atoms, thus, corresponds to an enhancement of hydrophilicity of fabrics due to an increase of polar groups in the surface of the mulberry silk fabric. It also helps to interact with more number of dye sites and hence higher dye exhaustion.

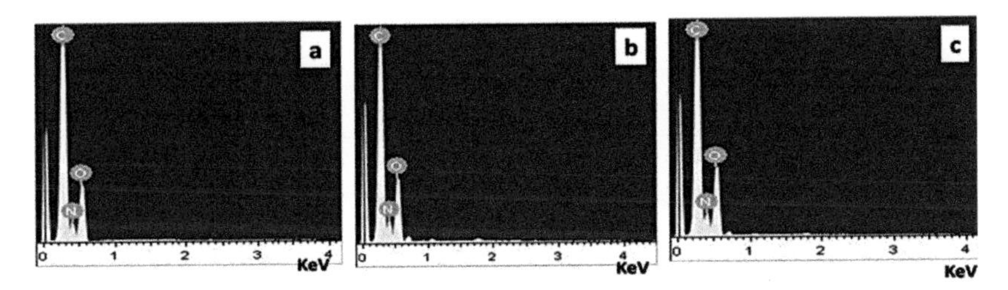

FIGURE 9.3 Energy dispersive X-ray images of the (a) untreated (b) He plasma-treated (c) He/N_2 plasma-treated silk.

TABLE 9.7 Results of Elemental Analysis for Untreated and Plasma-Treated Silk.

Different elements	Elements atomic (%)		
Elements	**Untreated**	**He plasma**	**He/N_2 plasma**
C	50.66	48.22	47.50
N	21.03	21.17	21.53
O	28.31	30.25	31.32

9.3.8 CHEMICAL ANALYSIS OF UNTREATED AND PLASMA-TREATED SILK FABRIC USING SECONDARY ION MASS SPECTROMETER

Time of flight secondary ion mass spectrometer was also used to analyse the surface chemistry of the silk fabric treated with plasma at 2-nm level. It can be seen from chemical images (molecular image) that OH^-/CH^- ratio is 0.88 in the untreated sample and it significantly increased to 2.22 on plasma treatment. A similar result was also observed for O^-/C^-. Due to the formation of more amine and amide groups, there was an increase in the NH^-/CH^- ratio from 0.04 to 0.11 in the untreated to plasma-treated samples, respectively (refer Table 9.8).

TABLE 9.8 The Molecular Ratio in the Different Samples Obtained from SIMS Chemical Images Data.

Molecular ratio	Untreated	He/N_2 Plasma treated
O^-/C^-	3.61	52.3
OH^-/CH^-	0.88	2.22
CN^-/CH^-	0.11	1.20
NH^-/CH^-	0.04	0.11

9.4 CONCLUSION

In general, conventional processes of optimization are time-consuming and expensive. Single-variable optimization methods are monotonous, and they can also lead to misinterpretation of results that are used to select the precise factors that influenced the process. The single-factor optimization cannot explain the actual interactions of the parameters of the experimental data because the interaction between different factors is overlooked, leading to a misinterpretation of the results. The results obtained for the silk fabrics showed higher dyeability compared with normal silk fabric due to the combined effect of plasma treatment and modified process conditions. Optimum condition for K/S of 4.46, fixation of 94.95% and strength loss of 7.65% were achieved with Reynolds Stress Model using Design-Expert 6 software at treatment time of 4 min, nitrogen gas flow rate of 50 ml/min, helium gas flow rate of 450 ml/min and dyeing temperature at 60°C for 60 min. The fit of the model was checked by the determination of R^2. The predicted values of the regression polynomial equation showed no significant difference from the experimental value, which shows the validity of the predicted model. Research is going on with other class and colour of suitable dyes on these optimized parameters.

Hence, within the range of independent variables studied here, any set of nitrogen (N_2) gas flow rate (with fixed He gas flow rate of 450 ml/min), plasma time and dyeing temperature, performance of the silk-dyed fabric in terms of K/S, fixation and strength loss can be predicted accurately. Atmospheric pressure He/N_2 plasma can be suitably used for colouration of silk fabric with acid dye at a significantly lower temperature.

9.5 ACKNOWLEDGEMENT

One of the authors **Mr. Pintu Pandit** acknowledges the scholarship support from the UGC-SAP, Govt. of India during his study period.

KEYWORDS

- **Atmospheric plasma**
- **mulberry silk**
- **Box–Behnken design**
- **EDX**
- **low-temperature dyeing**

REFERENCES

1. Basak, S.; Samanta, K. K.; Chattopadhyay, S. K.; Pandit, P.; Maiti, S. Green Fire Retardant Finishing and Combined Dyeing of Proteinous Wool Fabric. *Colour. Technol.* **2016,** *132*(2), 135–143.
2. Bhat, N. V.; Netravali, A. N.; Gore, A. V.; Sathianarayanan, M. P.; Arolkar, G. A.; Deshmukh, R. R. Surface Modification of Cotton Fabrics Using Plasma Technology. *Text. Res. J.* **2011,** *81*(10), 1014–1026.
3. Cireli, A.; Kutler, B.; Mutlu, M. Surface Modification of Polyester and Polyamide Fabrics by Low Frequency Plasma Polymerization of Acrylic Acid. *J. Appl. Polym. Sci.* **2007,** *104*, 2318–2322.
4. Gulrajani M. L.; Agarwal R.; Grover A.; Suri M. Degumming of Silk with Lipase and Protease. *Indian J. Fibre Text. Res.* **2000,** *25*(1), 69–74.
5. Hocker, H. Plasma Treatment of Textile Fibres. *Pure Appl. Chem.* **2002,** *74*,423–427.
6. Iriyama, Y.; Mochizuki, T.; Watamabe, M.; Utada, M. Preparation of Silk Film and Its Plasma Treatment for Better Dyeability. *J. Photopoly. Sci. Technol.* **2003,** *16*(1), 75–80.
7. Kiran, H. K.; Deasi, A. N. Atmospheric Pressure Plasma Treatment of Textile Using Non- Polymerising Gases. *Indian J. Fibre Text. Res.* **2011,** *36*, 289–299.
8. Pandey, R.; Patel, S.; Pandit, P.; Nachimuthu, S.; Jose, S. Colouration of Textiles Using Roasted Peanut Skin—An Agro Processing Residue. *J. Clean Prod.* **2018,** *172*, 1319–1326.
9. Prabhu, K. H.; M. D. Teli.; Waghmare, N. G. Eco-friendly Dyeing Using Natural Mordant Extracted from *Emblica officinalis* G. Fruit on Cotton and Silk Fabrics with Antibacterial Activity. *Fibers Polym.* **2011,** *12*, 753–759.
10. Sah, M. K.; Kumar, A.; Kumar, P. The Extraction of Fibroin Protein from Bombyx Mori Silk Cocoon: Optimization of Process Parameters. *Int. J. Bioinf. Res.* **2010,** *2*(2), 33–41.
11. Samanta, K. K.; Jassal, M.; Agrawal, A. K. Improvement in Water and Oil Absorbency of Textile Substrate by Atmospheric Pressure Cold Plasma Treatment. *Surf. Coat. Technol.* **2009,** *203*(10), 1336–1342.
12. Teli, M. D.; Rane V. M. Comparative Study of the Degumming of Mulberry, Muga, Tasar and Ericream Silk. *Fibres Text. East. Eur.* **2011,** *19*(2), 10–14.
13. Teli, M. D.; Pandit, P. Novel Method of Ecofriendly Single Bath Dyeing and Functional Finishing of Wool Protein with Coconut Shell Extract Biomolecules. *ACS Sustain. Chem. Eng.* **2017,** *5*(9), 8323–8333.
14. Teli, M. D.; Pandit, P. A Novel Natural Source Sterculia foetida Fruit Shell Waste as Colorant and Ultraviolet Protection for Linen. *J. Nat. Fibers.* **2017**. http://dx.doi.org/10.1080/15440478.2017.1328327.
15. Teli, M. D.; Pandit, P. Development of Thermally Stable and Hygienic Colored Cotton Fabric Made by Treatment with Natural Coconut Shell Extract. *J. Ind. Text.* **2017**. https://doi.org/10.1177/1528083717725113.
16. Teli, M. D.; Pandit, P. A Novel Natural Source *Sterculia foetida* Fruit Shell Waste as Colorant and Ultraviolet Protection for Linen. *J. Text. Assoc.* **2017,** *76*, 293–297.

17. Teli, M. D.; Samanta, K. K; Pandit, P.; Basak, S.; Chattopadhyay, S. K. Low-Temperature Dyeing of Silk Fabric Using Atmospheric Pressure Helium/Nitrogen Plasma. *Fibers Polym.* **2015,** *16*(11), 2375–2383.
18. Teli, M. D.; Pandit, P.; Samanta, K. K. Application of Atmospheric Pressure Plasma Technology on Textile. *J. Text. Assoc.* **2015,** *75*(6), 422–428.
19. Teli, M. D.; Pandit, P.; Basak, S. Coconut Shell Extract Imparting Multifunction Properties to Ligno-cellulosic Material. *J. Ind. Text.* **2017.** https://doi.org/10.1177/1528083716686937.
20. Tyagi G. K.; Singh A.; Gupta A.; Goyal A. Effect of Add-on Finish and Process Variables on Properties of Air-Jet Spun Polyester Yarns. *Indian J. Fibre Text. Res.* **2003,** *28*(6), 163–169.
21. Verschuren, J.; Kiekens, P.; Leys, C. Textile-Specific Properties that Influence Plasma Treatment Effect Creation and Effect Characterization. *Text. Res. J.* **2007,** *77*, 727–733.

CHAPTER 10

PARTIAL REPLACEMENT OF CARBON BLACK BY RICE HULLS AND SOY SHORT FIBRES AS FILLERS IN NATURAL RUBBER COMPOSITES: EFFECT ON MECHANICAL PROPERTIES

AJAY VASUDEO RANE[1], KRISHNAN KANNY[1], ABITHA V. K.[2], and SABU THOMAS[2]

[1]*Composites Research Group, Department of Mechanical Engineering, Durban University of Technology, Durban 4000, South Africa*

[2]*School of Chemical Sciences, Mahatma Gandhi University, Kottayam 686560, Kerala, India, E-mail: ajayrane2008@gmail.com*

CONTENTS

ABSTRACT

Short and long fibre filled polymer composites form a comparatively new family unit of materials, in spite of them being well established in numerous applications by now. There is an enormous range of materials in this category, some offering inimitable properties, some merely contending with other materials because of their moderately low cost. Their potential advantages are far from being fully realized, and we foresee continued growth in their use for many years to come. Research into these materials is essential to their development and exploitation and will be for many years to approach. A special feature of the chapter is that it includes significant discussion on natural rubber (NR)-matrix fibre composites, an important subclass of short fibre reinforced composites that is often neglected in reviews of polymer composites.

The aim of the paper is to report the partial replacement of carbon black with green fibre (rice husk and soy) and the reinforcing effect of green fibre (rice husk and soy) on NR. Natural rubber/rice husk and NR/soy compound were prepared using a laboratory size two-roll mill. The curing characteristics of the mixes were studied using oscillating disc rheometer, and the samples were vulcanized at 160°C. The mechanical testing of the vulcanizates involves hardness, tensile strength, elongation at break, compression set, specific gravity, abrasion resistance and tear strength. Thermal stability properties were also investigated.

10.1 INTRODUCTION

In general, any material consisting of two or more components with different properties and distinct boundaries between the components can be referred to as a composite material. Moreover, the idea of combining several components to produce a material with properties that are not attainable with the individual components has been used by man for thousands of years. There are two groups of composites—the first group comprises composites that are known as 'filled materials'. The existence of some basic or matrix materials, whose properties are improved by filling it with some particles, is the main feature of these materials. The second group involves composites that are called 'reinforced materials'. The basic components of these materials are long or short but thin fibres possessing high strength and stiffness. Filler (particulate or thin fibres) is a solid material which is capable of changing the physical and chemical properties of materials by interacting it with the surface or its lack thereof and by its own physical characteristics. Carbon fibres were a natural step aiming at a rise of fibre's stiffness, the proper level of which was not exhibited by glass fibres. Modern high-modulus carbon fibres demonstrate modulus that is by the factor of about four higher than the modulus of steel, whereas the fibre density is by the same factor lower. Though carbon fibres had lower strength than glass fibres at the beginning, modern high strength fibres demonstrate tensile strength that is 40% higher than the strength of the best glass fibres, whereas the density of carbon fibres is 30% less. High stiffness means that material exhibits low deformation under loading. However, in spite of stiffness being an important property, we do not mean that it should be necessarily high. Ability of structure to have controlled deformation can be also important. Shortage of material strength results in uncontrolled compliance, that is in failure after which a structure does not exist anymore. Usually, we need to have as high strength as possible. Thus, the structure cannot exist without controlled stiffness and strength. On account of the small size of the parts, material costs tend to be less of a concern for electronic composites than for structural composites.

Rubber, natural or synthetic, has now become an invaluable part of our day-to-day lives. But without any reinforcement, it has limited applications because of its feeble mechanical properties. The reinforced rubber composite has 10 times more strength than that of the unreinforced rubber.

Today we have a wide variety of products made from rubber composites ranging from automobile tyres to seal, valves and gaskets because of the excellent mechanical properties possessed by rubber composites. But the majority of these composites contain considerable amount of synthetic fillers such as silica and carbon black which are not only expensive and toxic but are also derived from non-renewable sources of energy. To make the matters worse, fillers such as silica also contribute to large amount of ash when used tyres are disposed of by incineration. Moreover, one of the most significant sources of greenhouse gas emissions is the production of carbon black.

Three criteria on the basis of which filler should be selected for rubbers are as follows:

1. The material must be capable of sustainable supply and should have a stable cost, which generally means no significant competition with other uses.
2. The preparation process must be easy to control, be uncomplicated and environmental friendly.
3. The material must be immediately available and with no or limited need to modify them.

To make eco-friendly products and to decrease global warming, the trend to use biomaterials has increased exponentially.

Green fibres, such as rice husk, soy filler, are abundant natural waste materials produced agriculturally and so are readily available. They are also ethically, economically and environmentally less undesirable as they do not compete with food products, are sustainably renewable and their use helps reduce the environmental and economic cost of waste disposal problems.

Green fibres are mainly composed of inorganic and organic parts (cellulose, protein etc.). The organic part in the green fibre helps to interweave the rubbery matrix and fibre particles and thus act as a compatibilizer. This helps to improve the reinforcing mechanism. The chemical composition of rubber is typically composed of a long hydrophobic hydrocarbon chain. Thus, the inclusion organic molecules from the green fibre may be useful for acting as a compatibilizer between the rubber matrix and the fibre and so make them more miscible.[8]

The raw polymer, natural rubber (NR), having considerable strength and appreciable elasticity and resilience at room temperature, is sensitive to hot and cold and is liable to oxidize to a sticky product. Today, apart from the adhesive solutions based on raw rubber, probably the only product made and used in the raw un-vulcanized state is the crepe sole for footwear.

Natural rubber is one of the most important elastomeric materials for innumerable commercial applications. However, NR by itself does not possess a high enough modulus and other mechanical properties, for which it needs to be reinforced.

Most of the synthetic rubbers used today are petrochemical derivatives. The petrochemical industry is based on non-renewable source of energy, also the ever dwindling fuel prices add to the already high cost of virgin polymer. Thus, addition of filler not only enhances the physical properties of the virgin rubber but also imparts cost-effectiveness to the final moulding material. The pollution-causing behaviour is the only demerit of using fillers, as it adds to the toxic emission caused due to the burning and recycling of filled rubber composite material. This can be overcome by replacing fillers with green fibre partly or completely. The aim of work on hand is to replace fillers, such as carbon black used in the rubber–filler composites partly by various green fibres, namely rice husk, soy filler and evaluate its performance. The use of green fibre not only lowers the emission of toxic gases but also helps in the agricultural waste management.

The objectives of this research work include development of a new class of biomaterial which is completely obsolete of synthetic fillers, namely rice husk and soy filler for rubber reinforcement, and also include development of NR-based green rubber composites using green fibres at variable loading from 0 to 50 phr by partly replacing the traditional carbon black filler. We shall also study the effects on rheological, mechanical and physical properties of the green rubber compounds when biomaterials are used as fillers.

In order to investigate the factors that can influence the properties of rubber composites, such as filler structure, filler–filler interaction and filler–rubber interaction, mechanical characterization of the green rubber composite was done. Based on the outcome of the characterizations of green rubber composites, the final aim was to develop green rubber for commercial applications in actual service conditions as well as to promote the use of green fibres for other applications.

The development of this research work will be of general interest and will benefit technologists in developing a new class of rubber products which are efficient and eco-friendly. It will also hugely benefit the agricultural fraternity as it will open up a huge market for waste and non-edible products.

10.2 GREEN FIBRES FOR COMPOSITE PRODUCTION

As a potential substitute for glass-filled composites over past few years, especially in the automotive sector, natural fibre-reinforced polymer composites have gained a worldwide acceptance. As natural fibres are lesser in weight, easier to handle, non-abrasive and cost effective, composites made from them are also sustainable and economical. Every natural fibre has its own surface morphology which decides the interfacial matrix–fibre adhesion. Some natural fibres such as piassava have fine protrusions containing silica on their surface which enhances interlocking at the fibre–matrix interface and thereby improving mechanical properties of the composites. Hydrophilic nature of natural fibres creates wetting issues with the resin matrix which are mostly improved by chemical, radiation and corona treatments. The mechanical properties of natural fibre composites (NFCs) generally increases with volume fraction of fibres up to an optimum level, and then tend to decrease.[2]

Natural fibres are generally classified in the writing as being derivative from plant, animal or mineral sources according to their origin. Plant fibres are composed of cellulose. Common examples include cotton, linen, jute, flax, ramie, sisal and hemp. These fibres are extracted from the fruits, seeds, leaves, stem and skin of plants. Hence, they are categorized as seed fibre (collected from seeds or seed cases, e.g. cotton and kapok), leaf fibre (collected from leaves, e.g. sisal and agave), bast fibre or skin fibre (collected from the skin or bast surrounding the stem, e.g. jute, kenaf, hemp, ramie, rattan, Soybean, vine and banana fibres), fruit fibre (collected from the fruit of the plant, e.g. coconut, coir fibre) and stalk fibre (stalks of the plant, e.g. straws of wheat, rice, barley and other crops including bamboo, grass and tree wood). The most widely used natural fibres are cotton, flax and hemp, although sisal, jute, kenaf and coir are equally popular, classification given in Table 10.1.

10.2.1 ADVANTAGES OF USING GREEN FIBRES AS REINFORCEMENT IN RUBBER

Nowadays, environmental concerns have resulted in rehabilitated interest in natural resources. For the introduction of materials and products, recyclability and environmental safety are becoming increasingly important.

TABLE 10.1 Classification of Fibres.

Bast	Leaf	Seed					Core	Grass/reeds	Other
		Fibre	Pod	Husk	Fruit	Hulls			
Hemp	Pineapple	Cotton	Kapok	Coir	Oil palm	Rice	Kenaf	Wheat	Wood
Ramie	Sisal		Loofah		Soy	Oat	Jute	Oat	Roots
Flax	Agava		Milk weed			Wheat	Hemp	Barley	Galmpi
Kenaf	Henequen					Rye	Flax	Rice	
Jute	Curaua							Bamboo	
Mesta	Banana							Bagase	
Urena	Abaca							Corn	
Roselle	Palm							Rape	
	Cabuja							Rye	
	Albardine							Esparto	
	Raphia							Sabai	
								Canary grass	

Natural fibres such as flax, hemp, banana, sisal, oil palm, and jute have a number of techno-economical and biological advantages over man-made fibres such as glass fibres.[12] The combination of interesting mechanical and physical properties together with their environmentally friendly character has triggered interest by a number of industrial sectors, notably the automobile industry.

The major advantages are the following:

1. Low specific weight, which results in a higher specific strength and stiffness when compared with glass-reinforced composites.
2. Renewable resource with production requiring little energy. CO_2 is used whereas oxygen is given back to the environment.
3. Processing atmosphere is friendly with better working conditions, and therefore there are reduced dermal and respiratory irritations.
4. High electrical resistance.
5. Thermal recycling is possible.
6. Good thermal and acoustic insulating properties.
7. Biodegradability.

10.2.2 DISADVANTAGES OF USING GREEN FIBRES AS REINFORCEMENT IN RUBBER

The disadvantages include

1. Enormous variability
2. Poor moisture resistance
3. Poor fire resistance
4. Lower durability

It is quite clear that the advantages outweigh the disadvantages, and most of the shortcomings have remedial measures in the form of chemical treatments. The natural fibres that are generally used in rubber composites are lignocellulosic fibres which have an advantage over synthetic ones as they buckle rather than break during processing and fabrication.

The fibrous reinforcing constituent of composites may consist of thin continuous fibres or relatively short fibre segments. When using short fibre segments, fibres with high aspect ratio (length to diameter ratio) are used. Continuous fibre reinforced composites are generally required for high performance structural applications. The specific strength (strength to density ratio) and specific stiffness (modulus to density ratio) of continuous carbon fibre-reinforced composites can be superior to conventional metal alloys.

For short fibre reinforcement, different types of elastomers have been used as matrices. Typically, the matrix has considerably lower density, stiffness and strength than that of the reinforcing fibre material, but the combination of matrix and fibre produces high strength and stiffness while still possessing a relatively low density.

In a composite, the matrix is required to fulfil the following functions:

1. To bind together the fibres by virtue of its cohesive and adhesive characteristics
2. To protect them from environments and handling
3. To disperse the fibres and maintain the desired fibre orientation and spacing
4. To transfer stresses to the fibres by adhesion and/or friction across the fibre matrix interface when the composite is under load, and thus to avoid any catastrophic propagation of cracks and subsequent failure of the composites
5. To be chemically and thermally compatible with the reinforcing fibres
6. To be compatible with the manufacturing methods which are available to fabricate the desired composite components

10.3 LONG PLANT FIBRE AS REINFORCEMENT IN RUBBER

Long fibre composites are the pinnacle of thermoplastic performance. They combine high levels of stiffness, strength and toughness together in a single material. No other method of reinforcing thermoplastics is able to match their performance trifecta of crucial properties. The high mechanical performance characteristics of long fibre composites is the reason they are often chosen as substitutes for metals, as a replacement for underperforming plastics or as alternatives to higher cost engineering polymers through up-engineering of lower cost plastics. Longer length, or higher aspect ratio, of reinforcing fibres provide long fibre composites with increased strength, which translates into the ability to resist deformation or creep under loads and fatigue endurance with minimal compression.[8] More fibre surface area provides the ductile polymer with more ability to grab onto and transfer stress to the stronger internal fibre skeleton formed during component moulding. It is important to maintain maximum fibre length through careful control of processing parameters. Fibre attrition can occur from shear in the injection moulding press or from tight runner radii and improper gating in the mould. Performance can be reduced by significant reduction in median fibre length. Orientation of reinforcing fibres within injection-moulded components also significantly influences composite strength. Although long fibres intertwine to form an internal structural skeleton, they also align in the direction of polymer flow as moulds fill. To obtain maximum performance, design moulds so fibres align perpendicular to the direction of stress forces during component use.

Typically, stiffer polymers are more brittle. However, with long fibre composites, the longer length of reinforcing fibres inverts that analogy. The higher aspect ratio of the reinforcement in long fibre composites facilitates more efficient energy transfer between the polymer and fibre upon impact and dissipates those forces throughout the composite structure instead of localizing them in one area. The high toughness of long fibre composites makes them one of the most durable of polymeric materials with structural characteristics. Moreover, it also makes them ideal for applications that experience repeated impact forces but need to retain their shape without permanent deformation or deterioration of function. Superior energy dissipation also increases their sound and vibration dampening capabilities. The inclusion of long fibre reinforcement helps composites resist cracking and impedes crack propagation by forming an internal fibre skeleton. Although long fibre composites offer excellent impact resistance, designs can permit failure at specified loads to prevent damaging forces from transferring to adjacent systems. Longer fibre length also minimizes fragmentation during failure. In addition, long fibre composites retain a significant amount of their durability at elevated and low temperatures making them desirable for devices exposed to varying climate.

10.4 SHORT PLANT FIBRE AS REINFORCEMENT IN RUBBER

Short fibre rubber composite can be defined as a compounded rubber matrix containing discontinuous fibres that are distributed within the rubber to form a reinforcement phase. Quite a number of researches have been carried out on natural and synthetic fibre-reinforced rubber composites. The reinforcement of an elastomer with fibres combines the elastic behaviour of rubber with the strength and stiffness of the reinforcing fibre. Short fibres are also used to improve or modify certain thermodynamic properties of the rubber for specific applications or to reduce the cost of the fabricated articles. According to O'Connor, the limitations to date of short fibres in rubber compounding applications have been due to difficulties in achieving uniform dispersion of fibres, fibre breakage during processing and difficulties in handling, incorporating and bonding the short fibres into the rubber matrix. To use a fibre-reinforced elastomeric composite most effectively, a basic understanding must be obtained of how the various properties depend on the fibre properties, the matrix properties and the processing methods. The presence of fibres in a rubber increases the modulus, but other improvements are also sought with fibre addition. The possibility of much improved properties brought about by the alignment of fibres, to give a composite with some measure of anisotropy, is of great interest.[4] In 1936, Guth et al. derived an equation for calculating the modulus of a fibre-reinforced matrix, applicable to fibre-reinforced rubbers, which has been much quoted. This equation is commonly referred to as the 'Guth Gold' equation and is expressed as:

$$G = G_o (1 + 0.67 fc + 1.62 f^2c^2)$$

Where G = modulus of composite material, G_o = modulus of matrix material, f = length to diameter ratio (aspect ratio) of fibre and c = volume concentration of fibre.

When the fibre aspect ratio is in the range 10–50, moduli ratios of 102−103 can be achieved if there is good adhesion between the fibre and the matrix.

A later modification taking into account the fibre orientation is the Boustany–Coran equation:

$$E_{comp} = E_r \{1 + K_f V_f [26 + 0.85 (l/d)]\}$$

Where E_{comp} = modulus of the composite, E_r = modulus of rubber, K = constant, f = a function of fibre orientation, l/d = fibre aspect ratio and V_f = fibre volume fraction.

The use of natural short-length cellulose fibre as rubber reinforcement claimed that this fibre, being naturally available in short lengths (from wood pulp), would circumvent the complicated route of making short fibre from synthetic polymer and thus would be more cost-effective. The authors reported that the oriented cellulose fibres increases tensile strength, but the most important changes were in increased modulus and reduced elongation at break. The modulus of the fibre-reinforced materials, as defined by Young's modulus, showed a steady increase with increasing fibre concentration. Tensile strength, however, first decreased with increasing fibre level and then increased beyond that of the matrix material at higher fibre loadings. This was explained by the fact that at low concentrations, the matrix is not restrained by enough fibres, and high strains occur in the matrix at low stresses, causing debonding. The matrix strength is thus reduced by the debonded fibres. When enough fibres are present, as concentration increases, the matrix is restrained. Eventually, at high concentrations, fibre orientation becomes difficult because the rubber matrix cannot easily flow around the fibres. This further causes a reduction in tensile strength again.

10.5 NATURAL RUBBER AND POLYISOPRENE

10.5.1 NATURAL RUBBER

Natural rubber (Fig. 10.1) can be isolated from more than 200 different species of plant, including surprising examples such as dandelions. *Hevea Brasiliensis* is the only tree source that is commercially significant. Latex is an aqueous colloid of NR, and is obtained from the tree by 'tapping' into the inner bark and collecting the latex in cups. The latex typically contains 30–40% dry rubber by weight, and 10–20% of the collected latex is concentrated by creaming, or centrifuging, and used in its latex form. Historically, such latex would be exported to consumer countries, but as it is expensive to ship a product with a high percentage of water, consumer companies are increasingly setting their latex processing plants in the producer countries, where the cheaper labour rates are an additional incentive. The remaining latex is processed into dry rubber as sheets, crepes and bales.

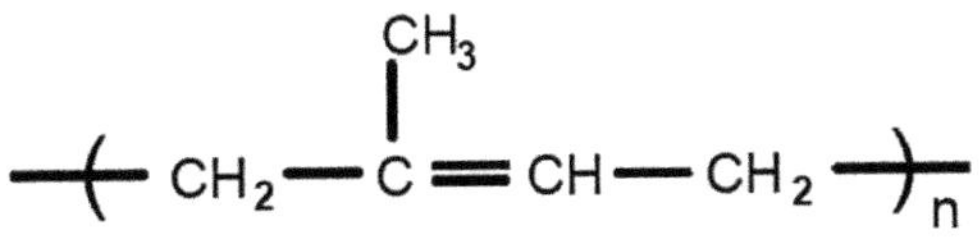

$$\left(CH_2 - \overset{\displaystyle CH_3}{\overset{|}{C}} = CH - CH_2 \right)_n$$

FIGURE 10.1 Structure of natural rubber.

1. Ribbed smoke sheets
2. White and pale crepes
3. Estate brown crepes
4. Compo crepes
5. Thin brown crepes
6. Thick blanket crepes
7. Flat bark crepes
8. Pure smoked crepe

Under each category, there are generally up to 5 divisions, for example 1RSS, 2RSS, 3RSS, 4RSS, 5RSS for ribbed smoked sheets (RSSs); the higher the number, the more inferior the quality. The Malaysian rubber industry has, however, played a pioneering role in producing NR grades to technical specifications, and this system is being followed by other producer countries. Currently, the following countries sell technically specified grades:

1. SMR – Standard Malaysian Rubber
2. SIR – Standard Indonesian Rubber
3. SSR – Specified Singapore Rubber
4. SLR – Standard Lanka Rubber
5. TTR – Thai Tested Rubber
6. NSR – Nigerian Standard Rubber

Natural rubber can be cross-linked by the use of sulphur, sulphur donor systems, peroxides, isocyanate cures and radiation, although the use of sulphur is the most common method. The sulphur vulcanization of NR generally requires higher added amounts of sulphur, and lower levels of accelerators than the synthetic rubbers. Sulphur contents of 2–3 phr and accelerator levels of 0.2–1.0 phr are considered to be conventional cure systems.[10]

Natural rubber can yield a hard rigid thermoplastic with excellent chemical resistance when cured with over 30 phr of sulphur. Such a product is termed ebonite. Natural rubber requires a certain degree of mastication (reduction in molecular weight) to facilitate processing, although the advent of constant viscosity grades and oil extended grades has substantially reduced the need for mastication. Peptisers are often used to facilitate breakdown of the rubber during mixing, although quantities of greater than 0.6 phr can cause a reduction in the final level of physical properties.

10.5.2 USES

The uses of NR are myriad, and a complete summary is not really possible. Its unique and excellent properties are utilized in tyres, shock mounts, seals, isolators, couplings, bridge bearings, building bearings, footwear, hose, conveyor belts, plant linings and many other moulding applications. Lattices and solutions are used to produce adhesives, carpet backings, upholstery foam, gloves, condoms and medical devices such as catheters. Natural rubber is also frequently used in blends with other elastomers.

10.6 RICE HULLS

The hard protecting coverings of grains of rice are called rice hulls (or rice husks). In addition to protecting rice during the growing season, rice hulls can be put to use as building material, fertilizer, insulation material or fuel. Rice hulls are the coatings of seeds, or grains, of rice. To protect the seed during the growing season, the hull is formed from hard materials, including opaline silica and lignin. The hull is mostly indigestible to humans. Winnowing, used to separate the rice from hulls, is to put the whole rice into a pan and throw it into the air while the wind blows. The light hulls are blown away, whereas the heavy rice falls back into the pan.[5]

Polymer composites with different biomasses are being used widely. Probably the most common and popular types are wood plastic composites (WPC). Although constant research has resulted in improvements in these composites, the end users have been limited to more or less decking and fencing.

Polymeric composites with rice hulls are the latest addition to the composites 'family', and their potential is very exciting. These virtually new composites have greater possibilities for end users than even WPC because silica in the rice hulls gives additional structural strength, versatility, durability along with lower costs. Moreover, it is possible to achieve aesthetically pleasing veneers that will be important as ideal substitutes for natural wood. With rice hulls available in abundance in most countries at very little cost and being cheaper than other biomasses, the possibilities are immense. The standard process for converting polymer composites to end products such as for decking and fencing has been extrusion, but now polymer composites with rice hulls in pellet form are spreading into injection moulding, which forms >60% of the plastic processing industry and will soon spread to compression moulding and other processes. As soon as newer resins with polymeric composites with rice hulls in pellet form are made available by resin manufacturers using modified polymers and combinations of additives, this will turn into a reality. Technology of polymer composites with rice hulls is virtually new, and we present valuable information to producers and end users from the raw materials to the production processes of these composites resins to the manufacturing technology of the end products and recommendations for end applications.[6]

10.7 SOY

Dietary fibre is a class of compounds that includes a mixture of plant carbohydrate polymers, both oligosaccharides and polysaccharides, for example cellulose, hemicelluloses, pectic substances and gums that may be associated with lignin and other non-carbohydrate components (e.g. polyphenols, waxes, saponins, cutin, phytates and resistant protein). Fibres extracted from some grains and seeds present physical and functional properties that make them useful for the food industry, which encourages researchers to search for novel raw materials that meet the needs of these areas, with a particular focus on food industry residues, such as malt bagasse, oat and rice hulls, or on residues from agriculture, such as the fibrous residue of banana pseudo-stems. Every year, billions of pounds of rice hulls are generated by rice-producing countries and most are thrown away as a waste by-product. Rice hulls represent approximately 20% of the dry weight of the rice harvest. Rice hulls consists 36–40 g/100 g cellulose and 12–19 g/100 g hemicelluloses, and they also contain fats, gums, alkaloids, resins, essential oils and other cytoplasmic components (extractives), and with an ash composition of approximately 12 g/100 g, which are made primarily of silica (80–90 g/100 g). Due to the high silica content present, rice hulls have not yet been exploited in the food and feed industry. Hence we propose to use in NR composite, but the removal of silica can be an alternative to convert this residue in a suitable fibre-rich food ingredient.[11]

Soy protein, a protein isolated from soybean, is a widely used food ingredient owing to its emulsification and texturizing properties. It is generally considered as the storage protein held in discrete particles called *protein bodies*. Soy protein can aid in preventing heart problems and therefore, of late, its popularity has increased at a fast rate. Soy proteins can be classified into different categories, such as soy protein isolate, soy protein concentrate and textured soy protein (TSP), depending on the method of production.[11] Soy protein isolate contains about 90% protein and is the most refined form of soy protein. Soy protein concentrate is basically soybean without the water-soluble carbohydrates, containing about 70% protein. Textured soy protein is made from soy protein concentrate by giving it some texture. The major proteins stored in soybean are globulins. Due to their hydrophilic nature, soy protein films do not have good mechanical and barrier properties as do most protein films, and they are used to produce flexible and edible films. Soy protein has been regarded as a practical substitute to the petroleum polymers in the manufacture of plastics, adhesives and packaging materials. Water sensitivity and poor mechanical performance, which limit the applications of soy protein materials, could be resolved, for example, by blending with polymer matrix.

10.8 REQUIREMENTS OF GREEN FIBRE COMPOSITES

Green composites still need to pass certain requirements to reach a large-scale of utilization. Fully biodegradable polymers and their composites must have the ability to satisfy urgent market needs. Nearly 30% of the plastics in municipal waste originate from goods that have been less than one year in use and tend to be heavily soiled by food and organic residues. Biodegradable alternatives could replace a large part of this voluminous part of plastic waste, which is difficult to dispose of or to recycle. The waste could be diverted from landfills and incineration to composting sites near the end user. Bio-composites and biopolymers are suitable for the packaging industry, especially for producing goods that are not subject to durability and are likely to end up soiled with organic matter.[2] As far as quality and processing performance are concerned, green composites must compete with the present plastics. Bacterial polyesters, for example, meet various quality and processing performance requirements. These materials can be processed by all types of thermal manufacturing. However, the esters are not flexible enough to form films or foils. They also tend to become brittle and to lose their vapour barrier properties. It is expected that these limits will be overcome by improving blend formulations.

10.8.1 WHY GREEN COMPOSITES?

Often when pursuing research into green composites, we say that we are protecting the environment and that we are working for nature. We may as well stop kidding ourselves—nature will be fine, nature will work out OK and adapt to the changes. If we do not change the way we are at present, it's humans that will cease to exist. Some scientists and engineers have realized that they need to take responsibility for the outcome of their work. Researching ways of creating faster machines and bigger toys, without due consideration of the effects on the environment or on people, is irresponsible. In this context, we are defining green composites as composites that are designed with the lowest environmental 'footprint' possible. Furthermore, we are focusing on fibre-reinforced polymer composites in this chapter as these are the most abundant material group of the composite family in use. All the usual selection criteria, such as ultimate mechanical properties, interfacial adhesion with the matrix (high or low for strength or toughness applications respectively), cost, availability of resources, chemical properties, resistance to moisture and so on, determine the choice of fibre. It is also determined by the 'green' criteria—whatever those might be. It is necessary to consider all issues impacting our 'footprint' if we are to take ourselves and our research seriously when we say we are working on 'green composites'.[2]

10.8.2 REUSE, RECYCLING AND DEGRADATION OF GREEN FIBRES COMPOSITES

The concern for environment and sustainable growth has created more awareness among the researchers to develop composites based on recycled materials and materials from nature. Reuse and recycling extends the useful life of the raw material resources in instances where a market exists for the recycled products, and it is economical to carry out collection and reprocessing. Recycling and reuse of materials have been getting more interest nowadays.

Green fibre composites are defined as materials that contain a reinforcement (such as natural plant fibres), supported by a polymeric binder or matrix material. Composites are generally based on a thermoset matrix such as elastomers, unsaturated polyesters, epoxies or phenolics, although thermoplastic matrices are also used. The use of composites has been growing steadily since their introduction in the 1960s. Applications of composites include automotive hoods, storage tanks, aircraft parts, bath tubs and vanities. Polymer-based composites will be increasingly used in auto body applications, because they do not rust, are lightweight, exhibit negligible fatigue and can better handle minor impacts with less damage, compared with traditional metal panels.

The majority of polymer-based composite materials are used in sheet moulding compound or sprayed, chopped, fibre applications, whereas the dominant composite used in the aerospace industry is carbon-epoxy composite. A particular feature of elastomeric and thermoset composite manufacturing which is distinctive from that of thermoplastics is the high scrap rate and significant amounts of off-cuts and rejects that are generated. This is in part due to the manual lay-up often used in composite production.[1]

Although there are now sufficient quantities of elastomeric and thermoset composites available for recycling processes, there exists a number of barriers such as elastomeric and thermoset polymers which cannot be remoulded or reprocessed by remelting. Elastomeric and thermoset majority of material in many thermoset composites is inorganic glass reinforcement and mineral fillers. A wide range of reinforcements and fillers are used in elastomeric and thermoset composite materials, and these are present in varying proportions. Thermoset composite waste is likely to be contaminated and often contains metal inserts or fasteners, which make the recycling of thermoset composites problematic. To maximize the resale value of recycled constituents of polymer-based composites, it is necessary to recover the materials in a form as close as possible to their original form. In the case of fibres from continuous-fibre composites, they need to be recovered in a near-continuous form, to make closed-loop recycling of advanced composites feasible. Instead, most composite recycling operations recover the fibres in a short form (that is <3 cm), because these processes rely on simple size reduction technology. Such short fibres, however, can only be used in the preparation of short-fibre composites.

Chemical degradation of composites involves partial or selective degradation of polyester/styrene polymer network in the presence of water, ethanol, potassium hydroxide and various amides.[1] This process is rather inferior when compared with the quality of grinding recyclate, as potassium hydroxide causes detrimental effects on the recovered glass fibres. Moreover, a neutralization step is required, which generates large quantities of waste water and adds to the cost. Instead of aggressive substances, ethanol amines have been evaluated for fibre recovery and have resulted in improved quality of recycled glass fibres. The disadvantage of this process is that it yields fibres whose volume per weight is greater than that of the virgin constituents.

10.9 CASE STUDY

10.9.1 METHODOLOGY

10.9.1.1 RAW MATERIALS

10.9.1.1.1 Natural Rubber

Ribbed smoked sheets are obtained from fresh field latex sourced from well managed rubber plantations by coagulating and processed RSSs IV were used in the study, obtained from deviance information criterion. The Bureau of Indian Standards specifications for the grade of rubber are given in Table 10.2.

TABLE 10.2 Ribbed Smoked Sheets IV [RSS IV] Grades Specifications.

Parameters	Values
Appearance	Brown to dark brown ribbed sheets
Specific gravity	0.915–0.930
Density	0.92 g/cc
Copper content	8 ppm(max)
Manganese content	10 ppm(max)
Nitrogen content	5%(max)
Dielectric constant	2.37
Mooney viscosity	55–90

The raw rubber used was (from the same lot) to minimize the experimental errors due to variations of source of NR.

10.9.1.1.2 Carbon Black

Semi-reinforcing furnace black [SRF] (N774) was manufactured from Hi-Tech Carbon and supplied by AB Brothers, Mumbai specified in Table 10.3 was used.

TABLE 10.3 Semi-reinforcing furnace black [N774] Specification.

Parameters	Values
Appearance	Black
Iodine adsorption No(mg/g)	41.2
Oil(DBP) absorption No(Cc/100g)	122.71
Sieve residue on #325(%)	0.0457
Sieve residue on # 35(%)	0.000
Heat loss (%)	0.53
Pour density (Kg/m^3)	348.89
Avg. pellet hardness (Avg.)	28.9
Ash content (%)	0.25

10.9.1.1.3 Rice Hull

Rice hull was supplied by M/S Shree Guru Agrotech Pvt. Ltd. of mesh size 70–150 μ.

10.9.1.1.4 Soy Filler

Soy filler was supplied by M/S Future Agrovet Ltd. of mesh size 70–150 μ.

10.9.1.2 ACTIVATOR

10.9.1.2.1 Zinc Oxide

Zinc oxide supplied by RR Traders, Mumbai, bearing properties mentioned in Table 10.4.

TABLE 10.4 Specification of Zinc Oxide.

Parameters	Values
Zinc oxide (%)	99.45
Lead (Pb)(%)	0.031
Retention on 325 mesh (%)	0.012
Moisture at 105/2 hrs (%)	0.058

10.9.1.2.2 Stearic Acid (Co-activator)

Stearic acid was supplied by Rubosynth and had the following specification mentioned in Table 10.5.

TABLE 10.5 Specification of Stearic acid.

Parameters	Values
Appearance	Light to dark cream col free flow flakes
Odour	Characteristic Odour
Colour in Lovibond (1″)	8.3
Red max	
Colour in Lovibond (1″)	70.0
Yellow max	
Titre, Deg C	50.9
Acid value (mg KOH/g)	202.26
Iodine value(g/100 g) max	1.78
Saponification value (mg KOH/g)	203.67
Ester Value (mg KOH/g), max	1.41
% Ash, max	0.066
% M/V by mass, max	0.40
Copper (as Cu), PPM, max	<10.0

10.9.1.3 ANTIOXIDANTS

10.9.1.3.1 2, 2, 4-trimethyl-1, 2-dihydroquinoline

2, 2, 4-trimethyl-1, 2-dihydroquinoline was obtained from AB Brothers, Mumbai. It had the following specification as mentioned in Table 10.6.

TABLE 10.6 Specifications of 2, 2, 4-trimethyl-1, 2-dihydroquinoline (TDQ).

Parameters	Values
Appearance	Yellow brown beads
Ash content (%)	0.07
Volatile matter (%)	0.17
Alkalinity index	550.00
Softening point (°C)	87.60

10.9.1.3.2 Para-phenylene Diamine Derivatives.

6PPD [N-(1, 3-dimethyl buty1)–N′-pheny1-p-phenylenediamine] obtained from NOCIL, having specifications mentioned in Table 10.7, was used as antioxidant.

TABLE 10.7 Specifications of 6PPD.

Parameters	Values
Appearance	Dark purple pastilles
Ash (%)	0.01
Melting point (°C)	48.3

10.9.1.4 ACCELERATORS

10.9.1.4.1 N-Cyclohexyl-2-Benzothiazyl Sulfonamide

N-cyclohexyl-2-benzothiazyl sulfonamide supplied by AB Brothers, Mumbai, had the following specification mentioned in Table 10.8.

TABLE 10.8 Specification of N-Cyclohexyl-2-Benzothiazyl Sulphonamide (CBS).

Parameters	Values
Volatile matter 70°C (%)	0.12
Melting point (°C)	103.0
Sieve residue 0.063 mm (%)	0.10
Ash content (%)	0.12

10.9.1.4.2 Tetra Methyl Thiuram Disulphide

Tetra methyl thiuram disulphide supplied by AB Brothers, Mumbai, had the following specifications mentioned in Table 10.9, which was used as an accelerator for the R & D work.

TABLE 10.9 Specification of Tetra Methyl Thiuram Disulphide.

Parameters	Values
Melting point °C	151.70
Volatile matter %	0.27
Ash %	0.16
Sieve residue (%)	0.03
Oil %	1.88

10.9.1.5 VULCANIZING AGENT (CROSS LINKING AGENT)

10.9.1.5.1 Sulphur

Sulphur was supplied by Rubosynth. The specifications of sulphur are given in Table 10.10.

TABLE 10.10 Specifications of Sulphur.

Parameters	Values
Appearance	Yellow coloured powder
Melting Point (°C)	116°C
Purity (%)	99.68%
Fineness from 200 Mesh (25 μ) (%)	92.02%

10.9.2 FORMULATIONS OF COMPOUND BASED ON BIO-FILLERS-TABLES 10.11 AND 10.12

TABLE 10.11 Compounding Formulation using Rice Husk as Bio-filler in Natural Rubber Matrix.

Particulars	Sample code				
	NS-D1O	NS-D1A	NS-D1B	NS-D1C	NS-D1D
Natural rubber [RSS IV]	100	100	100	100	100
Zinc oxide	5	5	5	5	5
Stearic acid	2	2	2	2	2
Sulphur	2.5	2.5	2.5	2.5	2.5
CBS	1	1	1	1	1
TDQ	1.5	1.5	1.5	1.5	1.5
6 PPD	1.5	1.5	1.5	1.5	1.5
Carbon black (N774)	80	60	50	40	30
Rice husk	0	20	30	40	50
Aromatic oil	20	20	20	20	20
Total	213.5	213.5	213.5	213.5	213.5

TABLE 10.12 Compounding Formulation using Soy Filler as Bio-filler in Natural Rubber Matrix.

Particulars	Sample code				
	NS-E1O	NS-E1A	NS-E1B	NS-E1C	NS-E1D
Natural rubber [RSS IV]	100	100	100	100	100
Zinc oxide	5	5	5	5	5
Stearic acid	2	2	2	2	2
Sulphur	2.5	2.5	2.5	2.5	2.5
CBS	1	1	1	1	1
TDQ	1.5	1.5	1.5	1.5	1.5
6 PPD	1.5	1.5	1.5	1.5	1.5
Carbon black (N774)	80	60	50	40	30
Soy filler	0	20	30	40	50
Aromatic oil	20	20	20	20	20
Total	213.5	213.5	213.5	213.5	213.5

10.9.3 MIXING AND HOMOGENIZATION OF GREEN FIBRE

Mixes were prepared on a laboratory size two roll mixing mill (6 × 12 in.) as per ASTM D 3182–89. The mixing was carried out at a friction ratio of 1:1.25. The mill opening was set at 0.2 mm, and the elastomer was passed through the rolls twice without banding. Natural rubber was then banded on the slow roll with mill opening at 1.4 mm and was increased to 1.9 mm as the band became smooth. The temperature of the rolls was maintained at 70±5°C. The compounding ingredients were added as per procedure given in ASTM D 3184–89 and ASTM D 3182–89 in the following order: activator, filler, accelerator and curing agents. Before the addition of accelerator and sulphur, the batch was thoroughly cooled. After mixing all the ingredients, homogenization of the compound was carried out by passing the rolled stock end wise six times at a mill opening of 0.8 mm. The mill was opened to give a minimum stock thickness of 6 mm, and the stock was passed through the rolls four times folding it back on itself each time.

10.9.4 RHEOLOGICAL STUDIES

Cure characteristics of the mixes were determined as per ASTM D 2084 using oscillating disc rheometer (Fig. 10.2). It uses two directly heated, opposed bi-conical dies that are designed to achieve a constant shear gradient over the entire sample chamber. The sample of approximately 5 g was placed in the lower die that is oscillated through a small deformation angle (0–2°) at a frequency of 50 cpm. The torque transducer on the upper die senses the forces being transmitted through the rubber. The torque is plotted as a function of time, and the curve is called a cure graph. Minimum torque (ML), maximum torque (MH), scorch time (ts2) and optimum cure time [T90] are the important data that could be taken from the torque–time curve.

FIGURE 10.2 Monsanto R-100 ODR.

10.9.4.1 MOULDING

It is a process of curing the compound by means of heat and pressure. Here for our studies, we have used 8″ × 8″ × 2 mm and 6″ × 6″ × 2 mm sheets, which are prepared from compression mould (Fig. 10.3) using a 14″ × 14″ twin double daylight hydraulic compression set. Buttons are also compression moulded in the same press. Curing is done at 150°C for NR. After 24 h setting time, the sheet cut into test sample shape for ASTM standard testing (Table 10.13).

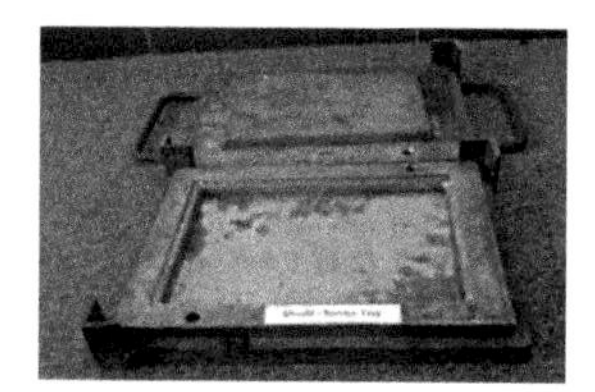

FIGURE 10.3 Mould of product.

Various properties of green rubber composites using various test condition of fillers are as given in Table 10.13.

TABLE 10.13 Outline of Testings.

Sl. no.	Test	Unit	Test specification
1	Hardness	Shore A	ASTM D 2240
2	Tensile strength	Kg/cm^2	ASTM D 412
2	Elongation at break	%	ASTM D 412
3	Compression set	%	ASTM D 395 Method B
4	Specific gravity	-----	IS 3400 part 9
5	Abrasion resistance	____	IS -34000 (3)
6	Thermal stability by thermal gravimetric analyser (TGA)	°C	ASTM E 1131
7	Tear strength	Kg/cm	ASTM D 624
8	Dynamic properties @RT and above for constant frequency and strain		ASTM D 5992
8.1	Loss modulus		
8.2	Storage modulus		
8.3	Tan delta		

Mechanical Properties – Hardness

Aim of testing To Measure the hardness using Shore 'A' Durometer

Test standard ASTM D 2240

Instrument used Durometer – Shore A

Test Procedure

1. Condition the test specimen 23±2°C for 1 h.
2. The test specimen of at least 6 mm thickness and 30 mm diameter with flat surface is made by compression moulding.
3. Place the test specimen on a non-conducting hard flat horizontal surface.
4. Hold the shore-A Durometer in a vertical position with the point of inventor at least 12 mm from any edge.
5. Take the maximum scale reading within the least possible time after applying the pressure foot to the specimen without any shock.
6. Check the hardness at three different points and calculate the mean reading.

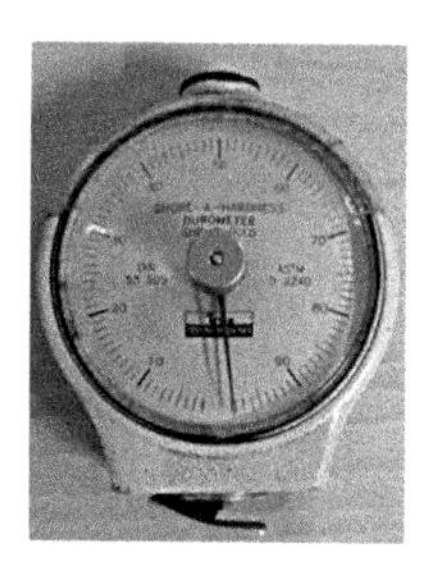

FIGURE 10.4 Shore A, Durometer.

Calculation

Average hardness = $[(S_1 + S_2 + \ldots S_I)/n]$ Shore A

Where S = sample

I = no. of samples viz., 1, 2, 3… n

n = total no. of samples

Tensile Strength, Elongation at Break

Aim of Testing	To Measure the tensile strength and elongation at break
Test standard	ASTM D 412 Die – C
Instrument Used	Universal testing machine (UTM)

FIGURE 10.5 Universal testing machine.

Test Procedure

1. Cure the slab in die as per specified time and temperature.
2. Make three dumb-bell pieces having dimension as mentioned below.

Width at the centre = 6 mm
Thickness = 2±0.2 mm
Calculation

1. T.S. = Breaking load in (N)/width × thickness (in mm)
2. Elongation at break = (Final length − initial length/initial length)

Compression set

Aim of Testing	To Measure permanent set properties of the rubber
Test standard	ASTM D 395 Method B
Instrument used	Steel plate, spacer and moulded button specimens

FIGURE 10.6 Steel plate, spacer for compression set apparatus.

Test Procedure

1. Test specimen is prepared from rubber compound by moulding.
2. Thickness is measured for each test specimen at the central portion.
3. Placed test specimen at centre place between the plates and spacers at a few distance away from specimen with thickness equal to 25% compression.
4. Tightened the bolt so that the plates were drawn together until they uniformly contact the surface of spacer.
5. Placed the whole assembly at air circulating oven at 70°C for 24 h.
6. After completion of specified period, the assembly is removed from oven and buttons are removed from the test assembly.
7. Transferred the test pieces quickly to a wooden bench and allowed them to recover at room temperature for 30–35 min.
8. Measured the thickness again and recorded.

Test Specimen

Before **After**

FIGURE 10.7 Moulded button test specimens.

Data Calculations and Results

$$\text{Compression set }(\%) = \frac{T_O - T_F}{T_O - T_S} \times 100$$

where

T_O = Initial thickness of test piece
T_F = Final thickness of test piece after recovery
T_S = Thickness of the spacer

Specific Gravity

Aim of testing: To measure the specific gravity of the sample
Test standard: IS 3400 part 9
Instrument used: Density determination kit

FIGURE 10.8 Density determination kit.

Test Procedure

1. A small piece was cut from the moulded sheet specimen.
2. Weight of the sample was taken in air medium and value recorded.
3. Weight of the sample was taken by placing it in water medium and value is recorded.
4. Specific gravity of the sample was calculated as per the formula given below.

Specific gravity = weight in air/(weight in air – weight in water)

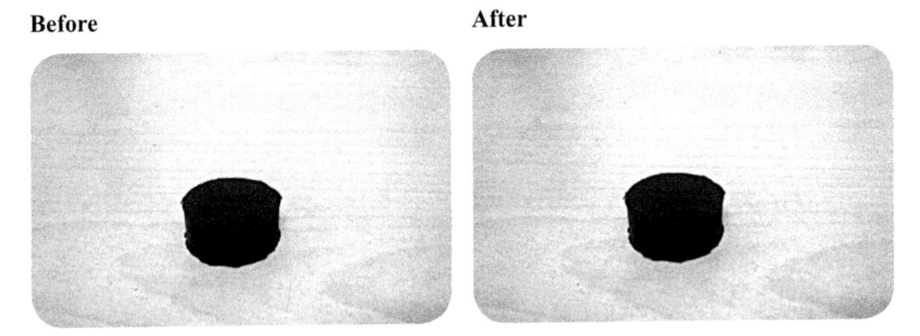

FIGURE 10.9 Moulded button test specimens for specific gravity measurements.

Abrasion Resistance

Aim of testing	To measure the abrasion resistance index (ARI) for the specimen
Test standard	IS-3400 part 3
Instrument used	Rotary drum abrasion tester

FIGURE 10.10 Rotary drum abrasion tester.

Test Procedure

1. The test samples were moulded as per specification.
2. The test piece was cylindrical in shape, of diameter 16±0.2 mm, with a minimum height of 6 mm.
3. Weighed the samples accurately to 0.1 mg.
4. Fitted the test piece into the test piece holder in such way that a length of 2±0.1 mm protrudes from the opening.
5. Put down the test piece holder on to the abrasive drum and test was started.
6. After completion of 84 revolutions of the drum, the samples were taken out.
7. Removed any rubber flash attached to the sample due to abrasion.
8. Reweighed the samples and calculate the Abrasion resistance index as per the formula given below.

$$\text{Abrasion resistance index} = (V_s/V_t) \times 100$$

Where
V_s = Volume loss of standard button.
V_t = Volume loss of test button.

Test Specimen

Before **After**

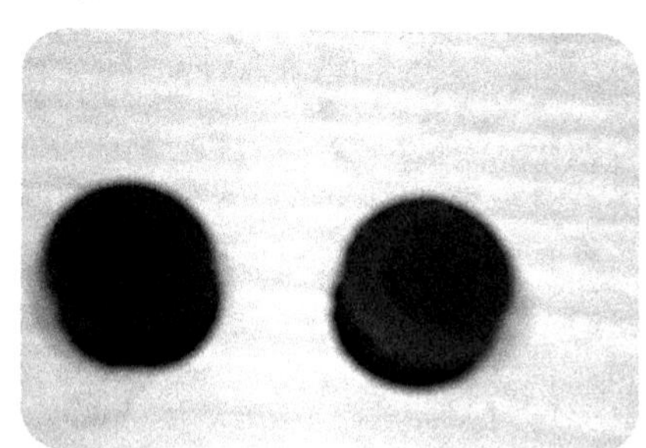

FIGURE 10.11 Moulded button test specimens for Abrasion test.

Thermal Stability

Aim of Testing	To study the thermal stability of the material under controlled thermal conditions.
Test Standard	ASTM E 1131
Instrument Used	Thermal gravimetric analyser (TGA)

FIGURE 10.12 Thermal Gravimetric analyser.

Test Procedure

1. Cut about 10–15 mg of sample in solid form and laced inside the pan.
2. The experiment is carried out from room temperature to 900°C at a heating rate of 20°C/min
3. Testing was done at Nitrogen atmosphere from RT to 570°C and then on oxygen atmosphere till 900°C.
4. After completion of run, the experiment is stopped and interpretation of the thermogram is done to calculate the wt. per cent of low boiling materials, polymers, carbon black if present and ash content/residue as applicable.

Test Specimen

Before **After**

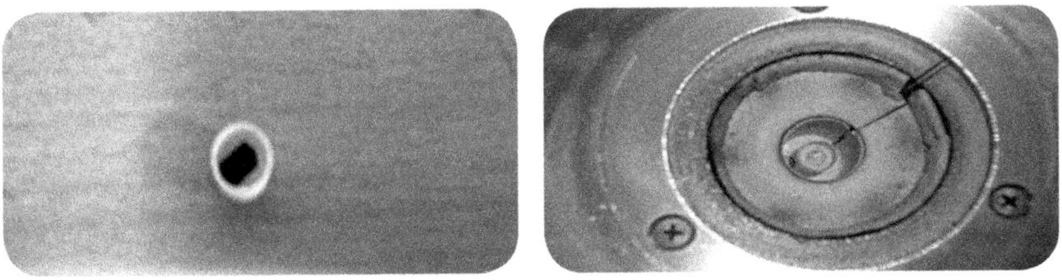

FIGURE 10.13 Test specimen before and after thermal analysis.

Tear Strength

Aim of testing	To Determine the tear strength of the rubber
Test standard	ASTM D 624
Instrument used	UTM

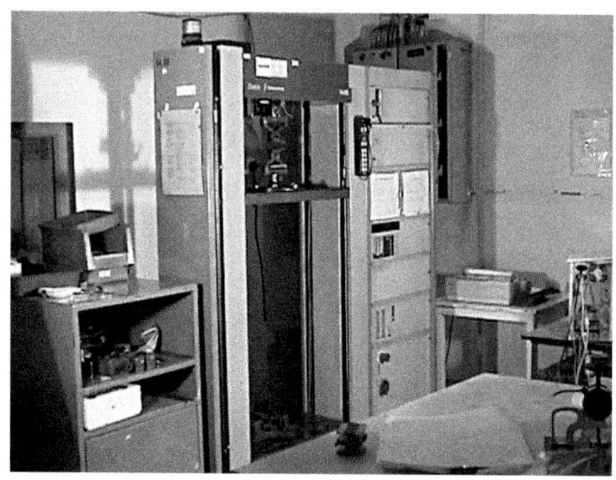

FIGURE 10.14 Universal Testing Machine for tear strength.

Test Specimen

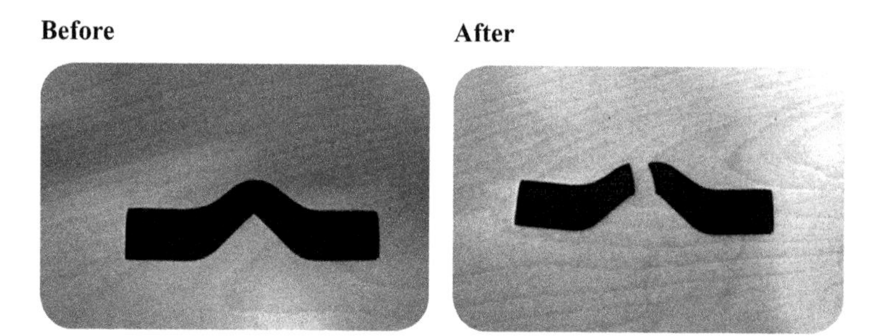

FIGURE 10.15 Tear Strength test specimen.

Test Procedure

1. Cure the slab in die as per specified time and temperature
2. Make three angel tear pieces having dimension as mentioned below

L = 102 mm
W = 19 mm
T = 2 mm
Angle = 90°
Calculation

Tear strength = Breaking load in (Kg)/thickness (in cm)

Dynamic Properties @ RT & Above

Aim of Testing	To study the behaviour of the samples at various dynamic and thermal environments
Test standard	ASTM D 5992
Instrument used	Dynamic mechanical analyser

FIGURE 10.16 Dynamic mechanical analyser.

Test Specimen

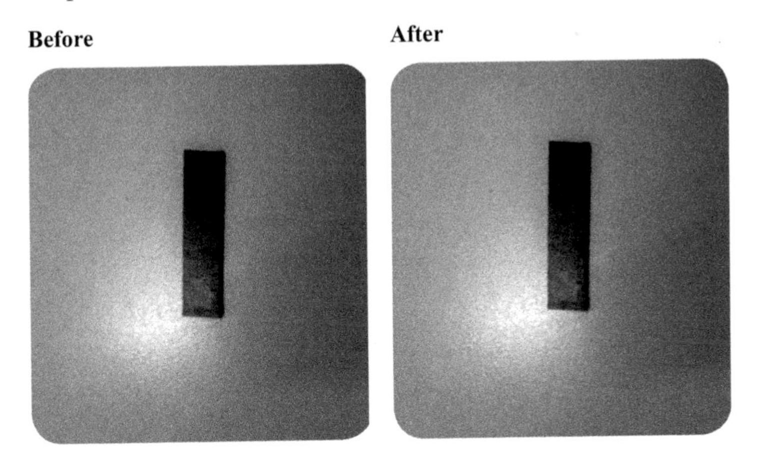

FIGURE 10.17 Test Specimen for dynamic properties.

Test Procedure

1. Test specimen was prepared as per ASTM standard.
2. The sample was placed between the sample holding jaws.
3. The chamber was closed for temperature conditions.
4. Frequency of 5 Hz was applied on the sample.
5. Dynamic Strain 0.1% and static strain 0.5% were applied on the sample.
6. Testing was carried out at 30°C, and values for loss modulus, storage modulus and tan delta were recorded.
7. Same procedure was repeated at 50 and 75°C, and values for loss modulus, storage modulus and tan delta were recorded.

10.9.5 INFLUENCE OF GREEN FIBRE

10.9.5.1 EFFECT OF GREEN FIBRE ON RHEOLOGICAL PROPERTIES

Table 10.15 indicates where optimum cure time of NR decrease by partial replacement of carbon black with soy filler which indicate some involvement of these filler with sulphur curing system, whereas Table 10.14 indicates increase in cure time of NR by partial replacement of carbon black with rice husk. In conclusion, the effects of particulate fillers differ from those of high aspect ratio fillers. Particulate fillers seem to increase cure rate in most cases, especially if they contain active groups on their surface which either may react with the resin or change reaction mechanism. High aspect ratio fillers seem to decrease the reaction rate due perhaps to their more localized influence.

TABLE 10.14 Rheological Studies at 150°C of Natural Rubber with Rice Husk.

Pre-vulcanized Properties	Unit	Compound code				
		NS-D1O	NS-D1A	NS-D1B	NS-D1C	NS-D1D
T90	min	6.26	6.5	6.75	6.98	7.25
Ts2	min	2.92	3.08	3.22	3.36	3.5
MH	lbs.in	84.12	74.58	70.38	67.13	61.93
ML	lbs.in	4.39	4.38	4.31	4.4	4.26
Mooney viscosity	MU	28	27	25	25	24

TABLE 10.15 Rheological Studies at 150°C of Natural Rubber with Soy Filler.

Pre-vulcanized Properties	unit	Compound code				
		NS-E1O	NS-E1A	NS-E1B	NS-E1C	NS-E1D
T90	min	7.5	6.97	6.83	6.85	6.69
Ts2	min	3.73	3.3	3.13	3.13	2.99
MH	lbs.in	80.03	70.26	64.94	59.95	56.25
ML	lbs.in	8.02	7.38	7.16	6.93	6.59
Mooney viscosity	MU	40	36	33	30	26

Tables 10.14 and 10.15 indicate reduction of MH value due to some non-reinforcement of these bio–filler; comparison with carbon black also shows decrease in Mooney viscosity due to softness of bio-material compare with carbon black.

10.9.5.2 EFFECT OF GREEN FIBRE ON HARDNESS

Fillers may either increase or decrease the hardness of a polymer depending on its interaction and particle size. Small particle size increases hardness, whereas coarse particles have little influence on the hardness of the composite. The effect depends on the interaction between polymer and filler. The general trend in filled material is that fillers increase hardness as the filler concentration is increased. In highly filled materials, especially those filled with silica flour, the hardness of the composite approaches the hardness of the filler. Natural rubbers have increasing tendency of hardness when carbon black is added. As in the case of rice husk and soy filler, there was some decreasing trend of hardness when bio-filler content was increasing, due to larger particle size dimension of green fibres than carbon black (see Fig. 10.18).

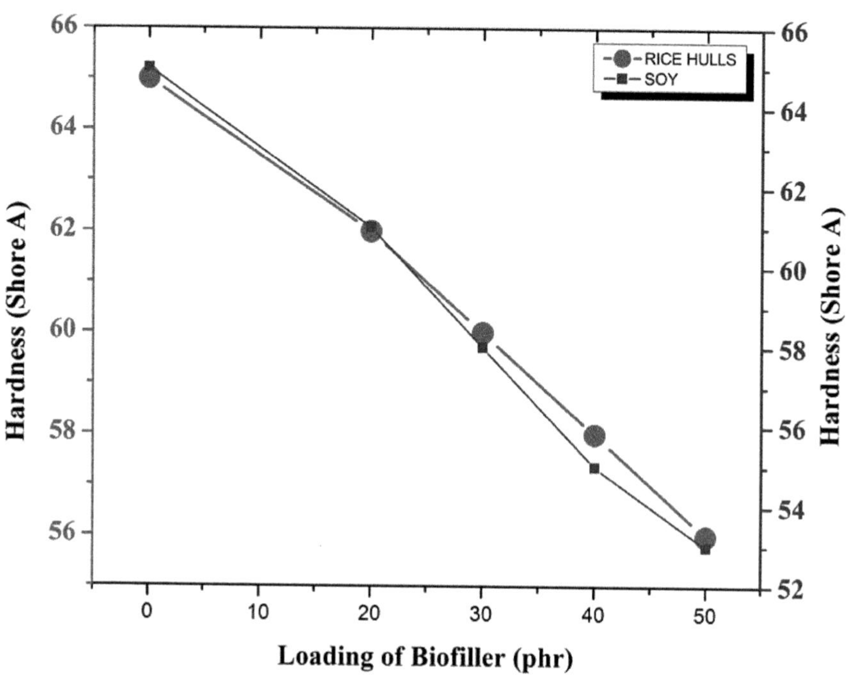

FIGURE 10.18 Effect of biofiller on hardness of natural rubber composites (Y axis on L-Side for Rice Hulls and R-Side for Soy)

10.9.5.3 EFFECT OF GREEN FIBRE ON SPECIFIC GRAVITY

Below the critical concentration of filler, some polymer is converted to the interphase layer in which the polymer has a higher density because of closer packing, therefore the density of the polymer increases. Above the critical concentration of filler, there are not enough polymers to cover the surface which increases the free volume, and the density of the composite decreases (see Fig. 10.19).

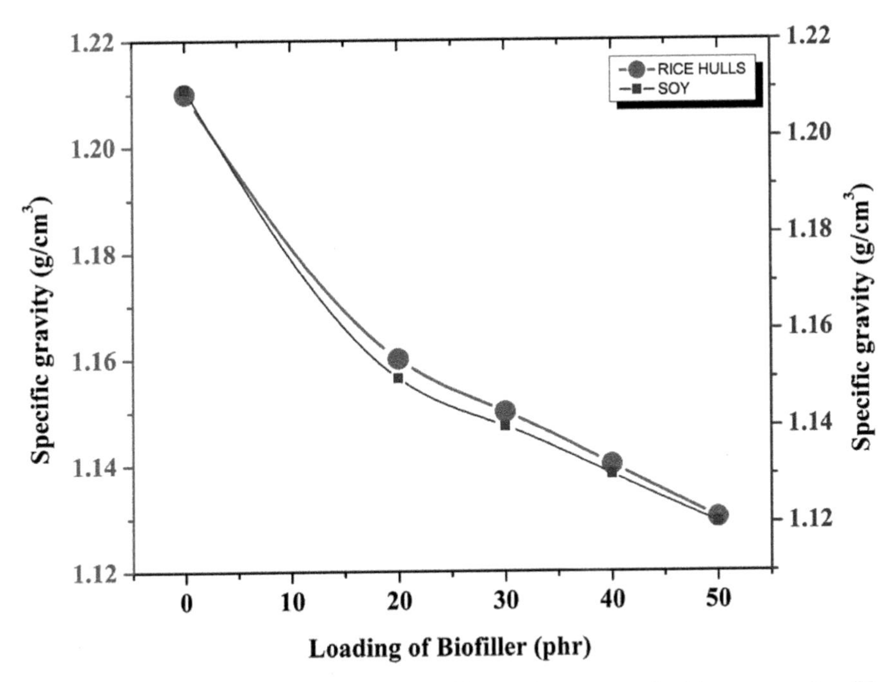

FIGURE 10.19 Effect of biofiller on specific gravity of natural rubber composites (Y axis on L-Side for Rice Hulls and R-Side for Soy)

10.9.5.4 EFFECT OF GREEN FIBRE ON MODULUS

Modulus is a convenient measure of composite stiffness. Fillers can contribute significantly to a stiffness increase, which reduces modulus. This decrease in modulus is due to poor interaction of rubber with filler matrix in presence of oil (see Figs. 10.20 and 10.21).

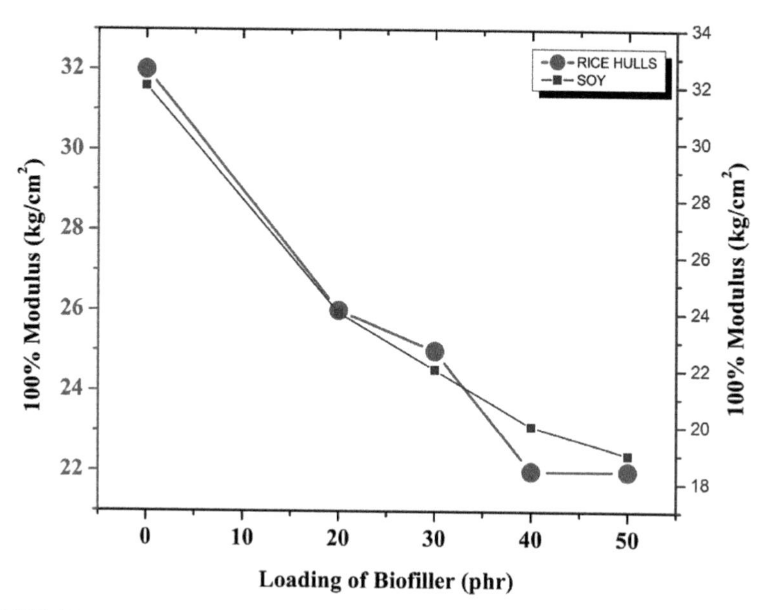

FIGURE 10.20 Effect of biofiller on 100% modulus of natural rubber composites (Y axis on L-Side for Rice Hulls and R-Side for Soy)

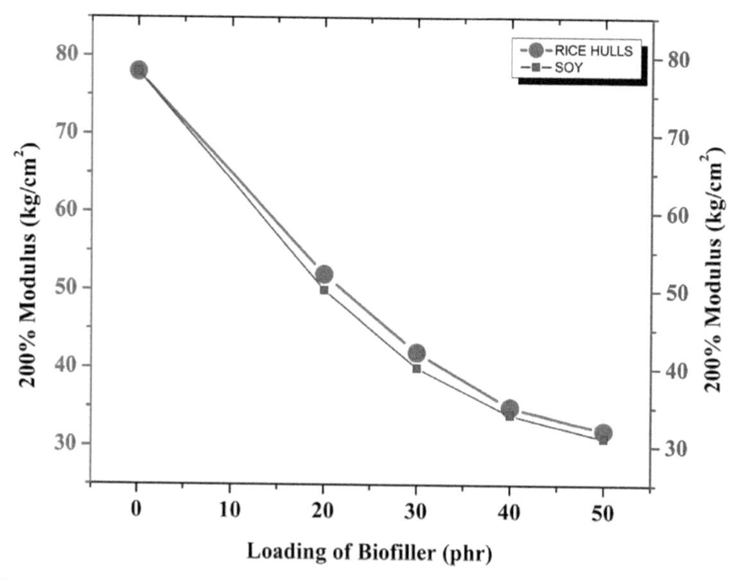

FIGURE 10.21 Effect of biofiller on 200% modulus of natural rubber composites (Y axis on L-Side for Rice Hulls and R-Side for Soy)

10.9.5.5 EFFECT OF GREEN FIBRE ON TENSILE STRENGTH

There is strong interaction between the filler and the matrix, which produces adverse results, that is increased interaction, increased interaction reduces tensile strength due to increasing material stiffness (see Fig. 10.22).

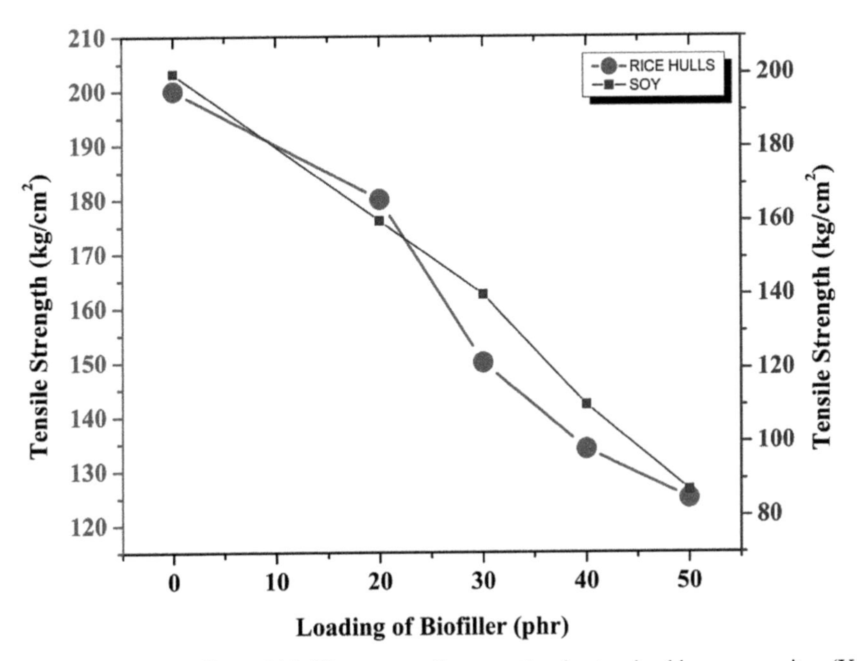

FIGURE 10.22 Effect of biofiller on tensile strength of natural rubber composites (Y axis on L-Side for Rice Hulls and R-Side for Soy)

10.9.5.6 EFFECT OF GREEN FIBRE ON TEAR STRENGTH

Fibres, due to their high aspect ratio, are the most efficient method of improving tear strength; fibres form large obstacles in the path of crack growth. Fibres with better adhesion to matrix are more efficient. To achieve optimum adhesion, the concentration of additive should coat the surface of carbon black with a monomolecular layer. Building additional layers on the carbon black surface reduces adhesion which also reduces tear strength. Due to more crystalline region of NR and larger particle size of bio-filler, deviation of tear path is less. So, less energy is required by bio-filler to tear (see Fig. 10.23).

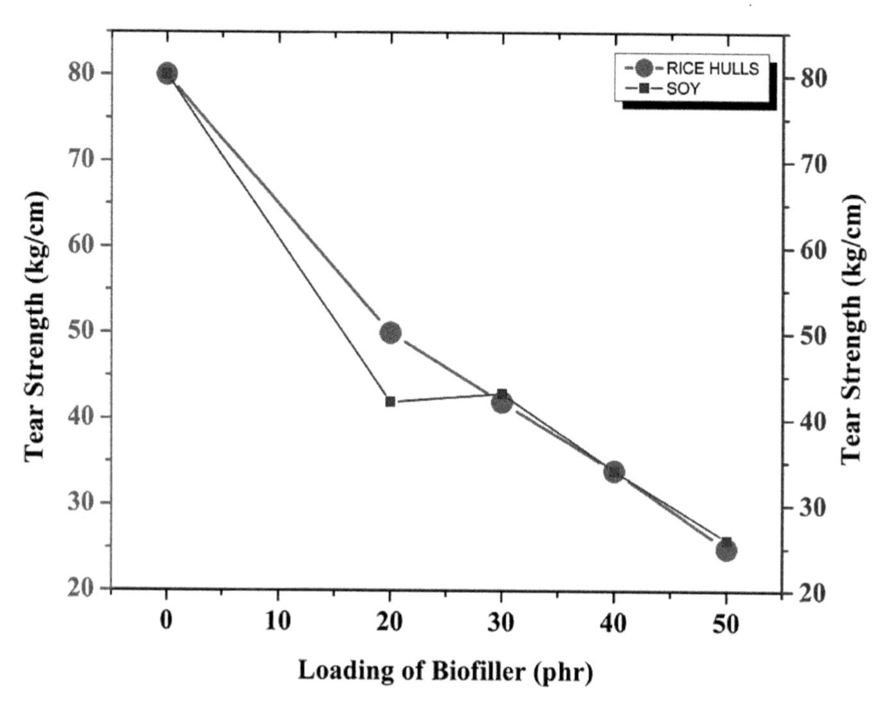

FIGURE 10.23 Effect of biofiller on tear strength of natural rubber composites (Y axis on L-Side for Rice Hulls and R-Side for Soy)

10.9.5.7 EFFECT OF GREEN FIBRE ON COMPRESSION SET

Compressive strength depends on the stiffness of the material; thus, all of the parameters which affect stiffness, including the effect of fillers, influence compressive strength. Compression set is a measure of the ability of rubber vulcanizes to retain their elastic properties after prolonged compression at constant strain under a specific set of condition and its permanent set of rubber vulcanizes. The poor performance of rubber compounds in terms of compression set is attributed to uncross-linked chains, which do not contribute to the permanent network and are able to relax during the compression stage. The compression set values are expressed as the percentage of rubber's permanent deflection. Figures 10.24 and 10.25 show increase in compression set value of NR with increment of green fibre content. Rubber shows lower percentage of compression set, will exhibit better elastic properties, which enhanced better resistance to deformation (see Fig. 10.24 and 10.25).

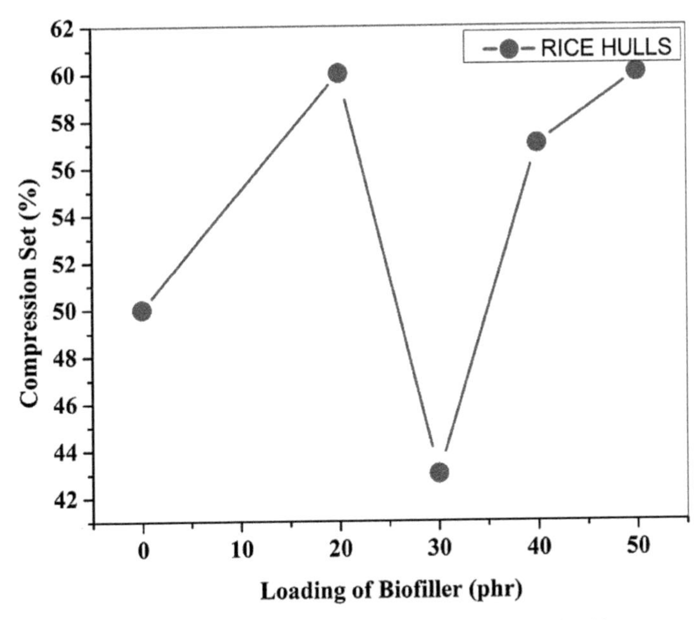

FIGURE 10.24 Effect of rice hulls on compression set of natural rubber composites

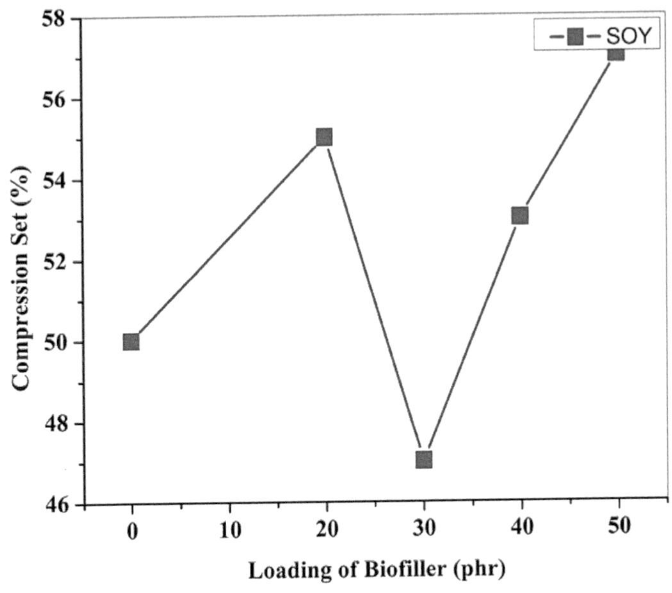

FIGURE 10.25 Effect of soy on compression set of natural rubber composites

10.9.5.8 EFFECT OF GREEN FIBRE ON DEGRADATION RATE

From the derivative of the thermo gravimetric curve, the maximum decomposition temperature was evaluated. Figure 10.26 indicates the temperature of maximum degradation of NR with rice husks initially becoming lower and then remaining constant to some extent. In case of soy filler, we find same trend. But when rice husk was used by partial replacement of carbon black, the degradation temperature was increased (see Fig. 10.26).

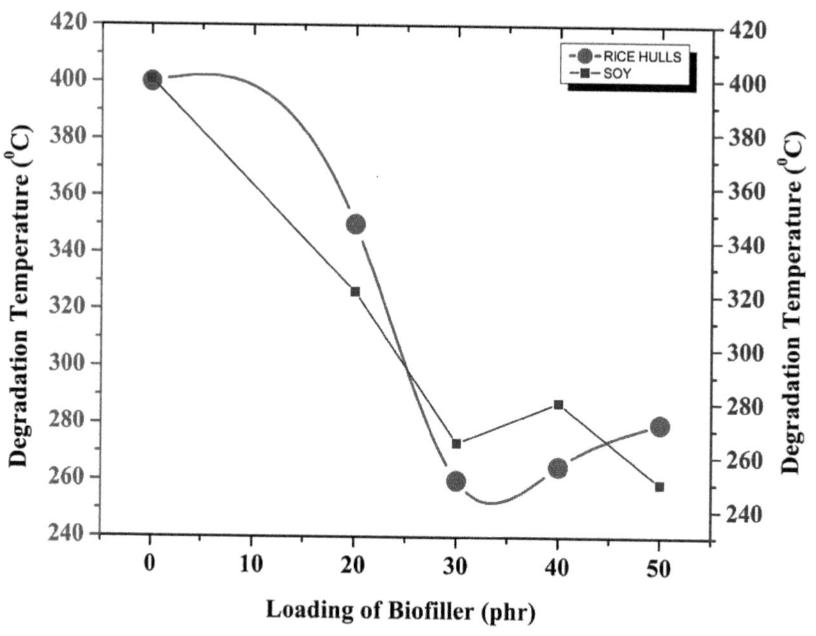

FIGURE 10.26 Effect of biofiller on thermal degradation of natural rubber composites (Y axis on L-Side for Rice Hulls and R-Side for Soy)

10.9.5.9 EFFECT OF GREEN FIBRE ON ABRASION RESISTANCE INDEX

The increase in bio-filler content in compound resulted in decreasing the ARI of rubber due to softness of bio-material. Two aspects of abrasion will be discussed, namely the abrasion resistance of filled materials and the use of fillers in the friction materials. Each has its own specificity and differs from other two.

Due to the elastomeric property of rubber, usually it has a low abrasion resistance. Fillers such as carbon black and silica can be added to impart abrasion resistance. Increased adhesion between the filler and the matrix contributes to an increase in abrasion resistance (see Fig. 10.27).

Abrasion resistance is the ability of a material to withstand mechanical action such as rubbing, scraping or erosion that tends progressively to remove material from its surface. Such ability helps to maintain the material's original appearance and structure.

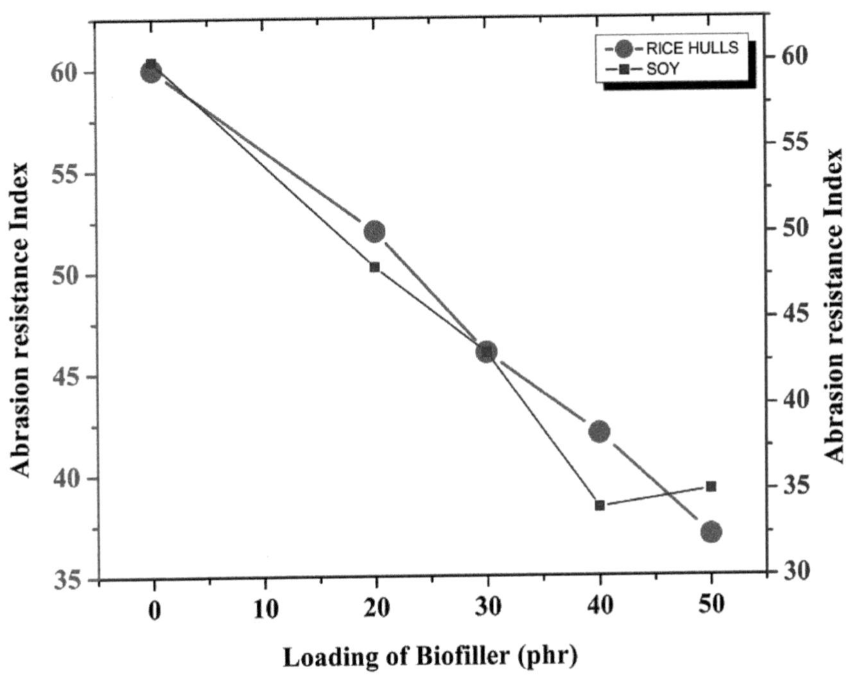

FIGURE 10.27 Effect of biofiller on wear of natural rubber composites (Y axis on L-Side for Rice Hulls and R-Side for Soy)

10.9.5.10 EFFECT OF GREEN FIBRE ON RESILIENCE

The ratio of energy returned to energy applied to produce the deformation is known as resilience. In addition, when this deformation results from single impact, the ratio between the returned and applied impact energy is called as rebound resilience.

Rebound resilience is directly proportional to degree of elasticity. Results from Figure 10.28 clearly show that the increasing proportion of bio-filler (rice husk, soy filler) increases the elasticity of NR composites; hence, the rebound resilience value increase when carbon black partially replaced by bio-filler.

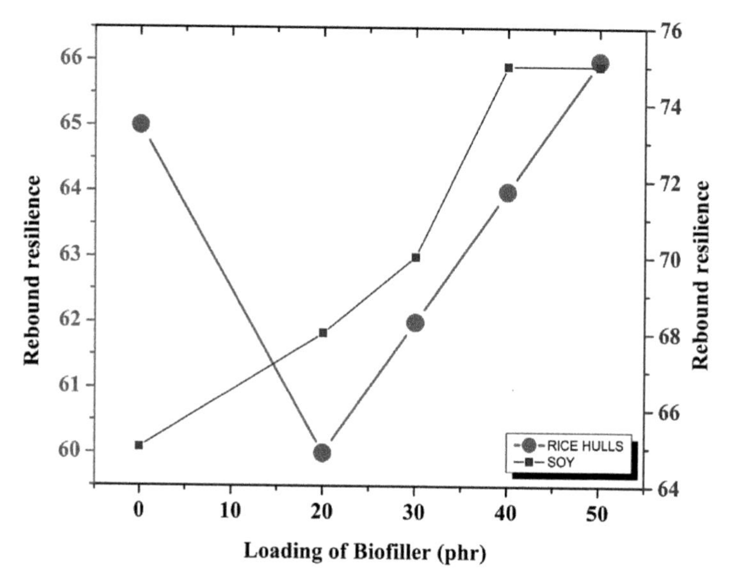

FIGURE 10.28 Effect of biofiller on rebound resilience of natural rubber composites (Y axis on L-Side for Rice Hulls and R-Side for Soy)

10.9.5.11 EFFECT OF GREEN FIBRE ON FLEXING

De-Mattia machine test method used to test rubber specimens for resistance to cracking produced either by extension or bending, depending on the relative adjustment of the stationary & movable grips and the distance to travel of the later.

Natural rubber shows poor crack initiation resistance but displays very good cut growth resistance, because NR crystallizes on staining and forms crystallites at the tip of the crack. But initially, when no partial replacement by bio-filler (carbon black: bio-filler =80:00) occurs, the value shows better [40,000 cycles] crack initiation. Again, in a similar fashion, the cut initiation value decrease when carbon black replaced by bio-fillers and among them, soy filler shows better comparative results (see Fig. 10.29).

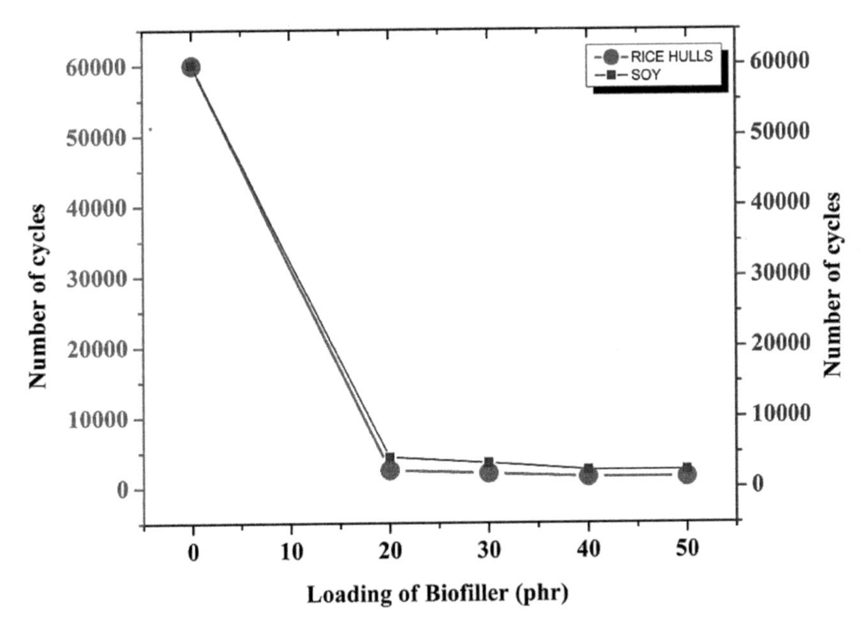

FIGURE 10.29 Effect of biofiller on flexing properties of natural rubber composites (Y axis on L-Side for Rice Hulls and R-Side for Soy)

10.10 END-USE APPLICATIONS OF GREEN FIBRE NATURAL RUBBER COMPOSITES

NFCs have competitive mechanical properties and reduced costs which made a widespread use of them in varied fields. NFCs have become more popular especially with automobile industry due to the advent of newer methodologies for processing and treatments of natural fibres. Several products, such as mobile phone cases, laptop cases, toys, snowboards, tennis rackets, building/construction materials, musical instrument parts, furniture and may more, have already developed from NFCs apart from automotive exterior and interior components.

Racing bicycles with NFCs have also been developed by Museeuw, with flax-carbon-epoxy prepegs. A Belgium-based bicycle company, named Maseeuw, claims that the flax content in the bicycle frames were up to 80%. Carbon-flax prepreg laminates have good stiffness and reduced vibration damping characteristics suiting them for dynamic applications.

Decking's, railing systems, window frames and fencings are also made from wood, flax, rice husk and bagasse-reinforced composites. Even though NFCs find applications in vast diverse fields, amongst all the automotive industry makes the most judicious use of them. In 1941, Henry Ford designed and built a car from compression moulded NFCs with 70% cellulosic fibres (including hemp) with a resinous binder. Now companies, such as Volkswagen, Audi, BMW, Daimler Chrysler, Opel, Peugeot and Renault, have already replaced many of their interior and exterior components with natural fibre-reinforced composites. They have replaced door panels, side panels, headliners, dashboard parts, back side of seats and many more with NFCs. Building/construction sector, especially with components that have more aesthetic value than the structural value, uses natural fibre-based composites. As they can't take up large loads, parts which require more strength cannot be fabricated from NFCs.

Table 10.16 shows some applications of NFCs in various sectors with emphasis to the fibres used and processing techniques employed.

TABLE 10.16 Applications of Natural Fibre Composites (NFC) in various sectors.

Technological areas	Automobile industry	Electrical and electronic industry	Sports good manufacturing	Civil industry – construction and building
Natural fibre used	Kenaf, Hemp, Abaca, Wood	Kenaf	Flax, Hemp	Bagasse, flax, wood, coir, stalk, rice husk
Processing method for NFCs	Injection moulding compression moulding	Injection moulding	Oven cure	Extrusion Injection moulding Compression moulding
End-use application	Door panels Seat backs Head liners Dash boards Truck liners	Mobile cases Laptop cases	Tennis racket Bicycle frames Snowboards	Decking Railing Door panels Window frames Boards for partition, ceiling and wall panel Bricks and blocks for walling

However, their impact strengths are worth commendable. Natural fibre composites have good thermal and acoustic insulation properties which suites them for many applications such as roofing's, flame retardant boards and floor coverings and others. For instance, the heat release rate of PE/

Flax composite is less than that of PE, which infers the flame retardency behaviour of these composites. The major potential of these composites lies in the building/construction sector, where glass fibres can be replaced from non-load bearing composites.

10.11 CONCLUSION

It has been confirmed from the above inspection that soy filler can be used as filler in rubber compounding as dry soy filler is combination of protein and carbohydrates, which are rigid and can form strong filler networks through hydrogen bonding and ionic interactions. They are also capable of interacting with polymers that possess ionic- and hydrogen-bonding groups. Through filler–filler and filler–rubber interactions, the rubber modulus is significantly increased. From the above observation, we found that replacing 10% carbon black with soy filler in a commercial NR-grade formulation, most of the specification properties, such as tensile strength, elongation, modulus, tear, hardness, were supposed to be met. But by increment of filler dose above 25%–30%, the physical properties become deteriorate. Natural rubber illustrates good result with soy filler.

Natural fibre composites have their own advantages and technical drawbacks. Hydrophilicity of natural fibres is of great concern, as it increases the moisture absorption, thereby reducing the dimensional stability of the composites. Due to hollow nature of the natural fibres, their composites show better thermal and acoustic insulation properties. But most of NFCs are not applicable for load bearing applications as they have low tensile strengths when compared with glass fibre composites. Natural fibres have excellent vibration damping property which can be made use in vehicle parts where there is excessive vibration and noise. These fibres absorb the waves and dampen it much faster than their synthetic counterparts. Hence, they are used in conjunction with carbon and aramid fibres so as to get both strength and damping effects. Potential areas of applications include underbodies of heavy trucks, cars, trains, bicycle frames and many more. In the building/construction sector, there lies a major potential for NFCs in the years ahead. As they have good thermal and acoustic insulating properties, they can be used in the interiors as well as in exterior applications where aesthetics have more importance. However, further studies on newer and existing natural fibres have to be promoted to discern the hidden potential of novel natural fibre composites.

Universal trends towards sustainable growth have brought to light natural, renewable, biodegradable raw materials, among them bast fibres. Science and technology continue in extending their use in textile and other industries.

Recent achievements and new applications of green fibres and associated products bast fibrous plants can provide a definite form to the background for the following conclusion.

Fast-growing population and eco- and health awareness creates large space for future expansion of other than cottons natural cellulosic fibres.

Present achievements in breading/production/processing extended the use of bast fibres in textiles. To make the way for these, difficult in processing fibres, into textile products being beyond their reach for centuries, it was necessary to develop new, refined bast fibre types, adapted to modern spinning systems; new, softer, finer 100% bast or bast blended yarns, amongst them knitting yarns; new crease resistant finishing treatments; new products, which could meet the needs of demanding apparel sector.

All the textile goods can boast high comfort/health properties and are ecological items which are made on the basis of green fibres. They can be labelled as 'NATURAL' which is the key to market success.

Green fibres/bast fibrous plants are used, thanks to research and development, in growing amounts for non-wood pulps.

Green fibres/bast fibrous plants will also be used in growing amounts in a wide spectrum of bio-composites materials. Being lignocellulosic, they can be combined with man-made or natural polymers to provide a wide range of useful composites in textiles (including geotextiles and non-woven).

Green fibrous plants provide valuable by-products such as seeds, waxes, fragrances and pigments. These may be used for food, fodder, pharmaceuticals, cosmetics and body-care items.

KEYWORDS

- **green fiber composites**
- **rice husk fiber**
- **soy fiber**
- **natural rubber**

REFERENCES

1. Akovali, G.; Bernardo, C. A.; Leidner, J.; Utracki, L. A.; Xanthos, M. *Frontiers in the Science and Technology of Polymer Recycling.* Proceedings of the NATO Advanced Study Institute on Frontiers in the Science and Technology of Polymer Recycling, Antalya, Turkey, 1997.
2. Caroline Baillie. *Green Composites.* Wood Head Publishing Limited: Cambridge, England, 2004.
3. Ciesielski, A. *An Introduction to Rubber Technology;* Rapra Technology Ltd.: Shawbury, Shrewsbury, Shropshire, United Kingdom, 1999.
4. De, S. K.; White, J. R. *Short Fibre – Polymer Composites;* Wood Head Publishers, 1996.
5. Defonseka, C. *Introduction to Polymeric Composites with Rice Hulls*; A Smithers Group Company: Shawbury, Shrewsbury, Shropshire, United Kingdom, 2014.
6. Jacometti, G. A.; Lea, R. P. F.; Mello; Nascimento, P. H. A.; Sueiro, A. C.; Yamashita, F.; Mali, S. The Physicochemical Properties of Fibrous Residues from the Agro Industry. *LWT—Food Sci. Technol.* **2015,** *62*, 138–143
7. Mark, J. E.; Erman, B.; Eirich, F. E. *The Science and Technology of Rubber,* 3rd ed.; Elsevier Inc, 2005.
8. Mohanty, A. K.; Misra, M.; Drzal, L. T. *Natural Fibers Biopolymers and Bionanocomposites*; Taylor and Francis Group, 2005.
9. Pai, A. R.; Jagtap, R. Surface Morphology and Mechanical Properties of Some Unique Natural Fiber Reinforced Polymer Composites—A Review. J. *Mater. Environ. Sci.* **2015,** *6*(3), 631–646.
10. Simpson, R. B. *Rubber Basics;* Rapra Technology Ltd.: Shawbury, Shrewsbury, Shropshire, United Kingdom, 2002.
11. Thomas, S.; Joseph, K.; Malhotra, S. K.; Goda, K.; Sreekala, M. S. *Polym. Compos.* **2014,** *2.* Wiley VCH.
12. Wallenberger, F. T.; Weston, N. *Natural Fibers Plastics and Composites*; Kluwer Academic Publishers, 2004.

CHAPTER 11

BIO-RESINS AND BIOPLASTICS SYNTHESIZED FROM AGRICULTURAL PRODUCTS FOR NOVEL APPLICATIONS: THE INDIAN PROJECTION

PADMANABHAN K.

Centre for Excellence in Nanocomposites, School of Mechanical and Building Sciences, VIT-University, Vellore 632014, India, E-mail: padmanabhan.k@vit.ac.in

CONTENTS

ABSTRACT

The quantum of bio-resins and bioplastics that can be produced from the conventional, organic and genetically modified (GM) plants is immense because India is a world class producer of sugarcane, sugar beet, other tubers such as potato and vegetables with starch, cashew and *badam*, castor oil and soya bean. As on date, advanced and state of the art plastics and composites are being used in low end applications as there is no incentive for farmers to produce plants and vegetables for the plastics and resins market exclusively. The use of advanced composites in low end applications escalates costs and shifts the material consumption that would deplete the natural resources through wrong usage at one end and lack of demand for natural resources at the other. This invited paper attempts to project the actual possibilities of the bio-resin and bioplastic market in this country and provides the knowhow for production of bio-polyethylene, bio-poly lactic acid, bio-polyester, phenolic cashew nut shell resin, castor oil-based plasticizers and soya bean-based epoxies. This chapter discusses their true potentialities in composites product applications involving structural, interior, electronic and chemical engineering markets. A novel working model with an economically feasible option is also provided for those concerned about their safe disposal, recycling, reuse and conversion into useable fuel with no impact to the environment.

11.1 INTRODUCTION

India is a world-class producer of sugarcane, sugar beet, other tubers such as potato and vegetables with starch, cashew and badam, castor oil and soya bean. The quantum of bio-resins and bioplastics that can be produced from these conventional, organic and GM plants is large. Cashew nut shell liquid (CNSL), which is the byproduct of cashew industry, is a unique resource of unsaturated long-chain phenolic distillate, cardanol.[1] India produces about 25,000 tons of cashew and 2500 tons of CNSL per annum. Most of the CNSL oil is consumed internally for furnace oil, paints, pharma, cosmetics, resins and adhesives, thus, only less than about 20% can be exported. Cardanol is a laminating resin, used in paints, coats, bonding resins and varnishes.[2] Cashew nut shell liquid and cashew friction dust are used in brake linings, pads, faces, discs and shoes. Cashew nut shell liquid comes with asbestos or non-asbestos containing brake applications. The natural meta-substituted alkylphenol can produce a series of phenolic resins by catalysed aldehydes or acids. Many have also characterized the polymerization of cardanol.[3] Some investigators have synthesized CNSL-based phenol-formaldehydes and studied their properties like thermal stability and compared them with standard phenol formaldehydes.[4] The thermal characterization and physical properties of CNSL were also studied. Tejas S Gandhi *et al.* characterized the Mannich base with Cardanol at a low viscosity and concluded that it can used as a polyol for synthesis of rigid polyurethanes.[5] A better variety of CNSL matrix materials for composite applications have been synthesized at VIT.[6,7] Novel mechanical properties were obtained, and the thermal characterization of different combinations of CNSL was done. This invited paper attempts to project the actual possibilities of the bio-resin and bioplastic market in this country[8] and provide the knowhow for production of bio-polyethylene, bio-poly lactic acid, bio-polyester, phenolic cashew nut shell resin.

11.2 THE INDIAN CONTRIBUTION

India produces about 350 million tons of sugarcane per annum, apart from CNSL, and all of it is used for the production of sugar. The demand for sugar is so high that even though the technology for producing polyethylene from ethylene and ethanol distilled from sugarcane is available, one has to depend on non-edible and GM sugarcane production for the synthesis of polyethylene from sugarcane. In order to assist in the production of less toxic epoxidized resins and not encroach upon the edible sector, the production of soya bean can also be GM. The farmers gain more from such GM cultivations as they generally yield more than conventional farming (say up to 94% as in GM crops), are more resistant to pests and assist in the synthesis of more environment friendly resins and bioplastics than their synthetic cousins. Besides, the fear of a biotechnologically modified edible crop does not come in at all. The farmers, however, need a guarantee of demands for the inspiration. As GM crops are basically resistant for any of the above applications, it would also mean that they can be organically produced without any pesticides.

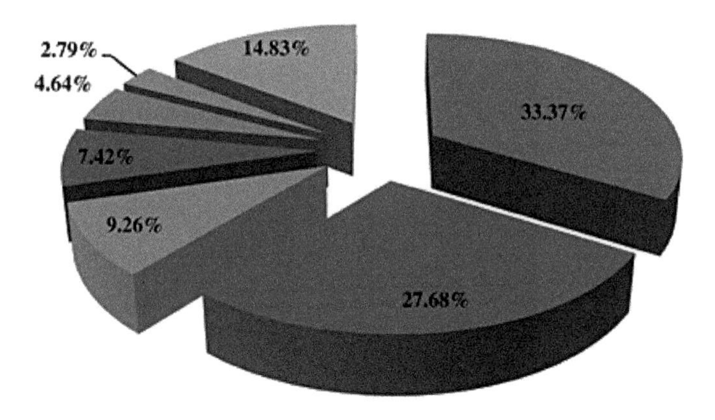

FIGURE 11.1 World cashew production.

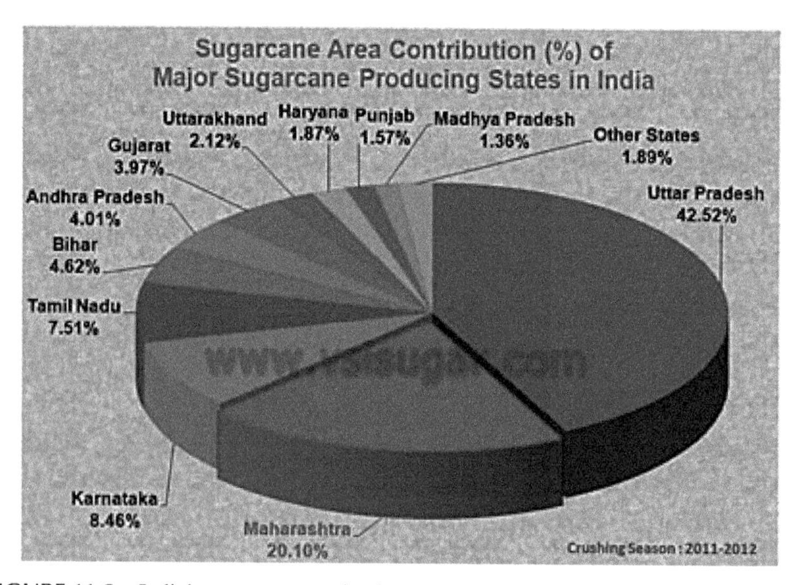

FIGURE 11.2 India's sugarcane production.

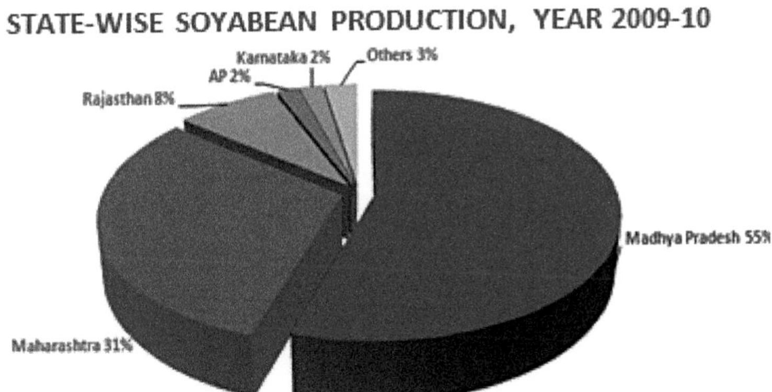

FIGURE 11.3 India's soya bean production.

Information about India's superiority in the production of GM crops is provided in Figures 11.1–11.3. Figure 11.4 shows a piece of the CNSL-based glass fabric composite produced at our laboratory. India's soya bean production data stand at 12 to 13 million tons. A portion of it could be used in producing epoxidized soya bean. It has been observed that most of the plant oil and resin production capabilities are confined to the southern states and Madhya Pradesh. Figure 11.4 illustrates that the CNSL composites exhibit a higher damping under vibrations than other synthetic epoxy-based composites.[9]

TABLE 11.1 Statewise Soya Bean Production in India, 2009–2010.

Sl. no.	Name of state	% Soya bean production
1	Madhya Pradesh	55
2	Maharashtra	31
3	Rajasthan	8
4	Andhra Pradesh (old)	2
5	Karnataka	2
6	Other states	3

One of the foremost applications of these bioplastic and bio-resin composites in their recycleability and ease of disposal through conversion of the bioplastics into useable fuel by depolymerization with the aid of a catalyst and condensing the pyrolysed gas into fuel oil[10] (Fig. 11.5). This is one of the ways to obtain bioplastic diesel and petrol.

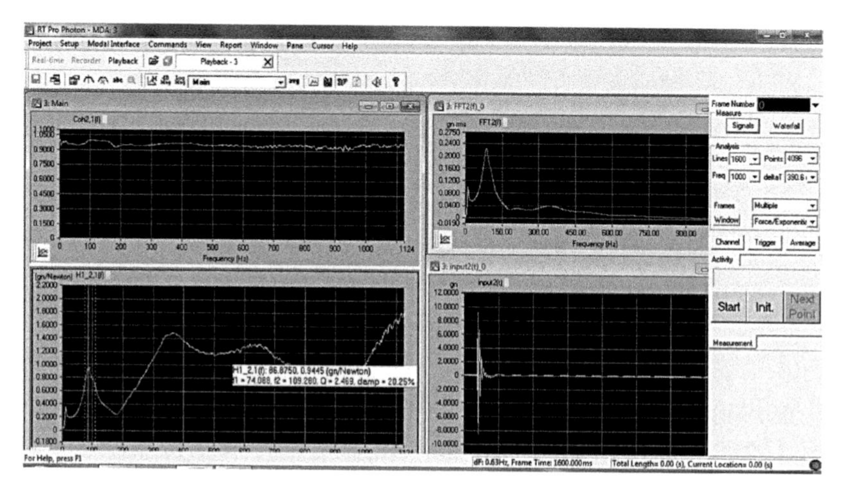

FIGURE 11.4 The damping properties of CNSL composites.

FIGURE 11.5 Bioplastic waste to fuel plant.

11.3 CONCLUSION

These conventional, organic and GM plants are sources of a discrete amount of bio-resins and bioplastics. This invited paper attempts to project the actual possibilities of the bio-resin and bioplastic market in this country. Their true potentialities in composites product applications, involving structural, interior, electronic and chemical engineering markets are discussed in this chapter. A novel working model with an economically feasible option is also provided for those concerned about their safe disposal, recycling, reuse and conversion into useable fuel with no impact to the environment.

KEYWORDS

- **bio-resins**
- **bioplastics**
- **CNSL oil**
- **genetic modification**
- **market potential**
- **bioplastic to fuel conversion**

REFERENCES

1. Gedam, P. H.; Sampathkumaran, P. S. Cashew Nut Shell Liquid: Extraction, Chemistry and Applications. *Progress Org. Coatings.* **1986,** *14*, 115.
2. Lubi, M. C.; Thachill, E. B. *Des. Monomers Polym.* **2000,** *3*,123.
3. Menon, A. R. R.; Sudha, J. D.; Pillai, C. K. S.; Mathew, A. G. *J. Sci. Ind. Res.* **1985,** *44*, 324.
4. Papadopoulou, E.; Chrysalis, K. *Thermochim. Acta.* **2011,** *512*, 105.
5. Gandhi, T. S.; Patel, M. R.; Dholakiya, B. Z. *Res. Chem. Intermed.* DOI: 10.1007/s11164-013-1034-2.
6. Harsha, S. N.; Padmanabhan, K. ICNP International Conference, Kottayam, 2015.
7. Sawant, A. V.; Takalkar, A. R.; Padmanabhan, K. ICNP International Conference, Kottayam, 2015.
8. Balamurugan, A.; Ramya, T.; Nagarajan, S. K. *Int. J. Manage.* **2011,** *2*(1), 84.
9. Harsha, S. N.; Padmanabhan, K.; Murugan, R. ICNP International Conference, Kottayam, 2015.
10. Padmanabhan, K.; T. Deepak Kumar, T.; Ganesh, P.; Haricharan, A. V.; Karthik, R. S.; Devasagayam, G. *J. Energy Storage Conversion.* **2012,** 3(2), 225.

CHAPTER 12

MECHANICAL PROPERTIES OF SANDWICH COMPOSITES MADE USING NATURAL FIBRES AND GLASS FIBRE MAT

R. GOPINATH[1,*], K. GANESAN[2], and R. POOPATHI[1]

[1]*Department of Civil Engineering, University College of Engineering, Tindivanam 604001, Tamil Nadu, India,*
[*]E-mail: ramanaocean@gmail.com

[2]*Department of Civil Engineering, Sudharshan Engineering College, Pudukkottai 622501, Tamil Nadu, India*

CONTENTS

ABSTRACT

Nowadays, due to the superior mechanical quality of the natural and synthetic fibres, composites made by using these fibres are extensively used. In the current study, mechanical properties of hybrid composites made using *Prosopis juliflora*, jute and glass fibres were experimentally investigated and discussed. Seven different types of composite laminates were made, with four of which were of sandwich type with *Prosopis* fibres placed in between the layers of glass and jute fibre mats, one of which was prepared using only *Prosopis* fibres and the remaining two were made using alternate layers of jute and glass fibre mats in 0, 90 and 45°C orientations. The laminates reinforced with jute and glass fibres in alternate 0 and 90°C orientation and with *Prosopis* sandwiched between two layers of glass fibre mats were found to exhibit better mechanical properties compared with other types. However, laminates reinforced only with *Prosopis* fibres were found to possess good impact and moisture resistance.

12.1 INTRODUCTION

The inherent advantages of natural fibres, such as abundant presence, cheap processing, ease biodegradability and high strength, make them extensively useful for various applications such as aerospace, automobile and infrastructure. Although natural fibres are obtained from various sources such as plants, animals and minerals, plant fibres are used in large extent for making composites mainly due to their inherent characteristics and abundant availability. Cellulose, hemicelluloses, lignin, pectin, wax and other impurities are the main composition of plant fibres. Cellulose consists of repeating units of D-anhydro glucose, with each unit comprising three hydroxyl groups.[1] Cellulose microfibrils embedded in the lignin matrix governs the strength, stiffness and stability of the fibres. The orientation of cellulose microfibrils to the fibre axis determines the stiffness of fibres. The fibres having spiral orientation shows ductile nature, and those with parallel orientation are rigid and inflexible. Moreover, they possess more tensile strength.[2,3] Some of the desirable properties of natural fibres are higher cellulose content, low microfibrillar angle, smaller fibre diameter and longer fibre.[4] Natural fibres are extracted from various parts of plants such as roots, seed, stem, fruits and leaves. Among the plant fibres, bast fibres commonly show the highest degree of polymerization.[5] Due to lower density, bast fibres offer specific strength comparable with glass fibres. Bast fibres alone accounts for nearly 4 million metric tons of global production annually. The tensile strength of bast fibres depends upon their diameter and location in the stems.[6] This is mainly due to the variation of fibre size along its longitudinal axis. High tensile strength is possessed by the bast fibres obtained from the mid region of the stem.[7] Although lignocellulosic fibres offer several advantages, they suffer from major drawbacks such as poor moisture resistance, poor adhesion with matrix and low processing temperatures.

The presence of the hydroxyl groups on the surface of natural fibres is mainly responsible for its hydrophilic nature which hinders effective reaction with matrix. Moreover, pectin and waxy substance covers the reactive functional groups of the fibre and acts as barrier to the adhesion between fibre and matrix.[8] Various chemical treatments, such as mercerization, acetylation, benzoylation, silane, permanganate, acrylation, cyanoethylation and isocyanate treatments and use of maleated anhydride grafted coupling agent (anhydride grafted polyethylene), are adopted to modify the surface of plant fibres so that better adhesion with the matrix is achieved.[9,10,11]

Prosopis juliflora is a woody weed which grows in abundance in tropical and sub-tropical regions of the world. The root of this species distributes both horizontally and vertically and taps the ground water from deep beneath the soil. It is a curse to the farmers as it causes rapid depletion of ground water.[12] Keeping aside these demerits, the wood from this species is used as fuel for household purposes in villages. In addition, good amount of high stiff fibres can be extracted from its bark. The characterization study on bark fibres of PJ revealed its promising characteristics such as high cellulose and lignin content, low microfibrillar angle and low wax content.[13] Although *P. juliflora* fibres possess the potentials required for reinforcing composites, studies on the mechanical properties of *Prosopis* reinforced composites have not been carried out till now. In the current chapter, fabrication and mechanical properties of seven different types of laminates made using *Prosopis*, jute and glass fibres were reported.

12.2 MATERIALS AND METHODOLOGY

12.2.1 MATERIALS USED

The fibres extracted from the bast of *P. juliflora* and subjected to alkali treatment using NaOH at 5% concentration were used after crushing it into finer form. Glass fibre bidirectional woven roving mat of 610 g and jute fibre mat of 367 g were used. Vinyl ester resin was used as matrix, and chemical compounds such as cobalt naphthanate, methyl ethyl ketone peroxide and dimethyl amine were used as acceralator, catalyst and promoter.

12.2.2 SPECIMEN PREPARATION

Seven different types of specimens were fabricated using hand layup technique and cold pressing method. Table 12.1 shows the details about the specimens. The representative samples of test specimens are shown in Figure 12.1(a)–(e). Out of seven types of laminates, four were of sandwich type with *Prosopis* fibres placed between layers of glass and jute fibre mats, one was prepared using only *Prosopis* fibres and the remaining two were made using alternate layers of jute and glass fibre mats in 0, 90 and 45°C orientations. After fabrication, the laminates were kept in an oven at a temperature of 60°C for a period of 2 h. As shown in Figure 12.1, the specimens required for performing tensile, impact, flexural and water absorption tests were cut from the laminate, which were followed by the appropriate ASTM standards.

TABLE 12.1 Details of the Specimens.

Specimen	Layup code	Glass (wt%)	Jute (wt%)	Prosopis (wt%)	Resin (wt%)
GPJG	Glass/PJ/glass	16		22	62
JPJJ	Jute/PJ/jute		20	16	64
PJ	PJ			32	68
GPJGPJG	Glass/PJ/glass/PJ/glass	26		18	56
JPJJPJJ	Jute/PJ/jute/PJ/jute		24	14	62
GJGJG	$Glass_0/jute_{90}/glass_0/jute_{90}/glass_0$	24	14		62
$[GJGJG]_{45}$	$Glass_{45}/jute_{45}/glass_{45}/jute_{45}/glass_{45}$	24	14		62

12.2.3 TESTING METHODS

The tensile test was carried out using TUN 400 universal testing machine by adopting the standards stipulated in ASTM 638. During the test, the cross head speed was maintained at 5 mm/min. In order to observe the change in length experienced by the specimen during the test, the extensometer was mounted at the centre of gauge length of the specimen. The flexural test was carried out according to ASTM 790. The span of the specimen was kept as 120 mm. The specimens were subjected to point load at the centre, and the corresponding deflection was measured using dial gauge of 0.01 mm accuracy. Both the impact and water absorption test were performed by adopting the specifications laid out in ASTM 256 and D 570. The test results of all the specimens are given in Table 12.2.

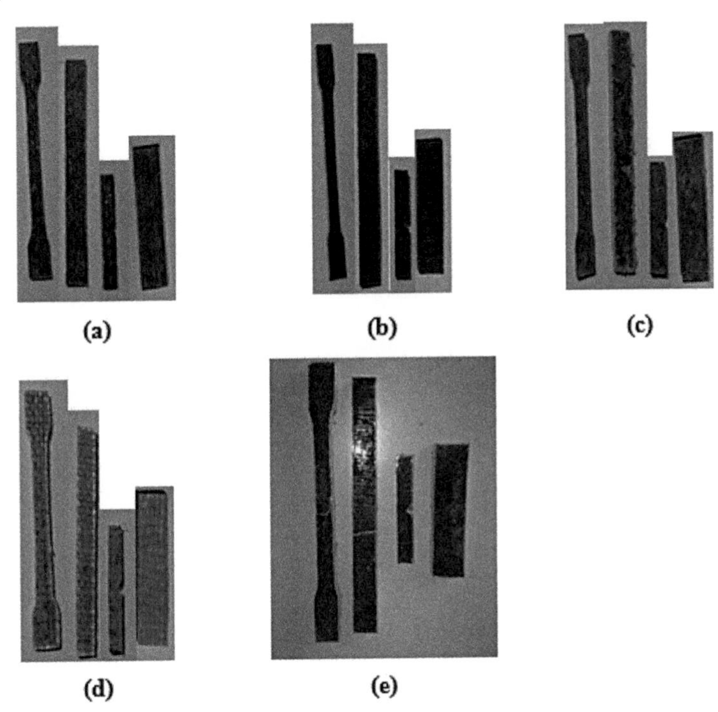

FIGURE 12.1 Representative samples of (a) GPJG, (b) JPJJ, (c) PJ, (d) GPJGPJG and (e) JPJJPJJ.

12.3 RESULTS AND DISCUSSION

12.3.1 TENSILE PROPERTIES

From Table 12.2 and Figure 12.2, it is evident that the specimen GJGJG with alternate layers of jute and glass fibre mats in 0 and 90°C orientations was found to have maximum tensile strength of 137.045 MPa. The presence of glass and jute fibres along the direction the tensile load has resulted in appreciable increase in tensile resistance. The next highest tensile strength was observed in specimen GPJG in which *Prosopis* fibres were sandwiched between two layers of glass fibre mats.

TABLE 12.2 Test Results of Specimens.

Specimen	Layup code	Tensile strength (N/mm^2)	Flexural strength (N/mm^2)	Energy absorbed (J)	Percentage of water absorption
GPJG	Glass/PJ/glass	70.65	185.38	7.30	0.78
JPJJ	Jute/PJ/jute	18.54	182.25	2.70	2.30
PJ	PJ	13.80	91.34	7.0	0.43
GPJGPJG	Glass/PJ/glass/PJ/glass	60.70	198.99	10	0.44
JPJJPJJ	Jute/PJ/jute/PJ/jute	38.54	176.50	6.80	1.80
GJGJG	$Glass_0/jute_{90}/glass_0/jute_{90}/hlass_0$	137.05	316.51	15.30	0.48
$[GJGJG]_{45}$	$Glass_{45}/jute_{45}/glass_{45}/jute_{45}/glass_{45}$	60.80	272.42	18	0.52

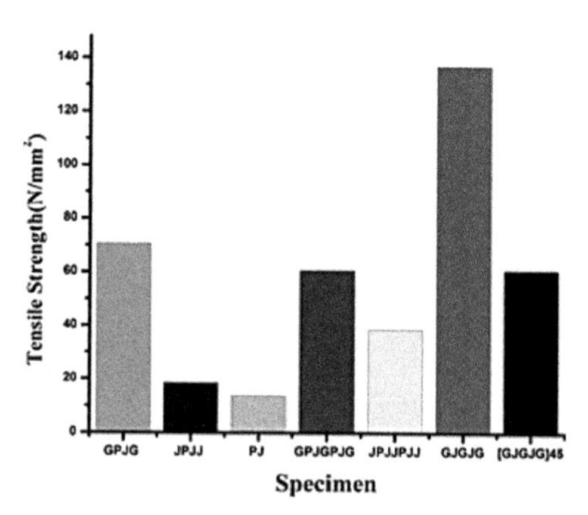

FIGURE 12.2 Plot of tensile strength.

The uniformity in distribution of *Prosopis* fibres in the core layer held between two layers of glass fibre mats has resulted in better stress transfer compared with specimen GPJGPJG wherein *Prosopis* fibres sandwiched between three layers of glass fibre mats. A similar trend was observed in the study wherein chopped banana and sisal fibres were used as sandwich between two and three layers of glass fibre mats. The tensile strength of banana fibre sandwiched and sisal fibre sandwiched two layer glass fibre specimens was found to be higher than three-layered glass fibre sandwiched specimens.[14] The lowest tensile strength was observed in specimens containing only *Prosopis* and *Prosopis* sandwiched bilayered jute fibre specimens. The failure of specimen PJ containing only *Prosopis* fibres was found to be in brittle mode.

12.3.2 FLEXURAL PROPERTIES

From Figure 12.3, it can be observed that specimens GJGJG and $[GJGJG]_{45}$ were found to have maximum flexural strength compared with other specimens. When the specimen is subjected to bending, stress transfer between the layers of fibre mats occurs through shear. The presence of glass and jute fibres in mat form enhances the flexural resistance substantially. However, all the sandwich specimens containing *Prosopis* were found to exhibit good flexural resistance. The fibre continuity in the core layer has resulted in better stress transfer from mid region to the supports. The lowest flexural strength was reported for specimen PJ containing only *Prosopis* fibres. The absence of fibres of high flexural strength such as glass at the periphery led to premature failure of the specimen PJ in brittle mode.

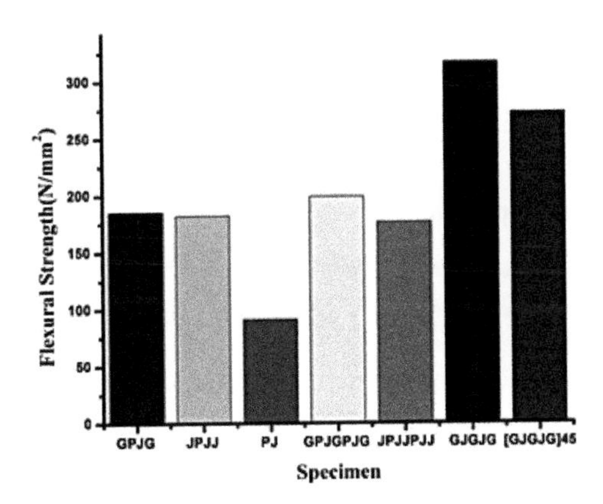

FIGURE 12.3 Plot of flexural strength.

12.3.3 IMPACT PROPERTIES

The fact that specimens GJGJG and $[GJGJG]_{45}$ tend to exhibit maximum energy absorption against impact load is evident from Table 12.2 and Figure 12.4. The specimen PJ containing only *Prosopis* fibres was found to possess high impact resistance. The interweaving of *Prosopis* fibres while mixing them intimately with the polymer matrix has abruptly improved its impact strength. Although the tensile and flexural strength of specimen PJ was considerably lower than other specimens, its impact strength was found to be good and comparable with glass fibre reinforced specimens. The specimen JPJJ with *Prosopis* sandwiched between two-layered jute fibre specimens was found to exhibit lowest impact resistance compared with other specimens. The absence of fibre mat in the mid region of the specimen has decreased its energy absorption capability, and this fact is proven from the result of *Prosopis* sandwiched three-layered jute specimen JPJJPJJ for which the impact resistance was found to increase moderately.

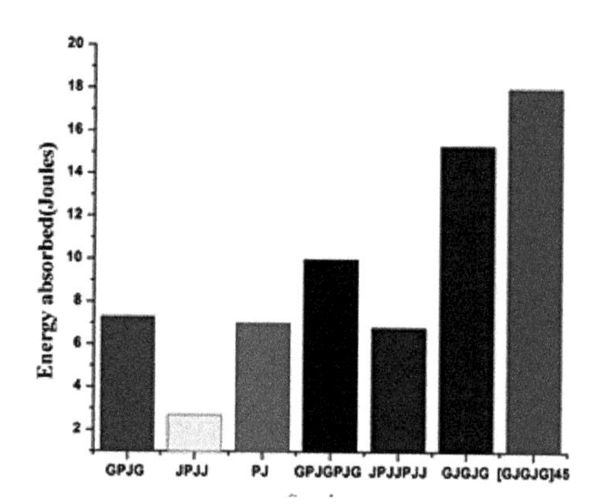

FIGURE 12.4 Plot of impact strength.

12.3.4 WATER ABSORPTION PROPERTIES

From Figure 12.5, it can be observed that specimens PJ and GPJGPJG reported lowest water absorption compared with other specimens. In specimen PJ, *Prosopis* fibres in mixed sizes enables voids filled up with fibres of next lower size. Intimate mixing of fibres with the resin before laying and alkali treatment of fibres at optimum concentration could be some of the other reasons for substantial improvement in water resistance. In specimen GPJGPJG, presence of glass fibres in the periphery as well as at the middle layer has reduced the moisture absorption appreciably. The specimen JPJJ exhibited highest water absorption compared with other specimens. Jute fibres offer least resistance against water absorption. Moreover, its presence at the periphery of specimen JPJJ has resulted in appreciable increase in water absorption.

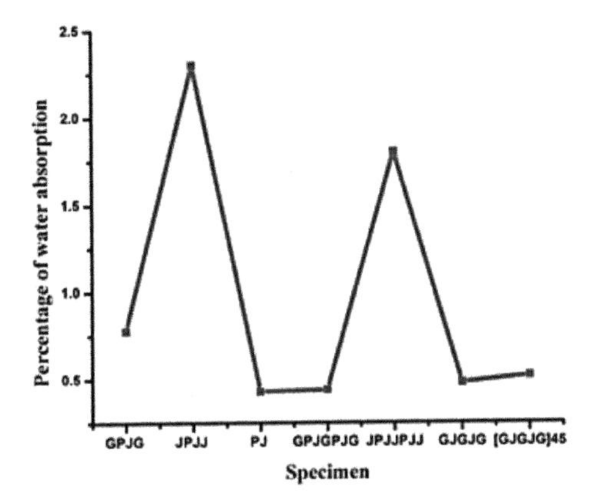

FIGURE 12.5 Plot of percentage of water absorption.

12.4 CONCLUSION

The mechanical properties of *Prosopis* sandwiched glass and jute fibre–reinforced specimens were studied and compared with specimens reinforced only with *Prosopis* and with glass/jute fibres in alternate layers. From the study, the following can be concluded:

1. The tensile strength of specimen reinforced with alternate layers of glass and jute was found to be highest among all other specimens. The orientation of glass and jute fibres along the direction of tensile load has enhanced the tensile resistance appreciably.
2. The tensile strength of two-layered glass fibre-reinforced *Prosopis* sandwiched specimens was found to be more than three-layered glass fibre-reinforced sandwich specimen. The efficiency of stress transfer between *Prosopis* fibres present in the core layer to the glass fibre at the periphery was found to decrease with increase in sandwich layers.

3. The tensile resistance of specimen containing only *Prosopis* fibres tends to be the lowest, and the failure was observed to be brittle.
4. The trend followed in the tensile strength was observed in the flexural resistance of specimens. However, specimen with *Prosopis* sandwiched between two layers of jute fibre mat was found to offer appreciably higher flexural resistance. The specimen reinforced only with *Prosopis* was found to exhibit the lowest flexural strength compared with other specimens.
5. The variation of impact resistance showed a different trend with specimen containing only *Prosopis* exhibited higher energy absorption when subjected to impact load. Compared with other specimens, specimen with *Prosopis* sandwiched between two layers of jute fibre mat was found to offer lowest impact resistance. The energy absorption capacity was drastically reduced because of the absence of jute fibre mat at the middle layer.
6. The water absorption of *Prosopis* sandwiched specimen with jute fibres placed at the periphery was found to be maximum, which proves that jute fibres have greater affinity towards moisture. The water absorption of specimen reinforced only with *Prosopis* fibres was found to be lowest. The uniform fibre distribution, complete wetting and alkali treatment of *Prosopis* fibres have improved the moisture-resistance appreciably. The water absorption of *Prosopis* sandwiched glass fibre-reinforced specimens tends to be lower which is mainly due to the presence of glass fibres at the periphery.

KEYWORDS

- ***Prosopis juliflora***
- **jute**
- **glass**
- **sandwich**

REFERENCES

1. Nevell, T. P.; Zeronian, S. H. *Cellulose Chemistry and Its Applications*; Wiley: New York, 1985.
2. Barkakaty, B. C. Some Structure Aspects of Sisal Fibers. *J. Appl. Polym. Sci.* **1976,** *20*, 2921–2940.
3. John, M. J.; Thomas, S. Biofibres and Biocomposites. *Carbohydr. Polym.* **2008,** *71*, 343–364.
4. Ishak, M. R.; Leman, Z.; Sapuan, S. M.; Edeerozey, A. M. M.; Othman, I. S. Mechanical Properties of Kenaf Bast and Core Fibre Reinforced Unsaturated Polyester Composites. *IOP Conf. Ser.: Mater. Sci. Eng.* **2010,** *11*, 1–6.
5. Jawaid, M.; Abdul Khalil, H. P. S. Cellulosic/Synthetic Fibre Reinforced Polymer Hybrid Composites: A Review. *Carbohydr. Polym.* **2011,** *86*, 1–18.
6. Charlet, K.; Baley, C.; Morvan, C.; Jernot, J. P.; Gomina, M.; Breard, J. Characteristics of Herme's Flax Fibres as a Function of Their Location in the Stem and Properties of the Derived Unidirectional Composites. *Comp. Part A Appl. Sci. Manuf.* **2007,** *38*, 1912–1921.
7. Bledzki, A. K.; Gassan, J. Composites Reinforced with Cellulose Based Fibres. *Prog. Polym. Sci.* **1999,** *24*(2), 221–274.
8. Kabir, M. M.; Wang, H.; Aravinthan, T.; Cardona, F.; Lau, K. T. Effects of Natural Fiber Surface on Composite Properties: A Review, eddBE2011. *Proceedings, Energy, Environment and Sustainability*, April 27–29, Queensland University of Technology: Brisbane, 2011; pp 94–99.
9. Mishra, S.; Mohanty, A. K.; Drzal, L. T.; Misra, M.; Parija, S.; Nayak, S. K.; Tripathy, S. S. Studies on Mechanical Performance of Biofibre/Glass Reinforced Polyester Hybrid Composites. *J. Comps. Sci. Tech.* **2003,** *63*, 1377–1385.
10. Sreekumar, P. A.; Thomas, S. P.; Saiter, J. M.; Joseph, K.; Unnikrishnan, G.; Thomas, S. P. Effect of Fiber Surface Modification on the Mechanical and Water Absorption Characteristics of Sisal/Polyester Composites Fabricated by Resin Transfer Molding. *Comp. Part A* **2009,** *40*, 1777–1784.
11. Faruk, O.; Bledzki, A. K.; Fink, H. P.; Sain, M. Biocomposites Reinforced with Natural Fibers: 2000–2010. *Prog. Polym. Sci.* **2012,** *37*, 1552–1596.
12. Pasiecznik, N. M.; Felker, P.; Harris, P. J. C. *The Prosopis Juliflora-Prosopis Pallida Complex: A Monograph*. HDRA: Coventry, UK, 2001; pp 162–168.
13. Saravanakumar, S. S.; Kumaravel, A.; Nagarajan, T.; Sudhakar, P.; Baskaran, R. Characterization of a Novel Natural Cellulosic Fiber from Prosopis Juliflora Bark. *Carbohydr. Polym.* **2013,** *92*, 1928–1933.
14. Arthanarieswaran, V. P.; Kumaravel, A.; Kathirselvam, M. Evaluation of Mechanical Properties of Banana and Sisal Fiber Reinforced Epoxy Composites: Influence of Glass Fiber Hybridization. *Mat. Des.* **2014**, *64*, 194.

CHAPTER 13

FABRICATION AND CHARACTERIZATION OF COMPOSITE MEMBRANE OF CROSS-LINKED POLYVINYL ALCOHOL AND HYDROXYAPATITE FOR TISSUE ENGINEERING

DEBAPRIYA BANERJEE[1], ANINDYA HALDAR[2], and PIYALI BASAK[3]

[1]*School of Bioscience and Engineering, Jadavpur University, 118 Raja S. C. Mullick Road, Kolkata 700032 West Bengal, India*

[2]*Department of Biotechnology, Heritage Institute of Technology, Kolkata, West Bengal, India*

[3]*Jadavpur University, 118 Raja S. C. Mullick Road, Kolkata 700032, West Bengal, India*

CONTENTS

ABSTRACT

Biocomposites are the composite materials which contain matrix phases formed by biopolymers. They are composed in such a manner that the final product can possess good biocompatibility. In the approach of implant coating and cellular recovery, the area of tissue engineering has resulted in a revolutionary change. The modern therapeutic technology has witnessed the biocomposites being successfully used for a variety of medical applications such as drug delivery, wound dressing and tissue regeneration. The highly porous polymer–ceramic composites can be applied as scaffold in tissue engineering. The major objective of this work is to fabricate a polyvinyl alcohol based composite with hydroxyapatite cross linked by glutaraldehyde. Characterization of the composites has been carried out by scanning electron microscopy, attenuated total reflection-Fourier transform infrared spectroscopy, swelling studies, X-ray diffraction and mechanical testing. The result of characterization denoted that the sample had a well-porous structure and could be applied for drug delivery.

13.1 INTRODUCTION

Tissue engineering is a field that combines technology with life science to repair, maintain or improve the function of tissue. There has always been an exiguity of tissue availability for the objective of tissue reconstruction. Various objections such as biocompatibility, tissue fatality, immune repudiation, correct dosage of cells, best timing for it and so on have been correlated to the mechanisms such as tissue grafting or fabricated biomaterial reinstatement. On the account of those limitations, there has been a revolutionary change in therapeutic field from classical tissue replacement to a rising integrative engineering access. In biomedical sector, tissue engineering is a developing field of importance. Tissue engineering mainly rotates around the formation of organic implant that can supplant, invigorate and alleviate affected tissue. An effective tissue engineering tool needs cultured cell in simulated matrix that could play an absolute reinstatement of the damage part. This kind of artificial matrix could be called scaffold. The materials which can be used to make scaffolds are polylactic-co-glycolic acid, polyvinyl alcohol (PVA), polylactic acid, hydroxyapatite (HA), gelatin, tricalcium phosphate and so on.

The materials that are formed of two or more than two definite parts can be described as composites. They subsist of a continuous phase which is termed as matrix phase. The other phase is discontinuous and is known as reinforcing agent. Here, PVA acts as the matrix, whereas HA acts as the reinforcing agent. The properties of the composites are motivated by the properties of their distinct components. The composites achieve the desired mechanical properties such as strength, fatigue defiance, toughness and so on.[1,5,10]

Hydrogels are hydrophilic polymeric networks having three-dimensional structures. They can absorb up to 90% of water. Hydrogels can be applied as scaffold as they can achieve the flexible mechanical characteristics of the anchor biological tissue. They use to imitate micro-environment for the cell growth inside the body. There are a few polymers which show the impulse to form the hydrogel. They are PVA, chitosan, gelatin and so on. Some hydrogels have area-specific applications; such gels find importance in drug delivery system. Hydrogels could also be used for controlled drug delivery as well as show drug evasion capacity.

Polyvinyl alcohol is a synthetic polymer having water solubility property. It has excellent adhesive ability that could be used for preparing hydrogels. It possesses well flexibility and tensile strength so that it is effective in case of fabricating scaffold. In addition, it is non-toxic and has a slow rate of degradation.

HA is a mineral of bone, thereby it is highly osteoconductive. The scaffold that is made by HA could help to invigorate the articular cartilage. In case of drug delivery system, it has the propensity to high burst release, which can be alleviated by polymer coating on it. The connection between bone and embed could be entrenched by HA.

Glutaraldehyde is a liquid organic compound which finds application as cross linker. It can be cross linked with synthetic polymers such as polyvinyl alcohol and gelatin. The rate of degradation of synthetic polymer film could be slower by cross linking with glutaraldehyde.

13.2 MATERIALS AND METHODS

13.2.1 MATERIALS

Polyvinyl alcohol with molecular weight 115,000 KDa was purchased from Lobacheme, India. Hydroxyapatite and deionized water were collected from Jadavpur University. Glutaraldehyde was purchased from Lobacheme, India. Simulated body fluid (SBF) has been prepared in laboratory of School of Bioscience and Engineering, Jadavpur University.

13.2.2 METHOD OF FABRICATION

By heating PVA in water, a clear solution was obtained. Hydroxyapatite solution was prepared by stirring HA in water in a magnetic stirrer at about 350 rpm. These two solutions are merged together by stirring for another 25 min. Diluted glutaraldehyde solution was added drop by drop to this mixture, and the whole solution was allowed to blend with each other by stirring it for 3 min. The resulting solution was kept in −20°C freezer for 24 h. The iced mixture was dried again in a freeze drier for 36 h. The composite was again dried in an oven for 5 h at about 60°C.[6,10]

13.2.3 CHARACTERIZATION TESTS

13.2.3.1 SCANNING ELECTRON MICROSCOPY

The microstructure of composite scaffold was evaluated by using scanning electron microscopy (SEM) (INSPECT F50, Fei, Netherlands). For the examination, the samples were coated with gold. The SEM micrographs helped to indicate the pores on the scaffold surface.

13.2.3.2 ATTENUATED TOTAL REFLECTANCE-FOURIER TRANSFORM INFRARED SPECTROSCOPY

The structural analysis of the samples using attenuated total reflectance-Fourier transform infrared (FTIR) (Thermo Nicolet, Nexus 870 spectrophotometer) were done from the peaks obtained for different vibrations (stretching vibrations, bending vibrations and so on) created due to the motion of molecular bonds inside the samples.[4]

13.2.3.3 SWELLING

The swelling test was carried out in water, SBF and pH media at 37°C. The dried samples were immersed in corresponding media followed by taking their weights. The samples were taken out and the weights were measured followed by removing the liquid from the surface of the swelled samples when swelling equilibrium had been reached after 24 h. The percentage of swelling was calculated by Equation 13.1.[2,9]

$$Swelling = \frac{(W_f - W_i)}{W_i} \times 100 \tag{13.1}$$

In Equation 13.1, W_i is the initial weights of samples before immersion, W_f is the final weights after swelling.

13.2.3.4 X-RAY DIFFRACTION

The phases present inside the samples were determined through X-ray Diffraction (XRD) analysis (Rigaku Miniflex X-Ray diffractometer, Ultima III). The crystalline HA information was also obtained by XRD analysis.[2,4]

13.2.3.5 MECHANICAL TESTING

To evaluate the mechanical strength of composite samples, tensile testing was carried out by using Instron machine (Instron, model 4204). The thickness of the sample, width of the sample and the length of the sample were measured by Vernier caliper. The load at failure was obtained from the Instron machine.

13.2.3.6 CONTACT ANGLE

To evaluate whether the samples were hydrophilic or hydrophobic in nature, Contact angle measurement (Rame Hart Goniometer, model 100-00-230) was carried out. A drop of fluid was placed on the surface of membrane. The angle between solid surface and liquid surface was measured by interfacial tensions between the liquid, solid and the medium in which the experiment was performed. In this work, the fluid medium taken was glycerol.

13.3 RESULT AND DISCUSSIONS

13.3.1 SCANNING ELECTRON MICROSCOPY

Figure 13.1(a)–(d) shows the SEM micrographs of the sample. Figure 13.1(a) shows some pores in sample 1. Figure 13.1(b) indicated the pores in a higher magnification and figure. Figure 13.1(c) shows the image of the pores as well as the uneven surface in a lower magnification of the uncross-linked sample in which Figure 13.1(d) is the close view of the uneven surface of the uncross-linked sample. The uneven surface is created due to the deposition of material. On account of various reasons, deposition could occur. The deposited material could be polymer or could be ceramic. The reason for the excess amount of deposition, though, could be attributed to the lack of proper drying. From the scanning electron micrographs of all the samples, it could be concluded that with the increase in the percentage of HA, there was a gradual decrease in the porosity within the surface of the membrane. It has been seen that the samples contain pores which are useful for supplying nutrition to the growing bone.

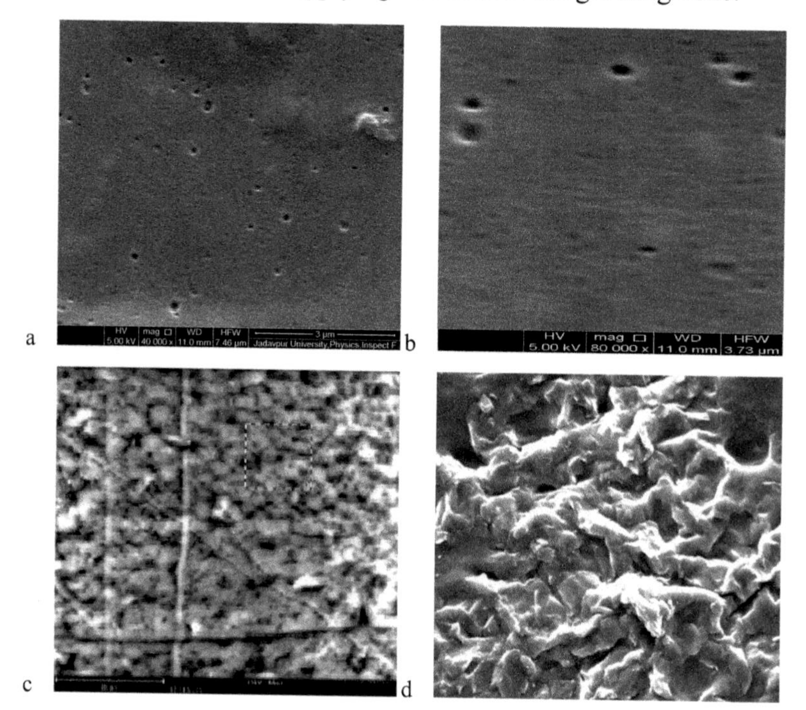

FIGURE 13.1 Microstructure of composite scaffold under SEM (INSPECT F50, Fei, Netherlands)

13.3.2 ATTENUATED TOTAL REFLECTANCE-FOURIER TRANSFORM INFRARED SPECTROSCOPY

Figure 13.2(a, b) shows the FTIR spectrum of the samples and its components. In Figure 13.2(a), two peaks were observed at 520.7 and 1157 cm^{-1}, which indicated the presence of $(Po_4)_3$ group of HA. A band was observed between 2962 and 3539 cm^{-1}, which was formed due to the stretching vibration of OH group of cross-linked PVA. A peak was obtained at 1720 cm^{-1}, which indicated the stretching vibration of carbonyl group of glutaraldehyde. This peak was obtained may be due to the presence of some unreacted glutaraldehyde in the sample. Two peaks were obtained at 1072 and 1157 cm^{-1}, which indicated the presence of C–O–C group of cross-linked compound. It verified the formation of strong acetal bridges between PVA and glutaraldehyde. Another peak was observed at 1467 cm^{-1}, which indicated the bending vibration of CH_2 group in PVA. In Figure 13.2(b), three peaks were observed at 559, 665 and 1143.7 cm^{-1}. They indicated the presence of $(Po_4)_3$ group of HA. A band was observed between 2960 and 3498 cm^{-1}, which was formed due to the stretching vibration of OH group of cross-linked PVA. Two peaks were obtained at 1024 and 1143 cm^{-1}, which indicated the presence of C–O–C group of cross-linked compound. It verified the formation of strong acetal bridges between PVA and glutaraldehyde. Another peak was observed at 1475.5 cm^{-1}, which indicated the bending vibration of CH_2 group in PVA. No peaks are obtained in the range of 1700 cm^{-1}, so it can be said that there is no unreacted glutaraldehyde portion in this sample.[3,8]

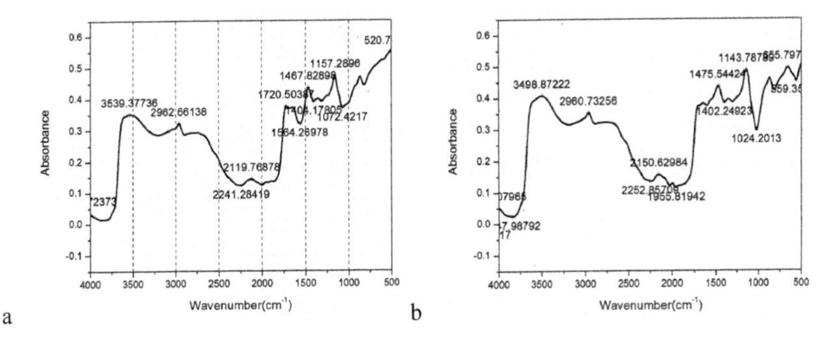

FIGURE 13.2 FTIR Spectrum for identification of functional group present in composite membrane of cross-linked PVA and Hydroxyapetite

13.3.3 SWELLING

Figure 13.3 represents the swelling behaviour of the samples in SBF, oxidative media and water. Two distinct factors were observed from the above two graphs. It is vividly observed that percentage of swelling is higher in water when compared with SBF. This is due to the existence of extinct salts of SBF, which retards the water to get in the pores of the sample. On the other hand, it is seen that the highest percentage of swelling is around 67% (in water), which is far smaller than the normal water absorption capacity (90%) of a HA scaffold. This could be due to the cross linking between PVA and glutaraldehyde.[2,9]

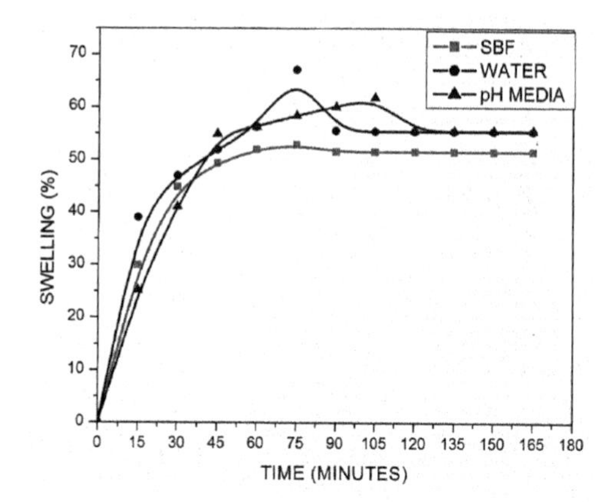

FIGURE 13.3 Swelling characteristics of Composite membrane of cross-linked PVA and Hydroxyapetite

13.3.4 X-RAY DIFFRACTION ANALYSIS

Figure 13.4 shows the XRD analysis of the sample. The graph consists of three peaks. First peak was obtained at $2\theta=20.2°$, which was formed due to the presence of the cross-linked PVA. One peak was observed at $2\theta=37°$, which was obtained because of crystalline HA material. Another peak was obtained at $2\theta=45°$ (see Fig. 13.4), which was also obtained because of the presence of the crystalline HA material in the sample. From Figure 13.4, it is clear that with increase in cross-linking, crystallinity decreases due to the breakage of H-bonding of PVA.[2,4]

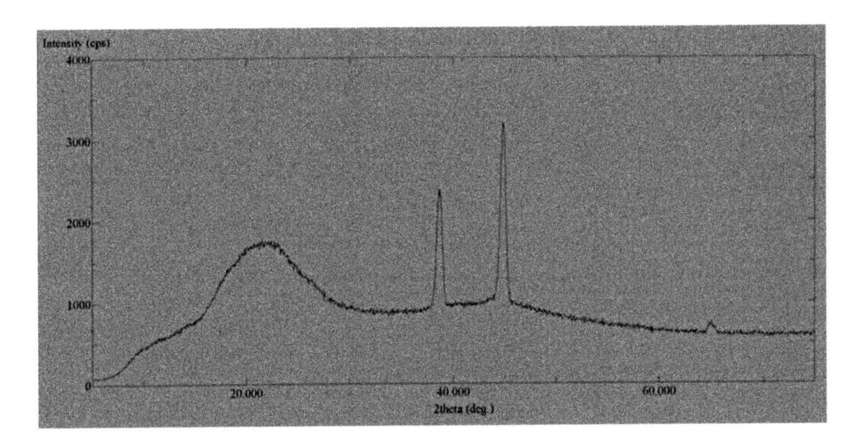

FIGURE 13.3 XRD analysis of Composite membrane of cross-linked PVA and Hydroxyapetite

13.3.5 CONTACT ANGLE

Table 13.1 shows the contact angle measurement of the sample.

When measured in the medium of glycerol, the contact angle of sample 1 was found to be 57.06°. Hence, the sample could be termed hydrophilic in nature. The contact angle of sample 2 was found to be 56.16° when measured in the medium of glycerol. Hence, the sample could be termed hydrophilic in nature. The contact angle of sample 3 was found to be 58.44° when measured in the medium of glycerol (see Table 13.1). Hence, the sample could be termed hydrophilic in nature. All of them were hydrophilic in nature. It is a desirable property of a scaffold.

13.3.6 TENSILE TESTING

Figure 13.5 shows the ultimate tensile strength of the samples. Table 13.2 shows the value of load at failure.

The ultimate tensile strength was calculated by Equation 13.2.

$$\text{Ultimate Tensile Strenght} = \left(\frac{\text{The load at failure}}{\text{Original cross sectional area}} \right) \quad (13.2)$$

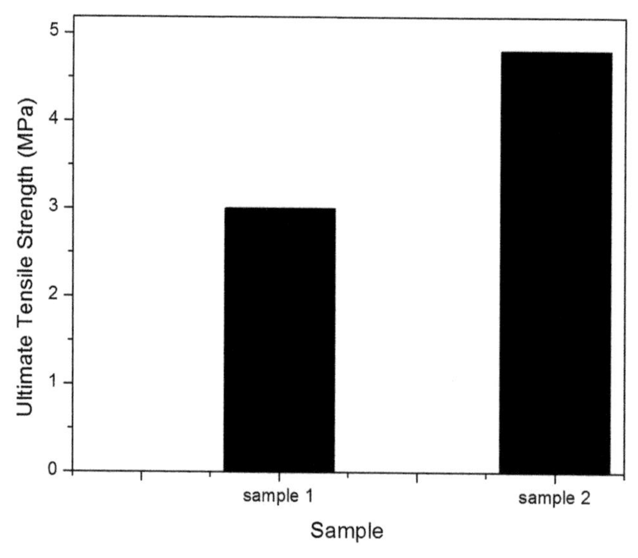

FIGURE 13.5 Shows the ultimate tensile strength of the samples.

For sample 1, the ultimate tensile strength is 3.03 MPa and for sample 2 it is 4.853 MPa (see Fig. 13.5). It shows the ultimate tensile strength of the samples. It suggested that sample 2 had a higher tensile strength than sample 1. This could be attributed to the higher percentage of HA content in the sample. Hence, better mechanical strength could be achieved by the increasing the percentage of HA in the sample. Here, the ultimate tensile strength depends on two factors—one is the amount of HA in the sample and another is the degree of cross linking in the sample. For high degree of cross linking, ultimate tensile strength is higher.

13.4 CONCLUSION

The purpose of this work is to fabricate a porous PVA-based HA composite scaffold cross linked by glutaraldehyde with the help of the process of freeze drying. The porous PVA/HA/glutaraldehyde membrane shows high affinity towards water. The sample undergoes a high percentage of swelling in water. The XRD, FTIR spectrum establishes that the sample has all the necessary functional groups as well as retains their vibrations. The sample shows various traits such as high porosity, completely hydrophilic in nature and good tensile strength, thus enabling it to be very useful for in vivo use. Further work will be conducted on the evaluation of the in vivo behaviour of the samples for application in bone regeneration. Future scope of this work lies in the enhancement of the rate of the drug release.[2,7,9]

KEYWORDS

- **polymer**
- **composite**
- **hydroxyapatite**
- **scaffold**
- **cross-linking**

REFERENCES

1. Asran-Ashraf, Sh; Henning S; Michler-Goerg H. Polyvinyl Alcohol–Collagen–Hydroxyapatite Biocomposite Nanofibrous Scaffold: Mimicking the Key Features of Natural Bone at the Nanoscale Level. *Polymer* **2010,** *51*(4), 868–876.
2. Basak, P; Adhikari, B; Sen, A. Surface Crosslinked Poly (Vinyl Alcohol) Hydrogel for Colon Targeted Drug Release. *Polym. Plast. Tech. Eng.* **2011,** *50*, 1357–1361 (accessed Sep 14, 2011).
3. Cheng-Ming, Tang; Yi-Hung, Tian; Shan-Hui, Hsu. Poly (Vinyl Alcohol) Nanocomposites Reinforced with Bamboo Charcoal Nanoparticles: Mineralization Behavior and Characterization. *Materials*. **2015**, *8*, 4895–4911.
4. Deville, S.; Saiz, E.; Tomsia, A. P. Freeze Casting of Hydroxyapatite Scaffolds for Bone Tissue Engineering. *Sci. Direct Biomater*. **2006,** *27*, 5480–5489 (accessed July 20, 2006).
5. Gang, W.; Bing, S.; Wenguang, Zhang; Chengtao, Wang. In vitro Behaviors of Hydroxyapatite Reinforced Polyvinyl Alcohol Hydrogel Composite. *Materials Chem. Physics*. **2008,** *107*, 364–369.
6. Hossan, Md. Yeathad; Islam, Molla Md. Ashraful; Islam, Md. Shahidul; Alam-Ashequl, Rana; Gafur, Md. Abdul; Mainul, Md. Karim; Fabrication and Characterization of Polyvinyl Alcohol-Hydroxyapatite Biomimetic Scaffold by Freeze Thawing in Situ Synthesized Hybrid Suspension for Bone Tissue Engineering. *Int. J. Emerging Technol. Adv. Eng.* **2012**, *2*(12), ISSN 2250-2459.

7. Lee-Jae, Young. Electrically Conducting Polymer-Based Nanofibrous Scaffolds for Tissue Engineering Applications. *Polym. Rev.* **2013,** *53*, 443–459.
8. Mansur-Herman, S; Sadahira-Carolina, M; Souza-Adriana, N; Mansur-Alexandra A. P. FTIR Spectroscopy Characterization of Poly(Vinyl Alcohol) Hydrogel with Different Hydrolysis Degree and Chemically Crosslinked with Glutaraldehyde. *Mater. Sci. Eng.* **2008,** *28*, 539–548.
9. Mingbo, Wang; Yubao, Li; Jiaqi, Wu; Fenglan, Xu; Yi, Zuo; Jansen, J. A. In Vitro and in Vivo Study to the Biocompatibility and Biodegradation of Hydroxyapatite/poly (vinyl alcohol)/Gelatin Composite. *Sci. Direct Biomate.* **2006**. DOI: 10.1002/jbm.a.3158 (accessed Aug 13, 2007).
10. Swetha, M.; Sahithi, K.; Moorthi, A.; Srinivasan, N.; Ramasamy, K.; Selvamurugan, K. Biocomposites Containing Natural Polymers and Hydroxyapatite for Bone Tissue Engineering. *Int. J. Biol. Macromol.* **2010,** *47*, 1–4.
11. Wu, X.; Liu, Y.; Li, X.; Wen, P.; Zhang, P.; Long, Y.; Wang, X.; Guo, Y.; Xing, F.; Gao, J. Preparation of Aligned Porous Gelatin Scaffolds by Unidirectional Freeze-Drying Method. *Acta. Biomater*. **2010,** *6*, 1167–1177.

CHAPTER 14

IN SITU NANOSILVER-IMMOBILIZED CHITOSAN/OXIDIZED CARBOXYMETHYLCELLULOSE BLEND DRESSINGS FOR WOUND MANAGEMENT

SHAKEEL AHMED[1,*] and SAIQA IKRAM[*]

*Department of Chemistry, Jamia Millia Islamia, New Delhi 110025, India, *E-mail: sikram@jmi.ac.in, shakeelchem11@gmail.com*

CONTENTS

ABSTRACT

The variety of wound types has resulted in a wide range of wound dressings with new products frequently introduced to target different aspects of the wound healing process. The ideal dressing should achieve rapid healing at a reasonable cost with minimal inconvenience to the patient. The present chapter offers the role of biopolymers in combination with synthetic ones, this opens up many phases towards the applications in medics including the tissue engineering and controlled drug delivery. One of the important aspects is the wound-care market where biopolymeric blend compositions as wound dressings are proving their potentials to enhance wound healing as these are capable of maintaining moist environment at the wound site. Synthetic biodegradable polymers are also increasingly used, although the use of natural polymers, such as cellulose and starches, is still common in biomedical research. Chitosan (CS) is a biopolymer having immense structural possibilities for modifications to generate novel properties and functions for its applications in wound care. The inherent antimicrobial and antifungal properties are the reason for it being one of the currently proposed and most promising polymers in wound dressings. The present chapter is focused on the synthesis of nanosilver-contained CS cellulose wound dressings for wound management. The introduction of green route for wound dressing loaded with silver nanoparticles has been explored.

14.1 INTRODUCTION

In the last 15–20 years, the research related to intelligent materials with specified nanoparticles has increased to a remarkable volume. This is related to the wide range of applications that offer these kinds of materials. In this context, the treatment of the wounds has evolved incredibly from the ancient times to the present. Wound healing is a complicated proceeding of cellular and biochemical events that can be accelerated by ideal wound dressing. Primarily, inhibition of bleeding and the protection of the wound from environmental irritants as well as water are done with dressing materials. There are three categories of wound dressing: biologic, synthetic and biologic–synthetic.[1] In order to maintain a balanced moist environment, during last few decades, researchers have focused on biologic–synthetic dressings to dry out the wound bed. The main function of traditional dressings is to absorb wound exudate and lead to formation of crust on the wound surface with remarkable scarring. modern dressings have widely replaced by aiming to improve healing by handling wound fluid in a way that prevents accumulation of excess exudate while maintaining a certain degree of moisture, and thereby enhancing the chance of obtaining new skin tissue without scarring.[2] Several types of wound dressings are available in which several bioactive compounds and materials have been added to accelerate wound healing. These dressings can deliver therapeutic agents such as vitamins, growth factors and mineral supplements to the wound site and also help in improving wound healing.[3] Several factors, such as the presence of underlying diseases, nutrition state, amounts of wound exudate and the microflora of the wound, will affect the performance of the chosen dressing.

Due to the beneficial biological and antimicrobial properties of chitosan (CS) and its high valuable potential for wound healing, it is very attractive for wound care.[4–6] CS is the second most abundant natural polymer after cellulose and a linear cationic homopolymer of β-(1,4)-linked N-acetyl-2-amino-2-deoxy-D-glucopyranose, obtained by deacetylation of chitin.[7,8] The role of CS as biomaterial is astonishing as demonstrated by the published scientific documents. Drawback of CS scaffolds is its low mechanical strength and high alkaline ionic product.[9] Cellulose is another natural polymer representing a renewable source and has strong mechanical strength.[10–12] The conversion of its hydroxyl group to dialdehyde by periodate oxidation is mostly used in derivation for further reactions.[13] Cellulose has similar structure as CS, providing possibility of producing a homogenous blend, and its availability makes it suitable for biomedical applications.

Silver has a long history of antimicrobial use. Moreover, it is a well-known antimicrobial agent against a wide range of over 650 microorganisms from different classes such as bacteria, fungi or viruses, and it is finding use in the form of silver nanoparticles nowadays[14] as a potent antimicrobial agent for use in wound dressing. Antibacterial efficacy of silver and its salts has been known for a long time, and in recent years, there has been extensive research done on the use of nanometric silver particles as antibacterial agents. Silver nanoparticles have been deployed in a wide variety of matrices and formulae.

In particular, silver nanoparticles have recently been used as antimicrobial agents for some Gram-positive/negative bacteria. To take advantage of this abovementioned feature and to combine some polymer for biomedical applications, such as drug delivery, with the antimicrobial effect is the focus of this work. For these reasons, a novel method has been developed for conversion of silver into silver nanoparticles within the film of CS or oxidized carboxymethylcellulose (OCMC). The present chapter reveals a simple and cost-effective chemical synthetic route to form nanosilver particles. Silver nanoparticles are formed in a simple oxidation–reduction reaction in CS and OCMC solution.

Hence, a very simple process of blending followed by a film-casting method through freeze drying in parallel to evaporation of solvent/s at room temperature were organized. For this purpose, the main component is CS, carboxymethylcellulose (CMC) Nanosized silver particles display unique physical and chemical properties and represent an increasingly important field in the development of nanomaterial. These could be used in numerous biomedical, electronic, catalytic, optical and quantum-size domain applications. A considerable amount of attention has been attracted by the bactericidal effect of noble-metal nanoparticles. Significant advantages of metal nanoparticles include the ease of fabrication and incorporation in different matrices, tunability of characteristics, possibility of functionalization and targeted delivery. Perhaps nanosilver is most extensively studied because of its low cost, ease of preparation of stable formulations and activity against a wide spectrum of prokaryotes while being relatively harmless to eukaryotes; though it may be noted that its cytotoxicity and genotoxicity in human cells have been discussed a lot. For the purpose, the main components, that is, chitosan, Carboxymethylcellulose (CMC) sodium salt (low viscosity) and including the simple reagents were gathered. CMC was purchased from Sigma Aldrich. Periodic acid, sulphuric acid, sodium bicarbonate, hydrochloric acid, potassium iodide and silver nitrate were of reagent grade. All the chemicals were used as such without any further purification. Deionized water was used throughout the experiments which are compiled in the current chapter.

The next target was to prepare the freeze-dried film containing nanosilver particles. Therefore, the oxidation of CMC was done as already reported earlier.[13] A mixture of CS and oxidized CMC with different weight ratios (95/5, 90/10 and 85/15) was dissolved in 2% (w/v) lactic acid solution and was stirred overnight at room temperature. The film-forming solutions, regardless of the CS/OCMC ratio, contained 1% (w/w) biopolymer solids in 2% (w/v) lactic acid and 0.02% (w/v) of silver nitrate was added into the solution and was gently stirred for 2 h. After stirring, solution was kept undisturbed for 2 h. Film-forming solution was then decanted into petri dishes and then kept in the deep freezer overnight at −80°C. Freeze drying was performed at −80°C for 48 h. The dried wound dressings were peeled out from the petri dishes and were kept in desiccators for further use.

The protocol may be depicted as follows for the ease of understanding:

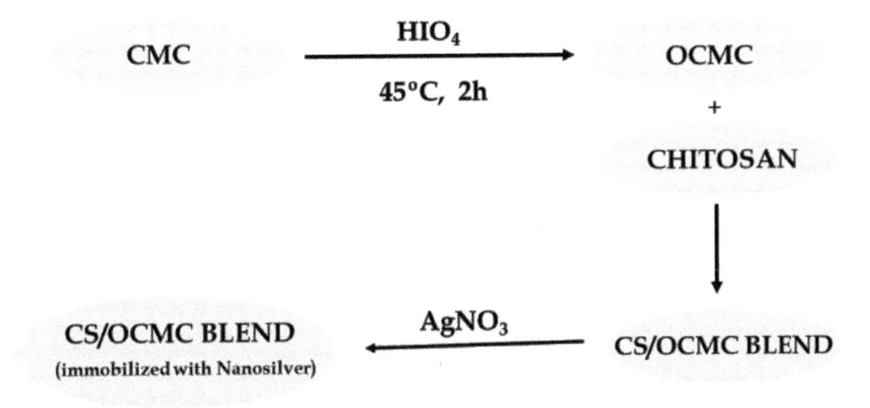

FIGURE 14.1 Nanosilver-contained chitosan (CS)/oxidized carboxymethylcellulose (OCMC) freeze-dried wound dressings.

The nanosilver containing CS/OCMC freeze-dried wound dressings obtained were opaque, reddish brown in colour (Fig. 14.1) and have very good flexibility and elasticity than pure CS wound dressings. In addition, when dried off, nanosilver containing wound dressings were easy to handle. The pure OCMC freeze-dried wound dressings were white in colour and very fragile so could not be handled easily. The oven-dried nanosilver containing wound dressings were transparent, shiny with same reddish brown colour with less flexibility and less porous as compared with freeze-dried wound dressings.

14.2 CHARACTERIZATION

To confirm the changes in the compositions of the individual polymers after certain modification like the oxidation of the CMC as well as in the blended polymers, the instrumental characterization is a must. Through these instrumental techniques, various parameters were obtained before and after the blending followed by film casting with and without nanosilver. The following characterizations were recorded:

14.2.1 ULTRAVIOLET (UV)-VISIBLE (UV-VIS) SPECTROSCOPY

Spectrophotometry is an important technique used in many biochemical experiments that involve DNA, RNA and protein isolation, enzyme kinetics and biochemical analyses. A brief explanation of the procedure of spectrophotometry includes comparing the absorbency of a blank sample that does not contain a coloured compound with a sample that contains a coloured compound. The spectrophotometer is used to measure coloured compounds in the visible region of light (between 350 and 800 nm); thus, it can be used to find more information about the substance being studied. In biochemical experiments, a chemical and/or physical property is chosen and the procedure that is used is specific to that property to derive more information about the sample, such as the quantity and purity. Spectrophotometry can be used as a method to create optical assays of a compound. As a spectrophotometer measures the wavelength of a compound through its colour, a dye-binding substance can be added so that it can undergo a colour change in order to be measured. This method is also convenient for use in laboratory experiments as it is inexpensive and relatively simple.

For the presented evaluation, UV-v spectral analysis was done by using Shimadzu UV-vis spectrophotometer (UV-1800, Japan). UV-vis absorption spectrophotometer with a resolution of 1 nm between 200 and 800 nm was used. Through the results of UV of the prepared aqueous solution of CS/OCMC containing silver nanoparticles, an absorption band at 352 nm was observed as shown in Figure 14.2, which is a typical absorption band of spherical silver nanoparticles due to their surface plasmon resonance,[15,16] confirming the formation of silver nanoparticles. By observing the solution's absorption spectra for 8 h, after every 20-min time interval, the stability of the solution was checked. The absorption spectrum of the solution depicts almost identical spectral features as the spectrum of the starting solution of silver nanoparticles. This confirms that the silver nanoparticles are not further Jimerized or agglomerated with many particles together. Here, in this solution of CS/OCMC, aldehyde groups of OCMC converting silver into nanosilver and CS work as stabilizers.

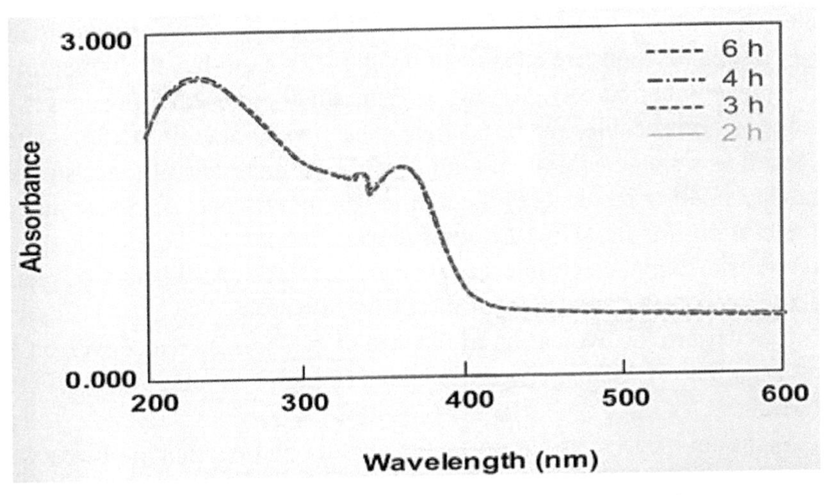

FIGURE 14.2 Ultraviolet-visible spectra of nanosilver-contained dressing.

14.2.2 *ELECTRON MICROSCOPY*

Microscopy is the technical field of using microscopes to view objects and areas of objects that cannot be seen with the naked eye (objects that are not within the resolution range of the normal eye). There are three well-known branches of microscopy, which are as follows: (i) optical, (ii) electron and (iii) scanning probe microscopy. Until the invention of sub-diffraction microscopy, the wavelength of the light limited the resolution of traditional microscopy to around 0.2 μm. To gain higher resolution, the use of an electron beam with a far smaller wavelength is used in electron microscopes.

14.2.3 *TRANSMISSION ELECTRON MICROSCOPY (TEM)*

Transmission electron microscopy forms a major analysis method in a range of scientific fields such as physical, chemical and biological sciences. Transmission electron microscopes find application in cancer research, virology, materials science as well as pollution, nanotechnology and semi-conductor research. In TEM, a beam of electrons is transmitted through an ultra-thin specimen, interacting with the specimen as it passes through it. From the interaction of the electrons transmitted through the specimen an image is formed; the image is magnified and focused onto an imaging device, such as a fluorescent screen, on a layer of photographic film or to be detected by a sensor such as a charge-coupled device camera. Transmission electron microscopes are capable of imaging at a significantly higher resolution than light microscopes, owing to the small de Broglie wavelength of electrons. This enables the instrument's user to examine fine detail—even as small as a single column of atoms, which is thousands of times smaller than the smallest resolvable object in a light microscope. Alternate modes of use allow for the TEM to observe modulations in chemical identity, crystal orientation, electronic structure and sample-induced electron phase shift as well as the regular absorption-based imaging.

In regard to the evaluation of the size of synthesized nanosilver, TEM was used. The TEM images were obtained by high-resolution TEM (JEOL, JEM-2000E7).

Shape and size distribution of the synthesized Ag nanoparticles was characterized by TEM study. The spherical shape of the particles can be seen by the shape and size distribution from Figure 14.3. The particle size distribution indicates that the particles formed fall within the narrow range of 0 5–0.51 nm.

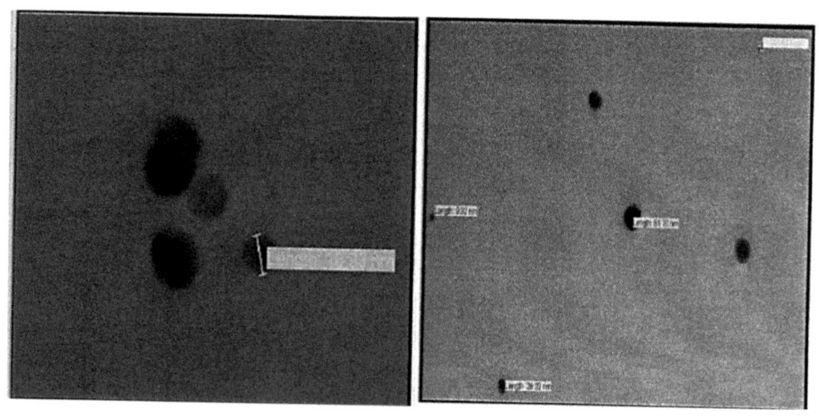

FIGURE 14.3 Transmission electron microscopy images of silver nanoparticles in CS/OCMC solution.

14.2.4 SCANNING ELECTRON MICROSCOPE (SEM)

The SEM uses a focused beam of high-energy electrons to generate a variety of signals at the surface of solid specimens. The signals that are derived from electron-sample interactions reveal information about the sample including external morphology (texture), chemical composition and crystalline structure and orientation of materials making up the sample. Data is collected over a selected area of the surface of the sample and a two-dimensional image is generated that displays spatial variations in these properties in most of the applications. Areas ranging from approximately 1 cm–5 μ in width can be imaged in a scanning mode by using conventional SEM techniques (magnification ranging from 20X to approximately 30,000X, spatial resolution of 50–100 nm). The SEM is also capable of performing analysis of selected point locations on the sample; this approach is especially useful in qualitatively or semi-quantitatively determining chemical compositions (using Energy-Dispersive Spectroscopy), crystalline structure and crystal orientations (using Electron backscatter diffraction).

Film-surface morphology of nanosilver-immobilized CS/OCMC blend wound dressings was examined by SEM. The dried film samples were mounted on a metal stub with double-sided adhesive tape. The morphological structures of the wound dressings were studied by a JSM-5600 LV SEM of JEOL, Tokyo, Japan and the images were taken at accelerating voltage 5 kV and a magnification 100 times of original specimen size.

Conversion of silver into nanosilver through aldehydic group of OCMC and stabilization by CS is expected to be highly effective as a polymeric chelating agent for silver ions. Thus, the generation of silver-impregnated CS film occurs spontaneously by a process involving complexation of silver ions with amine groups in CS. Scanning electron microscope measurements were carried out to further confirm the presence of silver nanoparticles in the CS film. As shown in Figure 14.4, the results of SEM of nanosilver-impregnated wound dressing indicate that the silver nanoparticles were spherical with particle size in the range of 5–51 nm. There is slight decrease in the porosity of resulting dressing as compared with the original CS/OCMC freeze-dried film. This is due to the formation of more complex structure by interaction of silver particles with amino groups of CS. As the amount of silver used is very small in quantity, it does not affect porosity to a large extent. Scanning electron microscope images of drug-containing CS/OCMC freeze-dried film were observed to be of decreased porosity and the pores were unevenly distributed throughout the membrane, as shown in Figure 14.5.

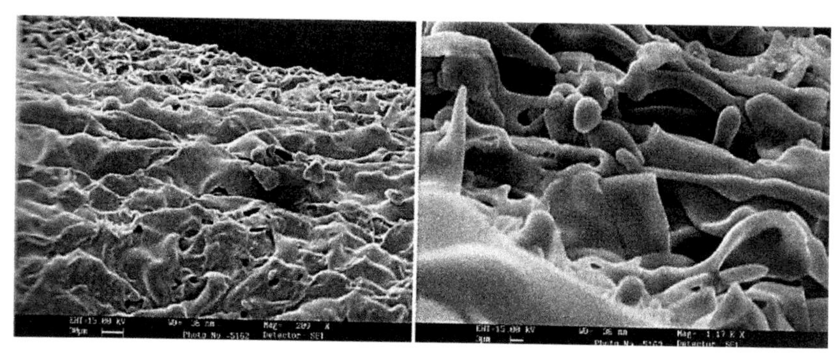

FIGURE 14.4 Scanning electron microscope images of the nanosilver-contained CS/OCMC wound dressing.

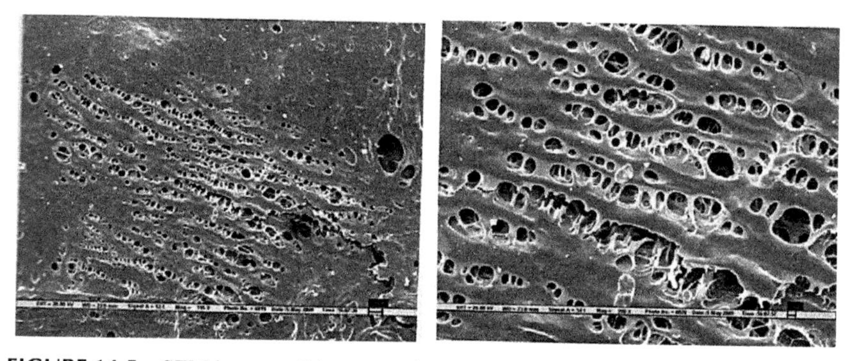

FIGURE 14.5 SEM images of drug-contained CS/OCMC freeze-dried film.

14.2.5 X-RAY DIFFRACTION (XRD)

X-ray diffraction or X-ray crystallography is a tool used for identifying the atomic and molecular structure of a crystal, in which the crystalline atoms cause a beam of incident X-rays to diffract into many specific directions. By measuring the angles and intensities of these diffracted beams, a crystallographer can produce a three-dimensional picture of the density of electrons within the crystal. From this electron density, the mean positions of the atoms in the crystal can be determined, as well as their chemical structure, their disorder and other information. X-ray diffraction by using X-ray optics has been applied to many different types of applications including thin-film analysis, sample texture evaluation, monitoring of crystalline phase and structure and investigation of sample stress and strain.

X-ray diffraction of the wound dressings was measured by the X-ray diffractometer Xpert PRO PANalytical, Holland. X-ray diffraction patterns were recorded with Cu-ά radiation ($\lambda = 1.54$ A). X-ray diffraction has been applied to mark the crystallinity in the structure of nanosilver with CS/OCMC freeze-dried wound dressings. X-ray diffraction (XRD) was further implied to examine the crystal structure of metal nanoparticles that confirm the formation of silver nanoparticles. The XRD pattern of freeze-dried silver nanoparticle-contained dressing (Fig. 14.6) shows characteristic peaks at 2θ values of 32.2, 38.3, 43.7, 66.4 and 78.1. These peaks for nanosilver-containing dressing are different from drug-contained CS/OCMC freeze-dried film. Here the sample loaded with model drug, showed that two peaks at 32.24 and 38.39 are absent.

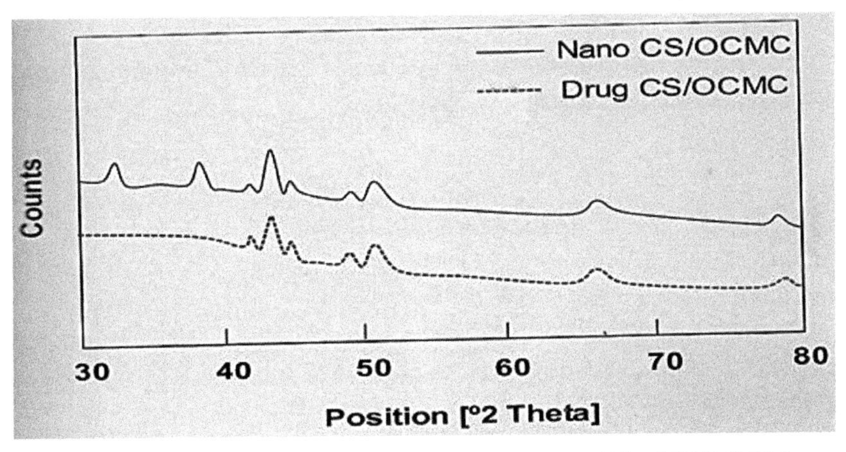

FIGURE 14.6 X-ray diffraction of nanosilver and drug-immobilized CS/OCMC freeze-dried wound dressings.

14.2.6 POROSITY MEASUREMENTS

Voids or pores are generally formed in thin wound dressings irrespective of the film preparation method (electro-deposition, evaporation or sputtering) as long as the deposition process involves a phase transformation from the vapour to the solid state. These voids can be extremely small (approximately 10 A) and high in density (about 1×10.7 cm-3). Advances in polymer synthesis followed by their wound dressing fabrication, including novel processing techniques, have led to a variety of nanoporous polymeric wound dressings. Lithography, pattern-transfer, track etching, solvent-based formation, layer-by-layer growth and so on are some of the well-known techniques. Pore size is one of the most important features that influence the application of the subjected wound dressings or wound dressings. As pore size plays an essential part in the porous wound dressings, some efforts have been made to regulate the pore size, but there's still no universal method to deal with this problem. The most widely used method to fabricate porous structures is the solvent evaporation and freeze-drying method. The porosity and specific surface area have a direct relationship with the film density for drug absorption amount, and a controllable porosity is preferable. Large specific surface area and low density are the key characteristics of the porous structures. Porous wound dressings have an excellent absorption capacity compared with traditional ones.

The porosity of nanosilver-immobilized CS/OCMC wound dressings was calculated as per the following relation:

$$\text{Porosity (\%)} = V_m - V_o \times 100 \tag{14.1}$$

Where, V_m and V_0 are the specific volumes of the freeze-dried CS/OCMC wound dressings and nanosilver-contained CS/OCMC wound dressings, respectively.

The trend of porosity is as follows: 5 wt % CS/OCMC film > nano-contained wound dressing as we can observe in Table 14.1, this is because in 5 wt% CS/OCMC more aldehyde groups are available to cross-link with amino groups of CS and formation of covalently cross linked Schiff's base takes place, which is a complex structure and in nanosilver-contained film silver ions interact with the amino groups of Cs and even ether oxygen and the hydroxyl group of unoxidized part of CMC anchor to silver ions tightly through ion-dipole interactions, this complex structure is responsible for the lowest porosity of nano-contained wound dressing as compared with the original sample.

14.2.7 DENSITY MEASUREMENTS

Many observations, encompassing both structural and dynamic behaviour, indicate that the properties of polymers in thin wound dressings deviate from what is generally observed in the bulk. The behaviour of polymers in thin wound dressings or close to interfaces is far from being understood. Questions which are much explored are usually about the density (irreversible and reversible changes), film stability and glass transition temperature and so on. The high surface-to-volume ratio, the influence of surface and interfacial properties become important if not dominant.

This interfacial sensitivity highlights the importance of the properties of the near surface region in polymer wound dressings—a property whose importance is beginning to be recognized. Hence, a complimentary property evaluation was done, which is helpful for a clear understanding of thin film properties and their applications in biomedics.

Density measurements of samples were carried out by taking into account the thickness of wound dressings of specific size and by measuring the weight of the samples. Weight per cubic centimetre was represented as the density of wound dressings.

TABLE 14.1 Density and Porosity of Chitosan (CS), 5 wt% CS/Oxidized Carboxymethylcellulose (OCMC) Freeze-Dried Film and NanoSilver-Contained CS/OCMC Wound Dressing.

Sample	OCMC (%)	Density (g/cm^3)	Porosity (g/cm^3)
Pure CS film	–	0.08	–
5 wt% CS/OCMC	5	0.06	86.1%
Nano-contained CS/OCMC	10	0.08	75.6%

14.2.8 SWELLING BEHAVIOUR

Solvent swelling of polymer wound dressings is important for applications such as food packaging, membrane separations and especially in controlled release of drugs. Although there are several methods for studying swelling of bulk samples by relatively good solvents, swelling of thin wound dressings or swelling by very poor solvents has been more difficult to measure.

Traditionally, swelling of polymeric wound dressings has been measured by weighing a sample before and after exposure to solvent by measuring the physical dimensions before and after exposure or by using a dyed solvent and examining a cross-section of the sample microscopically.

By placing the samples of known weight in water for 24 h, the swelling behaviour of the wound dressings was carried out in distilled water at 25±2°C. Subsequently, the samples were taken out and gently blotted with absorbent paper to remove the water adhering on the sample surface. The weight of the wet sample was measured. The swelling was determined by using following Equation 14.2:

$$\text{Equilibrium swelling} = \frac{W_i - W_f}{W_f} \times 100\% \quad (14.2)$$

Where, W_i and W_f are the weight of the wet and dry samples, respectively.

The swelling capacity of the antibacterial film and nanocomposites play an important role in the antibacterial activity, wound-healing process, and other biomedical applications such as drug release illustrating the swelling capacity of CS/CMC freeze-dried wound dressings and nanosilver-contained wound dressing. The order of the swelling capacity of wound dressings follows as CS/OCMC film > nanosilver-contained wound dressing (Table 14.2). The reason for lower swelling capacity of nanosilver-contained dressing as compared with CS/OCMC freeze-dried film is due to the complexation of silver ions with amine group of CS, or we can say that swelling capacity of film may be attributed to binding of silver nanoparticles with electron-rich O and N atoms of ether and amine group of CS. The swelling capacity of CS/OCMC freeze-dried film and nanosilver-contained dressing does not differ much from each other. Nanosilver-contained dressing shows enough swelling capacity, which makes the sample good for wound dressing.

TABLE 14.2 Swelling Behaviour of CS/OCMC Freeze-Dried Film and Nanosilver-Contained Wound Dressing.

Sample	Swelling (%)
CS/OCMC film	765
Nanosilver-contained CS/OCMC	741

14.2.9 WATER VAPOUR TRANSMISSION RATE (WVTR)

Moisture vapour transmission rate, also WVTR, is a measure of the passage of water vapour through a substance. It is also a measure of breathability which greatly contributes in the use of thin wound dressings as wound dressings. As the proper healing of the wounds is also related to the function of gaseous exchange between the wound and the external environment, proper permeation of the gases along the dressing and the wound area manages the moisture barrier, creating a dry atmosphere ensuring the prohibition of growth of bacterial and fungal infections around the wound.

Water vapour transmission rate tests were conducted by using LYSSY-80-5000. Flat film samples were clamped into the diffusion cell, which were then purged of residual water vapour by using moisture-free carrier gas. This moisture-free carrier gas is routed to the sensor until a stable WVTR has been established. Absorbent material saturated with distilled water provides an atmosphere of 100% RH. Molecules of water diffusing through the film to the inside chamber are conveyed to the sensor by the carrier gas. The computer monitors the increase in water vapour concentration in the carrier gas and reports that value on the screen as the WVTR.

As shown in Table 14.3 and Figure 14.7, WVTR was significantly affected by the ratio of the polymers. Significantly, wound dressings with more CS had high WVTR values compared with wound dressings with high OCMC content. 5–15% CS/OCMC film has 3753 g/m^2*dy and 3216 g/m^2*dy WVTR values, respectively. Water vapour transmission through a hydrophilic film depends on solubility and diffusivity of water molecules in the film matrix. However, although the OCMC fraction affects the solubility of the wound dressings, it did not cause higher permeability, which indicates that the main factor in water vapour permeability is diffusivity of water molecules through the film matrix. The impeded diffusivity could be due to promotion of strong intermolecular interactions between CS and OCMC molecules decreasing the intermolecular distance resulting in more compact wound dressings that caused the percolation of water molecules around the insoluble crystals resulting in longer diffusion paths. Nanosilver-containing wound dressings have low WVTR because amino groups in CS have strong tendencies to complex with metal ions. Moreover, CS is expected to be highly effective as a polymeric chelating agent for Ag^+ ions. Thus, complex of silver ions with amine groups of CS make the dressing less permeable as compared with CS/OCMC wound dressings but shows more WVTR as compared with drug-contained CS/OCMC wound dressings which has WVTR of 2183 g/m^2*dy. The WVTR of nanosilver-contained wound dressing is good, that is 2943 g/m^2*dy, which is in line with the commercially available dressings with high absorbing ability of wound exudate. It may be stated that the WVTR values of nano-contained wound dressings are still on the higher side of the dry and wet human skin values. Our results are in well agreement with the studies carried out by other workers.[17]

TABLE 14.3 Water Vapour Transmission Rate Values for CS/OCMC Freeze-Dried Film, Drug-Contained Cs/Ocmc Film and Nanosilver-Contained CS/OCMC Dressing.

5 wt% CS/OCMC	3753
Nanosilver-contained dressing	2943
Drug CS/OCMC (0.5%)	2183

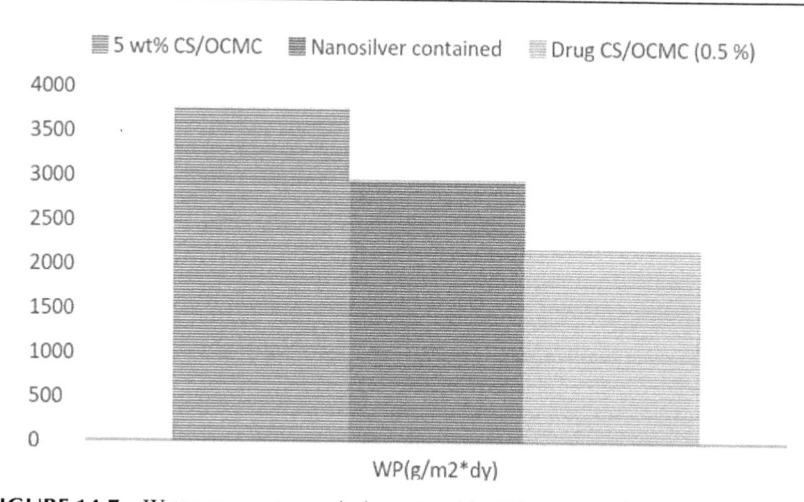

FIGURE 14.7 Water vapour transmission rate with different samples.

14.2.10 ANTIMICROBIAL EVALUATION

Antibiotics or antimicrobial agents are one of most important weapons in fighting bacterial infections and have greatly benefited the health-related quality of human life since their introduction. In recent years, there has been a growing interest in researching and developing new antimicrobial agents from various sources to combat microbial resistance. Therefore, antimicrobial activity screening and evaluating methods have received greater attention. Several bioassays such as disk diffusion, well diffusion and broth or agar dilution are well-known and commonly used.

Silver as antibacterial agent has been known to man for centuries. Therapeutic efficacy of indigenous silver metal for several disorders has been described by practitioners of traditional medicine, especially in the treatment of burns. Antimicrobial properties of silver nanoparticles in combination with CS towards enhancing its inherent antimicrobial activities are being increasingly reported from different parts of the world. The harmful microorganisms can be controlled with silver nanoparticles, and these result in the emergence of multiple drug-resistant bacteria. Moreover, it has created alarming clinical situations in the treatment of infections.

The antimicrobial activity of nanoparticles in CS/OCMC wound dressings was evaluated against *E. coli* through inhibition zone method which shows a very good result with very small amount of silver nanoparticles which was compared with the freeze-dried wound dressings of pure carboxymethyl cellulose, pure CS and CS/OCMC film with drug. Microbes were subcultured and incubated at 37°C for 24 h. In order to cultivate bacteria, fresh cultures were taken and spread on agar plates. Sterile wound dressings were placed in each agar plate and incubated again at 37°C for 24 h. The freeze-dried nanosilver-contained dressing exhibited a strong antimicrobial activity against *E. coli* (Gram-negative bacteria), which are general bacteria found on the contaminated wound. A recent study showed that impregnation, instead of coating the wound dressing with silver nanoparticles, improved the antimicrobial activity of the wound dressing and lowered possibility of the normal human tissue damage. This is probably due to the slow and continual release of silver nanoparticles followed by the slow conversion to silver ions in our physiological system and interaction with bacterial cells. Thus, silver ions will not be high enough to cause damage to normal human cells and can prolong the antimicrobial effect. The freeze-dried wound dressings containing silver nanoparticles were tested for antibacterial properties on *E. coli*. The effect of the silver nanoparticles on *E. coli* is shown in Figure 14.8. The zone of inhibition was observed for pure CS-based freeze-dried wound dressings due to inherent antimicrobial activity of CS, a slight zone of inhibition was also observed for wound dressings containing 0.5% tetracycline drug and a good zone of inhibition was observed for the freeze-dried wound dressings containing the silver nanoparticles thus confirming the antibacterial action, whereas there was no zone of inhibition for pure OCMC wound dressings, as shown in Figure 14.8. Good antibacterial properties of silver nanoparticles of the freeze-dried wound dressings in which a very small amount of silver nitrate (0.02%) was converted to silver nanoparticles in the presence of OCMC and CS was clearly indicated through these observations.

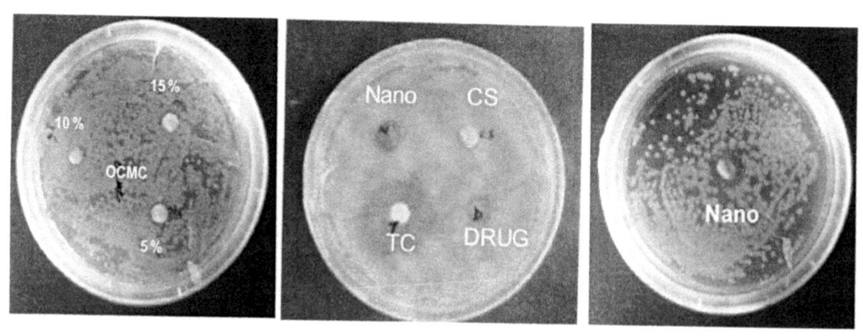

FIGURE 14.8 Zone of inhibition in pure CS, OCMC, TC, drug-containing wound dressings and nanosilver-contained CS/OCMC wound dressing.

14.2.11 SILVER RELEASE

Silver has antiseptic, antimicrobial and anti-inflammatory properties and is a broad-spectrum antibiotic. Silver is biologically active when it is in soluble form, that is as Ag^+ or Ag_0 clusters. Ag^+ is the ionic form present in silver nitrate, silversulphadiazine or other ionic silver compounds. Ag_0 is the uncharged form of metallic silver present in nanocrystalline silver. Free silver cations have a potent antimicrobial effect, which destroys microorganisms immediately by blocking the cellular respiration and disrupting the function of bacterial cell wound dressings. This occurs when silver cations bind to tissue proteins, causing structural changes in the bacterial cell wound dressings which in turn cause cell death. Silver cations also bind and denature the bacterial DNA and RNA, thus inhibiting cell replication.

Silver-containing dressings are reportedly used to assist with management of infected wounds and those at risk of infection. However, such dressings have varied responses in clinical use due to technological differences in the nature of their silver content and release and in properties of the dressings themselves. The relationship between silver content and rate of silver release in a simulated fluid model is examined by this study.

Figure 14.9 shows silver ion-releasing behaviour of the freeze-dried silver nanoparticle-impregnated wound dressing. The silver ions were released slowly and continuously from the freeze-dried silver nanoparticle-contained dressing that was prepared by mixing 0.02% (w/v) silver nitrate in 5% CS/OCMC solution (5% OCMC and 95% CS). Here, as wound dressings were prepared, silver ions were released gradually by mixing silver within the CS/OCMC solution, which causes the gradual release of silver ion.

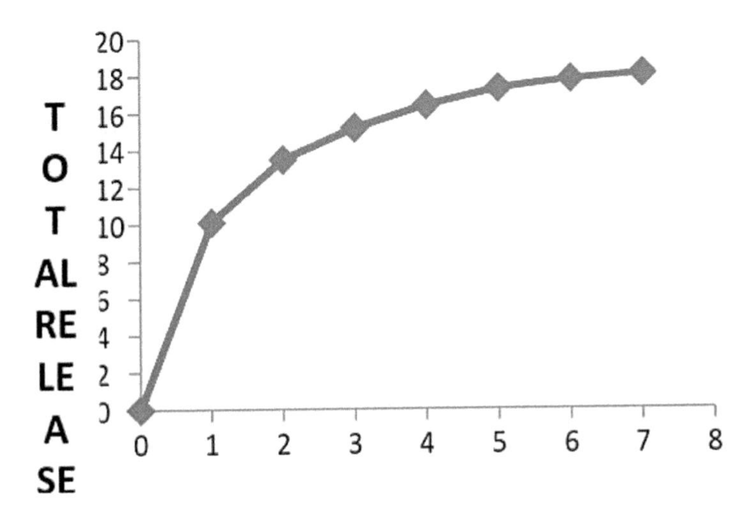

FIGURE 14.9 Silver ion releasing behaviour of nanosilver-contained CS/OCMC wound dressing.

14.2.12 DRUG RELEASE

In the present work, wound dressings, based on CS and oxidized carboxymethyl cellulose, were prepared by freeze drying. Tetracycline hydrochloride (TCH) was used as an antimicrobial drug. However, it has now been shown that having a continuous controlled release of drug around the wound environment achieves more rapid and successful wound healing. Hence, the behaviour of the drug during its release, for TCH in the CS/OCMC wound dressing was evaluated and shown in Figure 14.10. Drug release is very fast initially up to 10 h and then it becomes slow and almost stable for 24–48 h.

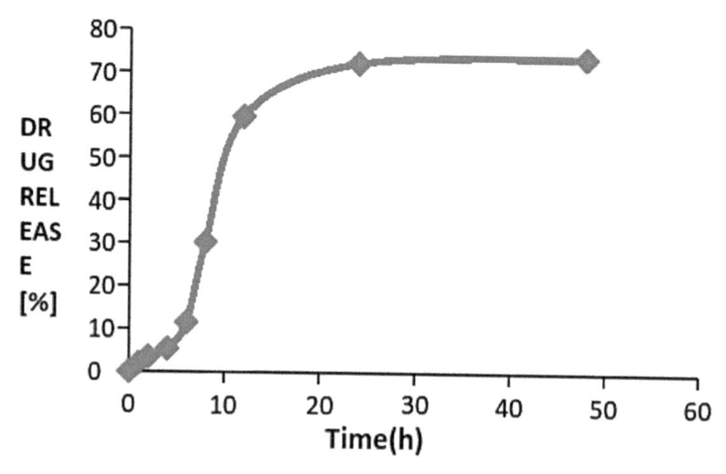

FIGURE 14.10 Release of TC from nanosilver-contained CS/OCMC dressing Conclusion.

Local wound infection causes tissue death, increase in wound size, wound hypoxia and vessel occlusion, all of which further delays the wound-healing process. Controlling microorganisms within a wound environment promotes wound healing. Microorganisms, that is, bacteria or fungi, are found in chronic wounds and if present in an acute wound can rapidly contaminate and infect, seriously impeding wound healing. High levels of bacteria, multiresistant organisms and bacterial BioWound dressings can have an impact on the wound healing process, especially in chronic wounds. Bacteria delay wound healing by competing with host cells for nutrients and oxygen; their waste products are also toxic to host cells. Bacterial wound infection causes raised blood cytokines, raised matrix metalloproteinase and decreased growth factors that can have adverse effects on wound healing.

The ultimate result of the study is the development of silver nanoparticle-incorporated CS/OCMC freeze-dried wound dressings possessing many beneficial features. The generation of silver-impregnated CS/OCMC film occurs spontaneously by a process in presence of CS and OCMC. Detailed characterization techniques by using UV-vis spectroscopy, SEM, TEM and XRD revealed the formation of nanosized silver particles. Water vapour transmission rate values of the drug-loaded dressings were similar to the ideal value for wound dressing, thus maintaining suitable moisture content at the wound surface. In this process, no additional chemical was added for the formation of silver nanoparticles so this CS–cellulose-based nanosilver containing freeze-dried wound dressings are very useful natural materials, which can be utilized for wound dressing in biomedical applications with mix properties of CS, cellulose and silver nanoparticles.

KEYWORDS

- **chitosan**
- **oxidized carboxymethylcellulose**
- **silver nanoparticles**

REFERENCES

1. Jayakumar, R.; Prabaharan, M.; Kumar, P. T. S. Novel Chitin and Chitosan Materials in Wound Dressing. In *Biomedical Engineering, Trends in Materials Science;* Anthony, N. Laskovski. Eds; 1990; pp 3–25. ISBN 978-953-307-513-6.
2. Nguyen, T. P. *Optimization of Antimicrobial Wound Dressings: Liposomal Hydrogels with Mupirocin.* University of Tromso, M.Ph., 2012.
3. Gupta, B.; Agarwal, R.; Sarwar, A. M. Antimicrobial and Release Study of Drug Loaded PVA/PEO/CMC Wound Dressings. *J. Mater. Sci.: Mater. Med.* **2014,** *25*, 1613–22. DOI: 10.1007/s10856-014-5184-6.
4. Younes, I.; Rinaudo, M. Chitin and Chitosan Preparation from Marine Sources. Structure, Properties and Applications. *Mr. Drugs*. **2015,** *13*, 1133–74. DOI: 10.3390/md13031133.
5. Croisier, F.; Jérôme, C. Chitosan-Based Biomaterials for Tissue Engineering. *Eur. Polym. J.* **2013,** *49*, 780–92.
6. Kamoun, E. A.; Chen, X.; Mohy, M. S.; Kenawy, E. S. Crosslinked Poly (vinyl alcohol) Hydrogels for Wound Dressing Applications: A Review of Remarkably Blended Polymers. *Arabian J. Chem*. **2015**, 1–14.
7. Ahmed, S.; Ahmad, M.; Ikram, S.; Chitosan: A Natural Antimicrobial Agent: A Review. *J. Appl. Chem.* **2014,** *3*, 493–503.
8. Ahmed, S.; Ikram, S.; Ahmad; Ikram. Chitosan & its Derivatives: A review in Recent Innovations. *J. Pharm. Sci. Res.* **2015,** *6*(1), 14–30. DOI: 10.13040/IJPSR.0975-8232.
9. Hasan, N.; Hashim, S. F. S.; Ariff, Z. M. Low-Frequency Sound Proof Cement-Based Syntactic Foam. *Int. J. Mater. Mech. Manuf.* **2013**, *1*, 384–7. DOI: 10.7763/IJMMM.2013.V1.83.

10. Vroman, I.; Tighzert, L.; Biodegradable Polymers. *Materials (Basel)* **2009,** *2*, 307–44. DOI: 10.3390/ma2020307.
11. Yu, L.; Dean, K.; Li, L. Polymer Blends and Composites from Renewable Resources. *Prog. Polym. Sci.* **2006,** *31*, 576–602. DOI: 10.1016/j.progpolymsci.2006.03.002.
12. Chandra, R. Biodegradable Polymers. *Prog. Polym. Sci.* **1998,** *23*, 1273–335. DOI: 10.1016/S0079-6700(97)00039-7.
13. Teotia, A.; Ikram, S.; Gupta, B. Structural Characterization of Chitosan and Oxidized Carboxymethyl Cellulose Based Freeze-Dried Wound Dressings. *Polym. Bull.* **2012,** *69*, 175–88. DOI: 10.1007/s00289-010-0415-6.
14. Ahmed, S.; Ahmad, M.; Swami, B. L.; Ikram, S. A Review on Plants Extract Mediated Synthesis of Silver Nanoparticles for Antimicrobial Applications: A Green Expertise.*J. Adv. Res.* **2015**. DOI: 10.1016/j.jare.2015.02.007.
15. Wei, D.; Sun, W.; Qian, W.; Ye, Y., Ma, X. The Synthesis of Chitosan-Based Silver Nanoparticles and Their Antibacterial Activity. *Carbohydr. Res.* **2009,** *344*, 2375–82. DOI: 10.1016/j.carres.2009.09.001.
16. Govindan, S.; Nivethaa, E. A. K.; Saravanan, R.; Narayanan, V.; Stephen, A. Synthesis and Characterization of Chitosan–Silver Nanocomposite. *Appl. Nanosci.* **2012,** *2*, 299–303. DOI: 10.1007/s13204-012-0109-5.
17. Mi, F. L.; Wu, Y. B.; Shyu, S. S.; Schoung, J. Y.; Huang, Y. B.; Tsai, Y. H.; et al. Control of Wound Infections Using a Bilayer Chitosan Wound Dressing with Sustainable Antibiotic Delivery. *J. Biomed. Mater. Res.* **2002,** *59*, 438–49. DOI: 10.1002/jbm.1260.

CHAPTER 15

SYNTHESIS AND CHARACTERIZATION OF BIODEGRADABLE POLYETHER URETHANE FOR THE PURPOSE OF CONTROLLED RELEASE OF ANTIBIOTICS

ANINDYA HALDAR[2], SOHINI SEN[1], DEBAPRIYA BANERJEE[1], NANDAN KUMAR JANA[2], and PIYALI BASAK[1*]

[1]*School of Bioscience and Engineering, Jadavpur University, Kolkata, India*

[2]*Department of Biotechnology, Heritage Institute of Technology, Kolkata, India*

E-mail: piyali_bioengg@school.jdvu.ac.in

CONTENTS

ABSTRACT

Polymeric material loaded with antibiotics is a promising area for prevention of implant infection by controlled release of antibiotics and for their biocompatibility. A polyether urethane (PEU) membrane has been synthesized using Isophorone diisocyanate (IPDI) and polyethylene glycol (PEG) using Dibutyltin dilaurate as a catalyst. Physical characterizations such as X-ray diffraction, Fourier transform infrared spectroscopy and field emission scanning electron microscopy have been done. Physical characterizations confirm the amorphous nature of the membrane and the presence of urethane linkage formed by the reaction of free hydroxyl group of PEG and free isocyanate group of IPDI and its microporous structure. In several solvents, swelling study of the polyurethane membrane has been done. In vitro degradation profile of the membrane has been studied. Antibiotic was loaded in the synthesized membrane and release of antibiotic from the membrane has been studied. The characterizations, degradation profile and drug release study confirm that the synthesized IPDI-based PEU can be used as biodegradable implant coating material to prevent bacterial infection.

15.1 INTRODUCTION

Polyurethanes have been widely used in several industries due to their versatile properties.[1] On account of their biodegradability, some of them are very useful for several medical applications. Biodegradability of the polymers depends on various factors such as chemical structure, molecular weight, degree of crystallinity and morphology.[2] Functional groups such as ether, ester, amide, urethane and so on in backbone of polymeric structures make them biodegradable.[3] It has been reported that microbial degradation of polyurethanes depends on diisocyanate and diol used (4). One important aspect regarding biodegradability of polyurethane is the ratio of soft and hard segment.[2,5] Incorporation of hydrophilic polyethylene glycol (PEG) may enhance biodegradability of polyurethane.[6] In the stability of polyether urethane (PEU), molecular weight of PEG plays a significant role.[7] As reported, a good polymeric coating matrix should have proper balance between hydrophobicity and hydrophilicity.[8] Controlled release of drug from the matrix is also an important aspect for medical application apart from biodegradability.[9] The release of drug from the matrix and the degradation rate should be controlled to achieve effective drug release for a long period.[6] As most of the major complications with implant devices arise from bacterial infection and biofilm formation, the best strategy could be coating of implant material, which is biodegradable, biocompatible and has the ability of releasing drug in controlled fashion.[10]

In this paper, we have synthesized a PEU using Isophorone diisocyanate (IPDI), which can be used as implant coating material. The PEU was characterized using X-ray diffraction (XRD), Fourier transform infrared (FTIR) spectroscopy, field emission scanning electron microscopy (FESEM), swelling study and degradation study. The release of antibiotic (Rifampicin and Streptomycin) from the IPDI-based PEU (IPDI-PEU) was studied in water and simulated body fluid (SBF).

15.2 EXPERIMENTAL DETAILS

15.2.1 *MATERIALS*

Polyethylene glycol 600, Dimethyl sulfoxide (DMSO), Di-chloromethane, Isopropanol Merck India Ltd. IPDI, Dibutyltin dilaurate (DBTDL) and Lipase from porcine pancreas Sigma-Aldrich. Streptomycin, rifampicin Central Drug House, India.

15.2.2 *METHOD*

15.2.2.1 *SYNTHESIS OF POLYURETHANE MEMBRANE*

The PEU was synthesized by reacting PEG (PEG 600) with IPDI in the presence of catalyst DBTDL in DMSO solvent at 60°C for 12 h with controlled and continuous stirring. Isopropanol was added to the reaction mixture after that, and the reaction was allowed to take place for another 1 h to block the free isocyanate groups. Film was casted by pouring the final solution in petri plates, and the membranes were kept for 8 h at 80°C in hot air oven. After that, the membranes were kept in dichloromethane for 30 min to remove unreacted DMSO and IPDI. Finally, membranes were washed in distilled water to remove water soluble impurities, if any, and finally oven-dried at 65°C.

15.2.2.2 PREPARATION OF SIMULATED BODY FLUID

Simulated body fluid has almost same ionic composition and pH similar to human blood plasma. Composition of Hank's balanced salt solution was used for this purpose.[11]

15.2.3 CHARACTERIZATIONS

15.2.3.1 SWELLING ANALYSIS

Dried and pre-weighed PEU membranes were immersed in SBF, DMSO and water. The membranes were taken out of solution, gently wiped and weighed at different time intervals. The total duration of swelling analysis was 2 h. The degree of swelling was calculated by the following equation:

$$\text{Degree of swelling}\,(\%) = \frac{(W_w - W_d)}{W_d}$$

where W_w and W_d are the wet and dry weights of the membranes, respectively.

15.2.3.2 X-RAY DIFFRACTION

X-ray Diffraction pattern was obtained for PEU membrane by using Rigaku miniflex X-ray diffractometer, Ultima III with a Cu radiation. The measurements were performed in the 2θ range of 3–60°C.

15.2.3.3 FOURIER TRANSFORM INFRARED ANALYSIS

Fourier transform infrared spectroscopy analysis of PEU membrane was performed by using Schimadzu FTIR instrument in the frequency range of

4000–500 cm^{-1}. Before analysis, membranes were cleaned using acetone to remove surface impurities, if present.

15.2.3.4 SURFACE MORPHOLOGY ANALYSIS

Field emission scanning electron microscopy of PEU membrane has been done by using Inspect F50 (FEI), after the use of gold coating of samples to investigate surface morphology.

15.2.3.5 DEGRADATION STUDY

Degradation study of the membrane was carried out for 28 days in hydrolytic (SBF), enzymatic and oxidative medium. Enzymatic medium was prepared by using porcine pancreatic lipase. Oxidative medium was prepared by mixing 0.1 M cobalt chloride solution in 20% hydrogen peroxide, prepared from 30% hydrogen peroxide solution by proper dilution with distilled water. In each medium, dried and pre-weighed pieces of PEU membranes were immersed and then incubated at 37°C. At different time intervals, the samples were taken out, wiped and oven-dried at 65°C. The weight of the samples was carefully noted after drying. The weight loss was calculated by the following equation:

$$\text{Weight Loss (\%)} = \frac{(W_w - W_d)}{W_d}$$

where W_w and W_d are wet and dry weight of the samples, respectively.

15.2.3.6 LOADING OF DRUG ONTO MEMBRANE

To prevent microbial infection, release of antibiotics at the site of implantation from a surface coating is desired. Streptomycin and rifampicin were used for drug-loading purpose. Pre-weighed PEU membranes were dipped in antibiotic solution for 48 h to facilitate drug loading onto the membrane. Water was used as solvent for both the antibiotics. Concentration of antibiotics used for drug loading was 200 and 1.2 mg/ml for streptomycin and rifampicin, respectively.

15.2.3.7 DRUG RELEASE STUDY

Drug release study was performed in both water and SBF. By using ultraviolet-visible spectrophotometer at regular intervals, the medium was assayed for drug release. The concentration of drug release was determined spectrophotometrically at 195 nm for streptomycin and 257 nm for rifampicin with replacement of fresh solvent. The total duration of drug release study was 30 days. The concentration of drug release was determined from the standard plot of absorbance versus concentration.

15.3 RESULTS AND DISCUSSION

15.3.1 SWELLING ANALYSIS

The swelling pattern of the IPDI-PEU membrane is elaborately presented in Figure 15.1. Highest swelling is displayed by the IPDI-PEU membrane in distilled water. This can be attributed to highly porous nature of IPDI-PEU. The swelling percentage of the membrane in SBF is little bit less when compared with water. Simulated body fluid contains several ions and the ions may block the micropores present in the membrane. Swelling percentage of membrane in DMSO is less compared with both water and SBF. Overall swelling percentages of membrane are 19.2–22.2%, 26–34.67% and 27–37.42% in DMSO, SBF and water, respectively.

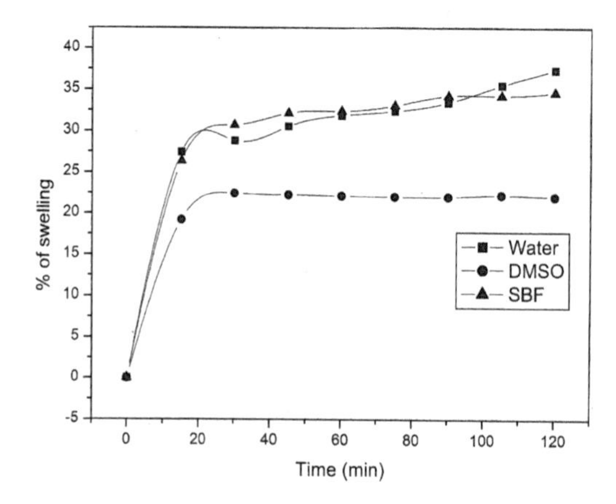

FIGURE 15.1 Swelling of isophorone diisocyanate-based polyether urethane (IPDI-PEU).

15.3.2 X-RAY DIFFRACTION ANALYSIS

X-ray Diffraction pattern of IPDI-PEU is shown in Figure 15.2. From XRD pattern of the IPDI-PEU, it is evident that the material is amorphous in nature, with a maximum peak at 2θ=20°C. The amorphous nature of the final material is facilitated by rapid cross linking between hard and soft segments in a disordered fashion.

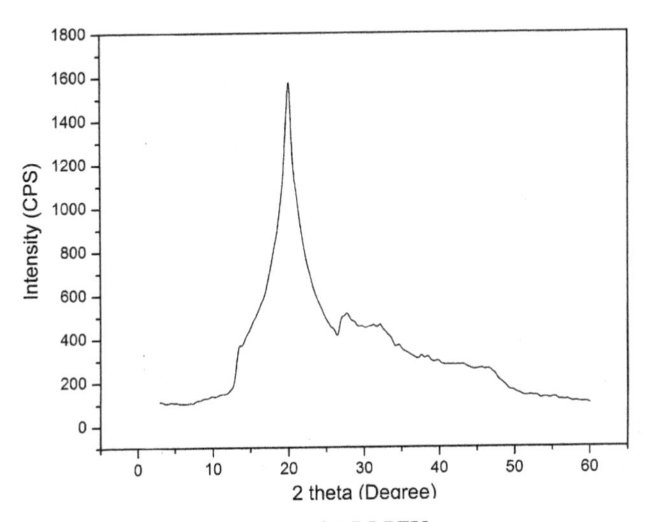

FIGURE 15.2 X-ray diffraction pattern of IPDI-PEU.

15.3.3 FOURIER TRANSFORM INFRARED ANALYSIS

Synthesis of PEU was confirmed by FTIR analysis, which is demonstrated in Figure 15.3. A strong peak at 1734 cm^{-1} indicates the presence of urethane linkage (–C=O vibration). A strong peak is observed at 1159 cm^{-1}, which is characteristic peak of aliphatic ether (C–O–C) linkage. For sp^2 C–N vibration, a peak nearly at 1260 cm^{-1} was observed. A peak at 1320 cm^{-1} confirms the presence of sp^2 C–O vibration. A strong peak at 2972 cm^{-1} represents the sp^3 C–H stretching vibration. Absence of significant peak in the range of 2270–2280 cm^{-1} indicates absence of free NCO group in the final material. Similarly, there is no significant peak observed in the range of 3530–3550 cm^{-1}, due to absence of –OH group. The presence of –NH vibration is indicated by a weak peak at around 3380 cm^{-1}.

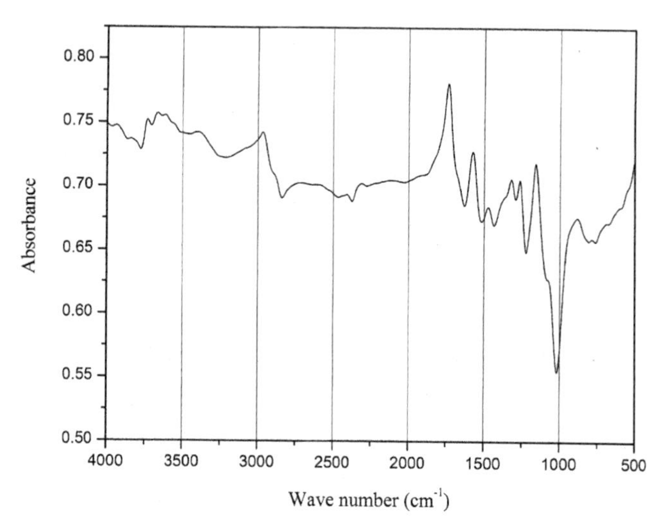

FIGURE 3.3 FTIR spectrum of IPDI-PEU.

15.3.4 SURFACE MORPHOLOGY ANALYSIS

Figure 15.4(a)–(d) represents the FESEM images of the IPDI-PEU membrane. Pores were found in 1200X zoom, but it was observed that pores were not equivalently distributed into the membrane. In general, micropores are formed as a result of microphase separation during solvent evaporation. But in this case, chance of solvent evaporation was minimum as DMSO has high boiling point (189°C). But during treatment of membrane in dichloromethane, the process may be the major step of microspore formation because of the removal of unreacted IPDI and DMSO. Removal of solvent from the membrane may assist in pore formation. Figure 15.4(b) and (c) represent presence of pores in the polyurethane structure. Figure 15.4(d) represents more magnified image of the porous membrane. The diameters of the pores are in the range of 1–20 μm. This porous morphology is desirable due to drug loading and controlled release of drug from membrane.

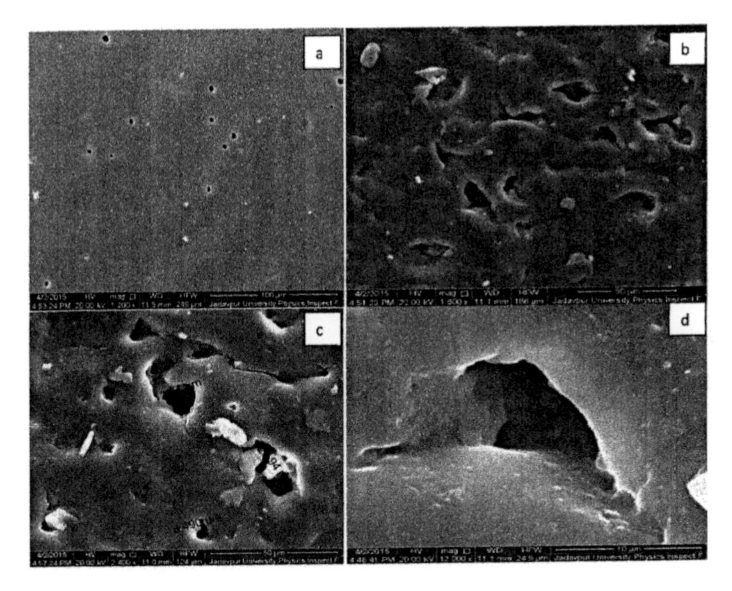

FIGURE 15.4 (a)–(d) Surface morphology of IPDI-PEU.

15.3.5 DEGRADATION STUDY

15.3.5.1 HYDROLYTIC DEGRADATION

Hydrolytic degradation profile of the PEU membrane is shown in Figure 15.5(a). A gradual increase in degradation percentage was observed. Maximum degradation obtained on 28th day is 13–16%. Degradation rate in SBF is comparatively low than that of oxidative and enzymatic medium. It may be due to presence of ions in SBF, which blocks the pores that are present in membrane.

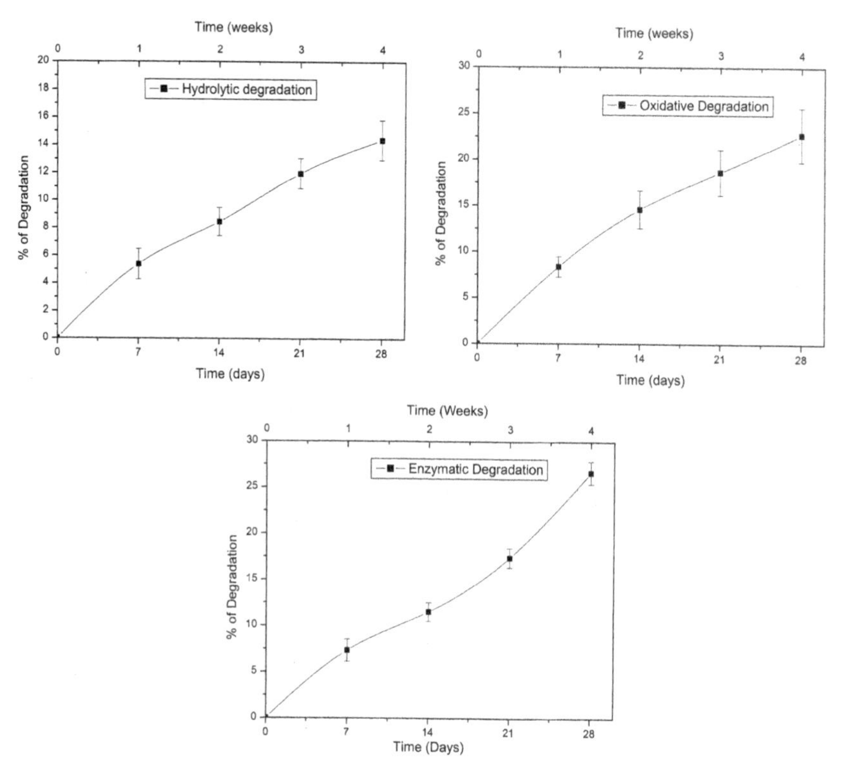

FIGURE 15.5 (a) Hydrolytic degradation of IPDI-PEU, (b) oxidative degradation of IPDI-PEU and (c) enzymatic degradation of IPDI-PEU.

15.3.5.2 OXIDATIVE DEGRADATION

The selected medium mimics the oxidizing power of in vivo environment. Figure 15.5(b) represents oxidative degradation of IPDI-PEU membrane. Degradation in oxidative medium is more than that in a hydrolytic one. Corrosive nature of cobalt chloride and hydrogen peroxide is the reason behind higher degradation rate. The maximum degradation percentage obtained was in the range of 19.6–23.7% on 28th day.

15.3.5.3 ENZYMATIC DEGRADATION

It was observed that IPDI-PEU membrane is much more sensitive to enzymatic medium compared with both hydrolytic and oxidative mediums

(Fig. 15.5(c)). One possible reason may be the attack of lipase on urethane linkage of PEU. We have got maximum degradation percentage in the range of 24.69–27.37% on 28th day.

Overall degradation pattern of the IPDI-PEU membrane confirms about its in vitro biodegradable nature.

15.3.6 IN VITRO DRUG RELEASE

15.3.6.1 RELEASE OF RIFAMPICIN

Figure 15.6(a) vividly displays release of rifampicin from membrane in both water and SBF. Release of rifampicin from IPDI-PEU was observed to be greater in case of water than that of SBF. The possible reason may be ions present in SBF block the pores and thus hinder release of antibiotics from membrane. Release of rifampicin into water was maximum on day 7, which was 0.0412 mg/ml and release of rifampicin in SBF was maximum on day 5, which was 0.0304 mg/ml. So, it is evident that the drug release pattern is controlled and as most of the implant infections occur during 1 week of implantation, it is very much desirable to release antibiotics in maximum concentration during that time period. Release of rifampicin was more than its minimum inhibitory concentration (MIC), which was 1 μg/ml.[12]

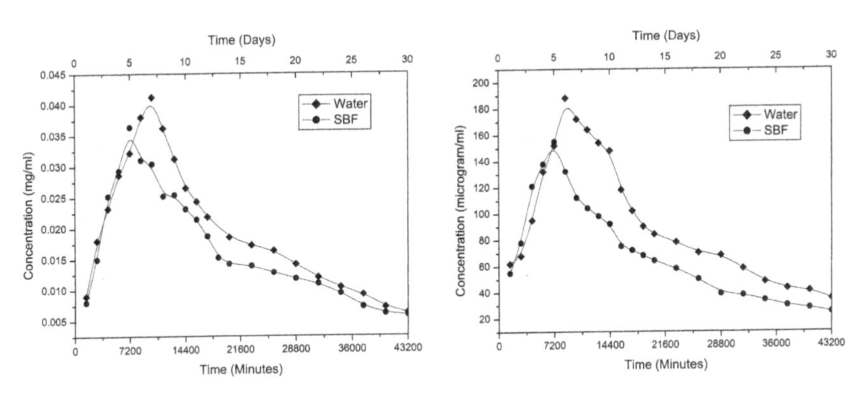

FIGURE 15.6 (a) Release of rifampicin from IPDI-PEU and (b) release of streptomycin from IPDI-PEU.

15.3.6.2 RELEASE OF STREPTOMYCIN

Release of streptomycin from membrane was observed to be greater in case of water than that of SBF. Release of streptomycin in water was maximum on day 6, which was 188 μg/ml, whereas release of streptomycin in SBF was maximum on day 5, which was 155 μg/ml. Release of streptomycin was more than its MIC (16 μg/ml) in both water and SBF (Fig. 15.6(b)).[13]

15.4 CONCLUSION

Successful synthesis of isophorone diisocyanate-based polyether urethane membrane was synthesized successfully. Swelling of membrane is higher in water. X-ray diffraction analysis reveals amorphous nature of membrane. Fourier transform infrared spectroscopy study confirms the presence of urethane linkage. Field emission scanning electron microscopy images provide proof of microporous structure of membrane, which were formed by removal of IPDI and DMSO during treatment with dichloromethane. Porous nature of IPDI-PEU facilitates better drug-loading capacity. In hydrolytic, oxidative and enzymatic mediums, the synthesized IPDI-PEU is degradable. Drug release study confirms sustained release of drug for longer time period. Thus, synthesized IPDI-PEU has good in vitro stability and sustained release characteristic. Overall, this degradable PEU can be used as implant coating material for sustained release of drug to prevent implant infection.

KEYWORDS

- **biocompatibility**
- **Microporous**
- **swelling study**
- **degradation profile**
- **drug release**

REFERENCES

1. Umare, S. S.; Chandure, A. S. Synthesis, Characterization and Biodegradation Studies of Poly(ester urethane)s. *Chem. Eng. J.* **2008,** *142*, 65–77.
2. Kim, Y.; Kim, S. Effect of Chemical Structure on the Biodegradation of Polyurethanes Under Composting Conditions. *Polym. Degrad. Stab.***1998,** *62*, 343–352.
3. Sarkar, S.; Basak, P.; Adhikari, B. Biodegradation of Polyethylene Glycol-Based Polyether Urethanes. *Polym. Plast. Technol. Eng.* **2011,** *50*, 80–88.
4. Darby, R. T.; Kaplan, A. M. Fungal Susceptibility of Polyurethanes. *Appl. Micobiol.* **1968,** *16*(6), 900–905.
5. Christenson, E. M.; Dadsetan, M.; Wiggins, M.; Anderson, J. M.; Hiltner, A. Poly(Carbonate urethane) and Poly(ether urethane) Biodegradation: In vivo Studies. *J. Biomed. Mater. Res.* **2004,** *69* (3), 407–416.
6. Basak, P.; Adhikari, B.; Banerjee, I.; Maiti, T. K. Sustained Release of Antibiotic from Polyurethane Coated Implant Materials. *J Mater. Sci. Mater. Med.* **2009,** *20*, S213–S221.
7. Sarkar, S.; Singha, P. K; Dey, S.; Mohanty, M.; Adhikari, B. Synthesis, Characterization and Cytotoxicity Analysis of Biodegradable Polyurethane. *Mater. Manufactur. Process*. **2006,** *21*, 291–296.
8. Shintani, S. Modification of Medical Device Surface to Attain Anti-infection. *Trends Biomater. Artif. Organs* **2004,** *18*(1), 1–8.
9. Basak, P.; Adhikari, B. Effect of the Solubility of Antibiotics on Their Release from Degradable Polyurethane. *Mater. Sci. Eng.* **2012,** *C32*, 2316–2322.
10. Arora, N.; Ali, A.; Sen, S.; Jana, N. K.; Basak, P. Synthesis and Characterization of Polyetherurethane Coatings for Preventing Implant Infection. *Compos. Interface.* **2013,** *21*, 51–58.
11. Hanks, J. H.; Wallace, R. E. Relation of Oxygen and Temperature in the Preservation of Tissues by Refrigeration. *Proc. Soc. Exp. Biol. Med.* **1949,** *71*, 196–200.
12. Sen, P. K.; Roy, B. N.; Chatterjee, R. Rifampicin: Biological Values for Action and Intolerance. *Indian J. Tub.* **1985,** *32*, 81–85.
13. Sunde, M.; Norstorm, M. The Genetic Background for Streptomycin Resistance in *Escherichia coli* Influences the Distribution of MICs. *Antimicrob. Chemother*. **2005,** *56*, 87–90.

CHAPTER 16

BIODEGRADATION STUDY OF POLYURETHANE FOR THERAPEUTIC APPLICATIONS

NEHA ARORA[2], ASIF ALI[1], SOHINI SEN[1,*], ANINDYA HALDAR[2], NANDAN KUMAR JANA[2], and PIYALI BASAK[1]

[1]*School of Bioscience and Engineering, Jadavpur University, Kolkata, India, *E-mail:sohinii904@gmail.com*

[2]*Department of Biotechnology, Heritage Institute of Technology, Kolkata, India*

CONTENTS

ABSTRACT

Polyurethane (PU) has been widely used for drug delivery applications such as patches, drug-eluting matrix and drug containing vehicle. Examination regarding biodegradation and tissue compatibility of PU and its potential as drug-delivery matrix was conducted in the study. Polyurethane membranes were set for in vitro biodegradation in simulated body fluid (SBF), esterase enzyme in phosphate buffered saline and in oxidative medium for few months. Results of the three types of degradation were compared. Characterization was performed with examining per cent weight loss, attenuated total reflectance Fourier transform infrared and X-ray Diffraction for the degraded membranes and the characterization result was compared with that of the membranes prior to degradation after biodegradation. For 1 month, implants were subjected to in vivo degradation in female albino mice for animal trial. After retrieval of the implant, tissue surrounding the polymer film was examined for histopathological compatibility. Tissues surrounding the polymer films were placed in 10% formalin for 48 h, dehydrated, paraffin-embedded and stained with eosin. Subcutaneous tissue response to PU implant was observed under light electron microscope at different intervals. Acute and subacute tissue showed minimum inflammatory reactions, and no necrosis or abscess formation was found indicating appreciable tissue compatibility to the implant. To employ the membranes in therapeutic applications, such as drug eluting medical implant coat, the drug release property was evaluated using rifampicin and gentamycin as model drugs. By immersing the film into drug solutions of SBF of known concentration, the membrane was loaded with drugs. Then drug release pattern was examined periodically by estimating optical density of different solutions of released drug at different intervals by ultraviolet-visible spectrophotometer. The drug release findings were compared with that of biodegradation study.

16.1 INTRODUCTION

Biodegradation is the breakdown of materials by the action of living organisms, which leads to alteration of the physical environment.[1] Biodegradation of polymers, such as polyurethane (PU), has been extensively analysed as this property makes it suitable for a variety of biomedical applications. Polyurethane has widespread biomedical applications, and one such application is its use as a coating to prevent infection of implant. There are a number of orthopaedic implants that are available and used for a variety of treatments; however, these implanted devices are a major source of implant-associated infection. To be used as effective implant coatings, it is beneficial if PU degrades with time when introduced in vivo. Hydrolysis, enzymatic attack, oxidation and environmental stress cracking are the most common pathways for PU degradation.[2] It is this segmented structure of PU that makes it susceptible to degradation. Polymers that possess in vivo applications should degrade with time so that there is no need for surgical eradication of the polymers out of the system. Therefore, the development of appropriate biodegradable polymer coating for medical devices has gained importance in recent research to entrap antibiotics.[3] Hydrolytic, oxidative and enzymatic degradation study of PU has gained importance in recent research. The conditions used to study in vitro degradation are such that mimic in vivo environment. Hypohalous and nitric oxide-based oxidants have been used to study the degradation of PUs.[4] In vivo oxidant environment is produced by a hydrogen peroxide (H_2O_2) and cobalt chloride ($CoCl_2$) solution at 37°C.[5] Monocyte-derived macrophages (MDM) are another source for the oxidants, believed to be involved in the degradation process.[6,7] Although MDM provide a significant source of several oxidative species, they also contain large amounts of enzymes with hydrolytic activity.[8,9] According to literature review, enzymes such as phosphatase, esterase, aminopeptidase and oxidoreductase are commonly found in the immediate surroundings of the implanted area.[10] In this study, hydrolytic, enzymatic and oxidative degradation of the PU membrane have been reported. Hydrolytic degradation is studied by using simulated body fluid (SBF) at 37°C. Oxidative degradation is studied by using 0.1 M $CoCl_2$ (cobalt chloride) in 20% H_2O_2 (hydrogen peroxide) oxidative solution at 37°C. Enzymatic degradation is studied using esterase enzyme from porcine liver. The degradation is characterized by X-ray Diffraction (XRD) and Fourier transform infrared (FTIR). The drug release study of PU membrane was also investigated by using antibiotics.

16.2 EXPERIMENT

16.2.1 MATERIALS

Hexamethylene diisocyanate (HDI) and esterase enzyme were procured from Sigma-Aldrich, India. Hydrogen peroxide and cobalt chloride were procured from Merck India Ltd. Rifampicin and gentamycin were procured from Central Drug house (CDH®), India.

16.2.2 METHOD

16.2.2.1 POLYURETHANE MEMBRANE SYNTHESIS

Polyurethane membrane preparation is done by a single-step solution polymerization method. Hexamethylene diisocyanate and polyethylene glycol (PEG 400) were mixed in a molar ratio of 2:1 in a two-necked, round-bottomed flask. 2.02 g of PEG and 30 ml benzene were mixed and 10 ml of benzene and 4.04 g of HDI were mixed simultaneously and separately. Here, PEG and HDI served to synthesize the hard segment and soft segment of the polymer, respectively. Each of the solutions was allowed to mix independently until they both become colourless. Then, the two solutions were mixed and stirred until the solution turned white. The solution was cast on a Petri plate and kept in 80°C for 5 h to obtain a white polymer. Polyurethane membranes were obtained by dissolving the white-coloured 2.5 g of PU into measured quantity that is 30 ml of dimethyl formamide and stirred till a colourless solution was obtained. This viscous polymer solution was recast and allowed to evaporate on a flat Petri dish overnight in a fume hood to obtain a film or membrane of PU membrane.

16.2.2.2 IN VITRO DEGRADATION STUDY

16.2.2.2.1 Hydrolytic Degradation Study

Hydrolytic study of the membrane was carried out for 1 month in 20 ml of SBF. The samples were incubated at 37°C. For formulation of SBF, we have used the composition of Hanks' balanced salt solution and pH was adjusted to 7.4.[11] The membranes were cut into dimensions of 2×2 cm. Dried and preweighed membrane samples were dipped in SBF and incubated at 37°C depending upon the different intervals. After appropriate time intervals such as starting from 15 min, 45 min, 3, 6, 24 h to 3, 7 and up to 30 days, the membranes were taken out, swelled in water and then oven-dried at 65°C. After drying, the final weight of the membrane was recorded. The weight loss of the membranes was calculated as follows:

$$\text{Weight loss} = \frac{(W_I - W_D)}{W_I} \times 100 \quad (16.1)$$

where W_I and W_D are the weights of the membrane before and after degradation, respectively.

16.2.2.2.2 *Oxidative Degradation Study*

Oxidative degradation of the PU membrane was carried out for a period of one month. The membranes were cut into dimensions of 2 × 2 cm. Dried and preweighed membrane samples were dipped in the oxidative medium and incubated at 37°C for different intervals like starting from 15 min and continuing up to 30 days. The oxidative medium was prepared by mixing 0.1 M cobalt chloride solution in 20% hydrogen peroxide prepared from 30% hydrogen peroxide solution by proper dilution with distilled water. The samples were taken out and percentage of degradation was determined at different time intervals. The weight loss was calculated as given in Equation 16.1.

16.2.2.2.3 *Enzymatic Degradation Study*

Enzymatic degradation study was carried out using esterase isolated from porcine liver and phosphate buffered saline (PBS) as the solvent for a period of one month. The pH was adjusted to 7.4. The membrane was cut in the dimensions 2 × 2 cm. Dried and preweighed membrane samples were dipped in 8 ml PBS-containing 2% (V/V) streptomycin. Enzyme stock solutions of 40 units/ml were prepared, from which 200 μl of enzyme stock solution was added to each of the sample.[12] The membrane samples were incubated at 37°C. At exactly similar time intervals like that of hydrolytic and oxidative degradation study, the membrane samples were taken out and percentage of degradation was determined. The weight loss was calculated as given in Equation 16.1.

16.2.2.3 DRUG LOADING

Perhaps the most direct approach for improving the efficacy of conventional antibiotics against implant-associated biofilms is to deliver the antibiotics in a controlled manner at the implant from a surface coating.[13] The antibiotics used were rifampicin and gentamycin. Concentration of antibiotics solution used for loading was 1.2 and 2.4 mg/ml for rifampicin and 2 and 4 mg/ml for gentamycin. Water was used as a solvent for gentamycin. Water and tetrahydrofuran (THF) in the ratio 1:1 was used for rifampicin. For loading of antibiotics, membrane was cut in dimension of 2 × 2 cm. The membranes were swelled in water and THF and finally were oven-dried at around 65°C. Preweighed membrane samples were dipped in antibiotic solutions and incubated at 37°C to facilitate the loading of drug onto the membrane by diffusion. The membranes were dipped in the antibiotic solution for 48 h. The membranes were dried and weighed after dipping. Increase in weight indicates the amount of drug loaded.

16.2.2.4 DRUG RELEASE STUDY

The antibiotic release study was carried out in SBF and water. The antibiotic-loaded membrane was dipped in SBF and water, and the samples were incubated at 37°C. At regular time intervals, the release medium was assayed by Ultraviolet-visible spectrometer. A release medium of 4 ml was collected and was replaced by fresh solvent. The optical density was recorded at 257 nm for rifampicin and at 333 nm for gentamycin. Drug release study was carried out for a period of one month. The concentration of drug released was determined from the standard plot of absorbance versus concentration for the respective antibiotics.

16.2.3 CHARACTERIZATIONS

16.2.3.1 X-RAY DIFFRACTION

X-ray Diffraction pattern was obtained for the degraded membrane using Rigaku miniflex X-ray diffractometer, Ultima III with a Cu radiation. The measurements were performed in the 2θ range of 10–60°C. Before XRD study, the membranes were swelled in water and oven-dried at 65°C.

16.2.3.2 FOURIER TRANSFORM INFRARED ANALYSIS

Fourier transform infrared analysis of the degraded membranes was performed by using Thermo Nicolet FTIR 380 instrument in the frequency range of 3000–500 cm^{-1}. Before FTIR study, the membranes were swelled in water and oven dried at 65°C.

16.3 RESULTS AND DISCUSSION

16.3.1 *IN VITRO* DEGRADATION STUDY

16.3.1.1 HYDROLYTIC DEGRADATION STUDY

Degradation of the membrane was performed in SBF at 37°C. The overall degradation was 14.48% as seen in Figure 16.1. As degradation was carried out in SBF, only hydrolytic degradation is responsible for the weight loss of the membrane. Slight alkaline pH of SBF has facilitated the hydrolytic degradation.

16.3.1.2 OXIDATIVE DEGRADATION STUDY

Oxidative degradation was carried out in $CoCl_2$ and H_2O_2 system at 37°C. The overall degradation was 7.68% for a period of one month as observed in Figure 16.1. The medium selected was such that it mimics the oxidizing power of *in vivo* environment.

16.3.1.3 ENZYMATIC DEGRADATION STUDY

Esterase enzyme was used to study enzymatic degradation of the poly (ether urethane) membrane, as it was found in the immediate surroundings of the implanted areas. Enzymatic degradation was carried out using esterase enzyme (40 units/ml) from porcine liver. The overall degradation was 11.40% for a period of one month as observed in Fig 16.1.

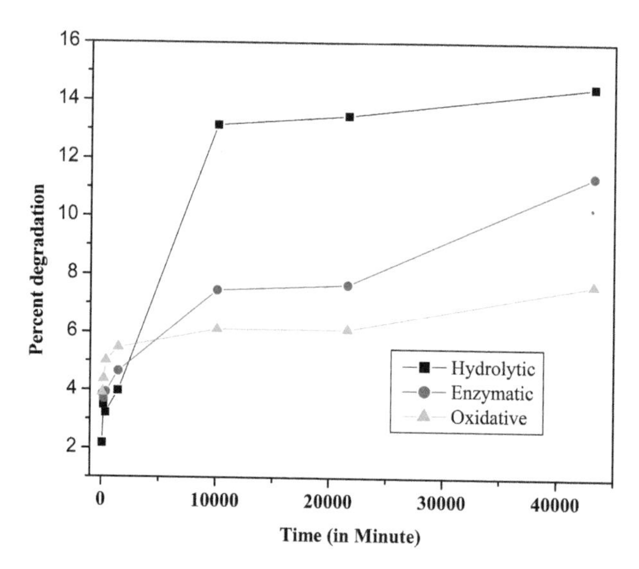

FIGURE 16.1 Comparative weight loss study for different types of degradation.

16.3.2 DRUG RELEASE STUDY

For both rifampicin and gentamycin, it was observed that the release of the antibiotics was more in SBF than in water. This might be due to the difference in ionic nature of the drug release medium. Simulated body fluid is ionic in nature, having slight alkaline pH. This nature may have facilitated the release of rifampicin in the medium.

16.3.2.1 RELEASE OF RIFAMPICIN AND GENTAMYCIN

Release of rifampicin is more in SBF than in water as reflected in Figure 16.2. The release of drug increases with increase in time and then gradually falls after a period of 15–20 days. This pattern appears for both the releases in SBF as well as water. The antibiotic concentration in the release medium is more than their minimum inhibitory concentration (MIC) (MIC for rifampicin is 1 μg/ml).[14] Release of gentamycin is more in SBF than in water as seen in Figure 16.3. The release of gentamycin in SBF is high during the first few hours of incubation and is maintained for next few days, after which the release gradually falls with time. As the first 6 h are very crucial for preventing implant infection, the effective release of drug during the first 6 h is essential.[13] The release in water gradually increases with time and then falls gradually. The antibiotic concentration in the release medium is more than their (MIC) (MIC for gentamycin is 0.5 μg/ml).[15]

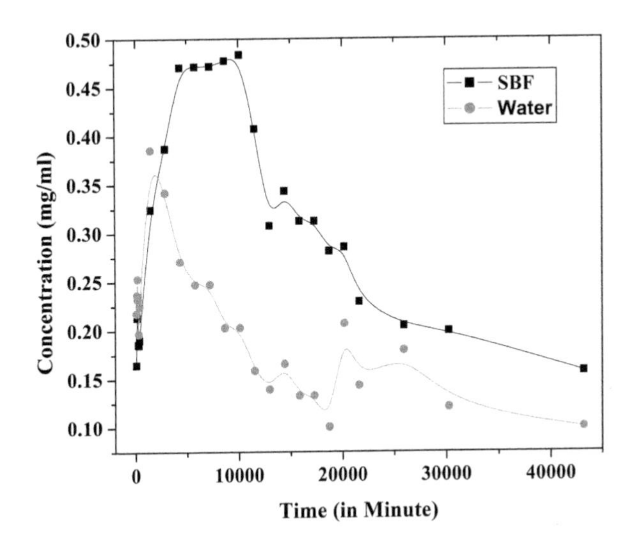

FIGURE 16.2 Release study of rifampicin.

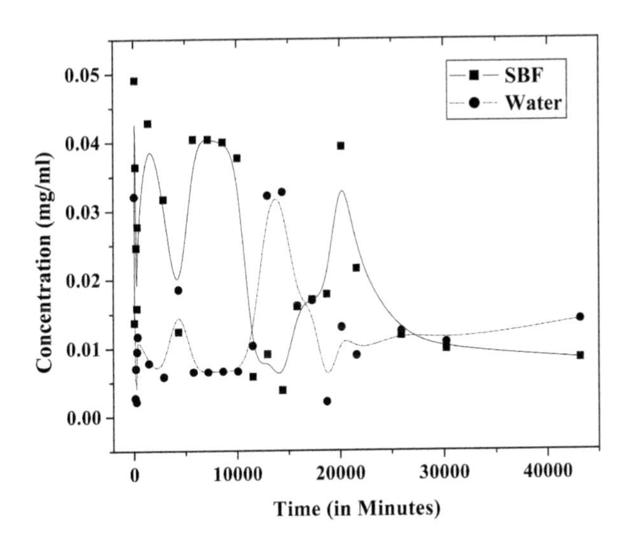

FIGURE 16.3 Release study of gentamycin.

16.3.3 FOURIER TRANSFORM INFRARED ANALYSIS

The peak present in FTIR spectra of nondegraded membrane at 1737 cm^{-1} is due to C=O stretching of urethane bond. The peaks at 1648 and 1589 cm^{-1} are due to N–H bend and C=C vibrations, respectively. The double peaks at 1284 and 1162 cm^{-1} are due to the presence of C–N bond and C–C stretch, respectively. In the case of the spectra of hydrolytically degraded membrane shown in Figure 16.4, these peaks show lower intensity, which is due to degradation of the membrane. In the case of oxidative degradation observed in Figure 16.5, intensity of the peak at 1737 cm^{-1} decreases, signifying degradation of urethane bond. The peak (1648 cm^{-1}) signifying N–H bend is not present, whereas a new peak at 1638 cm^{-1} signifying C=O stretch in amide bond is present. The oxidatively degraded membrane shows a peak at 1585 cm^{-1}, which is due to C=C stretch of an aromatic compound. This aromatic compound might have been formed due to degradation of the membrane. The peaks 1284 cm^{-1} and 1162 $cm^{-1}$are less in intensity in the case of oxidatively degraded membrane. In the case of enzymatically degraded membranes reflected in Figure 16.6, there is a decrease in the peak 1737 cm^{-1}, but the decrement is much less when compared with oxidative and hydrolytic degradation. The other peaks also show decrease in intensity, but the decrease is quite less when compared with oxidative and hydrolytic degradation.

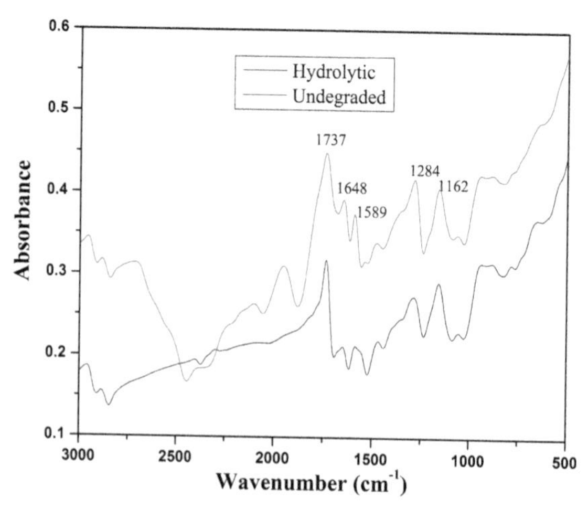

FIGURE 16.4 Comparative FTIR spectra for hydrolytic degradation.

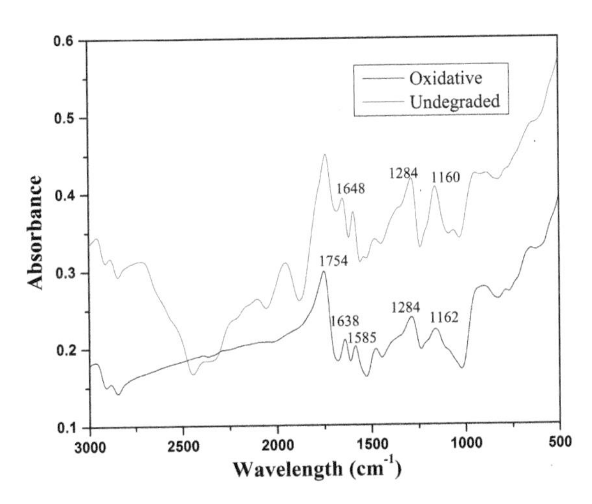

FIGURE 16.5 Comparative FTIR spectra for oxidative degradation.

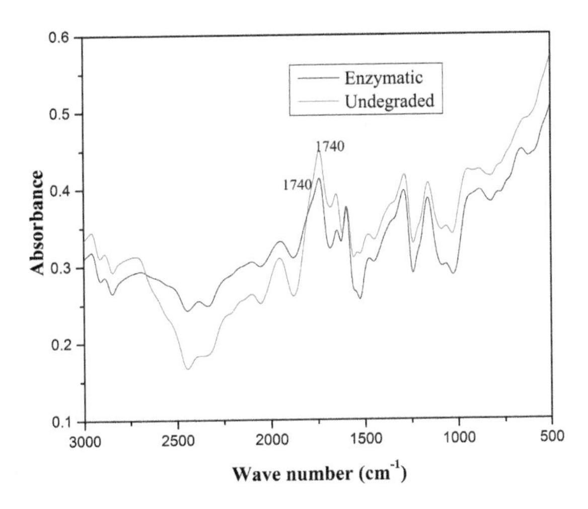

FIGURE 16.6 Comparative FTIR spectra for enzymatic degradation.

16.3.4 X-RAY DIFFRACTION STUDY

X-ray Diffraction analysis of the hydrolytically degraded PU membrane as in Figure 16.7.showed an increased intensity of the peak at $2\theta = 20°C$. This might be due to degradation of amorphous region of the membrane, resulting in an increase of the crystallinity of the membrane. The oxidatively degraded membranes also show an increase in peak intensity, confirming degradation of the membrane indicated in Figure 16.8. There is a slight increase in the peak, in the case of enzymatically degraded membranes shown in Figure 16.9. This can be attributed to less degradation of the membrane in enzymatic medium. From the XRD data, it can be concluded that there has been a less amount of degradation of the amorphous region in the case of enzymatic medium.

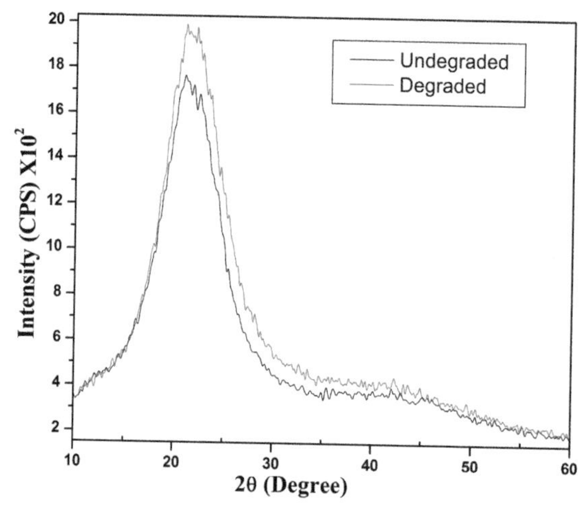

FIGURE 16.7 Comparative XRD pattern for hydrolytic degradation.

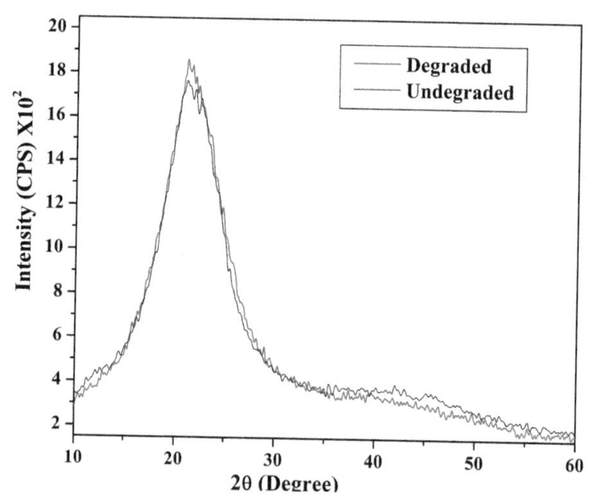

FIGURE 16.8 Comparative XRD pattern for oxidative degradation.

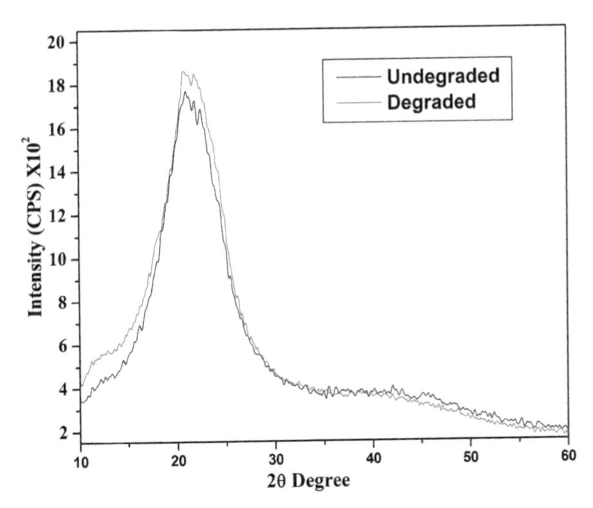

FIGURE 16.9 Comparative XRD pattern for enzymatic degradation.

16.4 CONCLUSION

The membranes synthesized from HDI show high potential for drug loading. The release of both rifampicin and gentamycin was above the MIC. Oxidative and hydrolytic media were found to be more effective for degradation than the enzymatic medium. The FTIR and XRD results confirm a higher degradation of the membrane in hydrolytic and oxidative media, although there is a higher weight loss percentage of the membrane in enzymatic medium than the oxidative medium. The histopathology study of the rat skin tissue confirms that the degraded products of the membrane show no harmful effect on the tissue. In addition, the membrane degrades completely inside the rat body without showing any toxicity. Hence, the membrane can be suitably used as a safe and biodegradable drug-loading matrix.

16.5 ACKNOWLEDGEMENT

The authors acknowledge Council of Scientific and Industrial Research for funding this project and for providing fellowship to the first author Sohini Sen.

KEYWORDS

- **polyurethane**
- **degradation**
- **drug release**

REFERENCES

1. Ratner, B. D.; Hoffman, A. S.; Schoen F. J.; Lemons, J. E. *An Introduction to Materials in Medicine,* 2nd Ed.; Elsevier Academic Press: London, 2004; pp 760.
2. Sarkar, S.; Basak, P.; Adhikari, B. Biodegradation of Polyethylene Glycol-Based Polyether Urethanes. *Polym. Plast. Technol. Eng.* **2011,** *50*, 80–88.
3. Basak, P.; Adhikari, B. Effect of the Solubility of Antibiotics on Their Release from Degradable Polyurethane. *Mater. Sci. Eng. C.* **2012,** *32*, 2316–2322.
4. Sutherland, K; Mahoney II, J. R.; Coury, A. J.; Eaton, J. W. Degradation of Biomaterials by Phagocyte Derived Oxidants. *J. Clin. Invest.* **1993,** *92*, 2360–2367.
5. Christenson, E. M.; Anderson, J. M.; Hiltner, A. Oxidative Mechanism of Poly(Carbonate Urethane) and Poly(Ether Urethane) Biodegradation: In Vivo and In Vitro Correlations. *J. Biomed. Mater. Res.* **2004,** *70*, 245–255.
6. Frautschiaq, J. R.; Chinn, J. A.; Phillips, R. E.; Zhaob, Q. H.; Andersonb, J. A.; Joshi, R.; Levy, R. J. Degradation of Polyurethanes In Vitro and In Vivo: Comparison of Different Models. *Colloids Surf. B.* **1993,** *1*, 305–313.
7. Marchant, R.; Hiltner, A.; Hamlin, C.; Rabinovitch, A.; Slocbodkin, R.; Anderson, J. M. In vivo Biocompatibility Studies. I. The Cage Implant System and a Biodegradable Hydrogel. *J. Biomed. Mater. Res.* **1983,** *17*, 301–325.
8. Miller, K. J.; Rose-Caprara, V.; Anderson, J. M. Generation of IL1like Activity in Response to Biomedical Polymer Implants: A Comparison of *In Vitro* and *In Vivo* Models. *J. Biomed. Mater. Res.* **1989,** *23*, 1007–1026.
9. Papadimitriou, J. M.; Ashman, R. B. Macrophages: Current Views on Their Differentiation Structure and Function. *Ultrastruct. Pathol.* **1989,** *13*, 343–372.
10. Schakenraad, J. M.; Hardonk, M. J.; Feijen, J.; Molenaar, I.; Nieuwenhuis, P. Enzymatic Activity Toward Poly(L-Lactic Acid) Implants. *J. Biomed. Mater. Res.* **1990,** *24*, 529–545.
11. Hanks, J. H.; Wallace, R. E. Relation of Oxygen and Temperature in the Preservation of Tissues by Refrigeration. *Proc. Soc. Exp. Biol. Med.* **1949,** *71*, 196–200.
12. Basak, P.; Sen, S. Degradation Study and Its Effect in Release of Ciprofloxacin from Polyetherurethane. *Adv. Mater. Res.* **2012,** *584*, 474–478.
13. Arora, N.; Ali, A.; Sen, S.; Jana, N. K.; Basak, P. Synthesis and Characterization of Polyether Urethane Coatings for Preventing Implant Infection. *Compos. Interfaces* **2014,** *21*(1), 51–58.
14. Evan, M.; Hetrick, H.; Schoenfisch, M. H. Reducing Implant-Related Infections: Active Release Strategies. *Chem. Soc. Rev.* **2006,** *35*, 780–789.
15. Sen, P. K.; Roy, B. N.; Chatterjee, R. Rifampicin: Biological Values for Action and Intolerance. *Indian J. Tub.* **1985,** *32*, 81–85.

CHAPTER 17

MICROSCOPY AND SPECTROSCOPY CHARACTERIZATION OF ECO-FRIENDLY COMPOSITES AND NANOCOMPOSITES

RAGHVENDRA KUMAR MISHRA[1,*], REMYA V. R.[2], MUFEEDA K.[3], SABU THOMAS[1], and NANDAKUMAR KALARIKKAL[1]

[1]*International and Interuniversity centre for Nanoscience and Nanotechnology, Mahatma Gandhi University, Kottayam, Kerala, India, *E-mail: raghvendramishra4489@gmail.com*

[2]*Department of Applied Chemistry, University of Johannesburg, Doornfontein Campus, South Africa*

[3]*Department of Engineering, Amity University, Dubai, United Arab Emirates*

CONTENTS

ABSTRACT

Petroleum-dependent thermoplastics are commonly used in a variety of platforms, especially in product and packaging. However, their application has led to increasing pollutant by-products. As a result, scientists are motivated to look for eco-friendly substitute for packaging materials which can be recycled as well as are biodegradable. Due to the outstanding mechanical characteristics of natural fibres, they have been widely employed to strengthen biopolymers to generate biodegradable composites. A detailed knowledge about the aspects of these kinds of composite materials is important for examining their usability to numerous items.

The main aim of the current chapter is to review microscopy and spectroscopy evaluation of such types of composites. Therefore, scanning electron microscopy, scanning probe microscopy, X-ray photoelectron spectroscopy and wide-angle X-ray diffraction techniques are considered in this chapter.

17.1 INTRODUCTION

The main purpose of microscopy and spectroscopy is to observe and understand materials with respect to their size and shapes at different length scale and reaction on light and so on. Considering the difference between the microscopy and spectroscopy, microscopy refers to the observation of materials which cannot be seen by naked eyes and it can also amplify the image, whereas spectroscopy refers to how materials behave with light. In the case of a microscope, acquiring the image is essential, but in the case of spectroscopy, just spectra would be able to tell all the information about matters; the spectrum helps to analyse the matter with respective frequency or wavelengths comprising a radiation. For example, in scanning electron microscope (SEM), there can be a different interaction of electron beam at different parts of a sample. Thus, the variations in the sample with respect to its position can be observed by the image at micro- or nanolevel. Most of the time, optical microscope is used; we cannot view the matter at the molecular level. This raises a need for electron microscopes, which utilizes the wave nature of the electrons.

However, matter can be identified with the help of passing light through the sample by taking the different spectroscopy characteristics into account. The sample characteristics depend on the intensity, wavelength, phase, full width at half maximum (FWHM) of a peak and so on.

For example, IR spectroscopy gives much more acquired molecular information by measuring the vibration of molecules on a surface, which acts similar to a molecular fingerprint. When the spectrum of sunlight is taken into consideration, it tells us the chemical composition of the sun because each element atom gives out its own spectral line. Every atomic or molecular radiation gives out a characteristic spectral line.

Recently, a lot of research efforts are geared to the development and application of environmental-friendly bio-reinforced composites and nanocomposites for sustainable development of our human society. For this purpose, these bio-reinforced composites and nanocomposites have been used in various fields including automotive, construction, packaging and medical fields and so on. It is also found that implementation of nanotechnology in these types of composites by using nanoscale materials into base polymers reduced the drawback of bio-reinforced composites and nanocomposites and made them a suitable material for extreme environment.[13,27]

In addition the term 'biocomposites' refers to the combination of matrix and fibres (biofibres) such as wood fibres (hardwood and softwood) or non-wood fibres (e.g. wheat, kenaf, hemp, jute, sisal and flax) from both renewable and nonrenewable resources. Biofibres are one of the main constituents of biocomposites. These bio fibrous materials are obtained from trees, plants or shrub sources and so on. These composites are eco-friendly and biocompatibility to the system is provided them. Natural fibres have lot advantages over synthetic fibres such as low cost, density, shape ratio and mechanical behaviour. Therefore, the application of natural fibre for the reinforcement of the composites has received increasing attention by the both academic sector and industrial sector. However, natural fibres based biocomposites have some drawbacks such as low modulus elasticity, high moisture absorption, decomposition in alkaline environments or in the biological attack and variability in mechanical and physical properties. It is necessary to study fibre properties precisely to solve this problem. For instance, dimensional stability, flammability, biodegradability and degradation are attributed to acids, bases and UV radiation that alter the biocomposites back into their basic building blocks (carbon dioxide and water). However, the properties of natural fibres that result from the chemistry of the cell wall components make some major problems in biocomposites. Therefore, more attention has been given to bio-nanocomposites nowadays.[23]

The properties of nanocomposites depend on their individual constituents, morphology and interfacial (matrix/nanofiller) characteristics. Recently, a lot of works have been done in the field of the biomedical application based on the bio-nanocomposites. The design and fabrication of nanocomposites for various biomedical applications requires a fundamental understanding of the interactions between polymers, nanostructures, growth factor, biocompatibility and the biochemical and biophysical rules that dictate cell biology. As the cell is complex and is formed by self-arrangement of macromolecules, it is very important. To employ this, biology offers the best models for strategies on how to rationally design high-performance biomaterials with the properties of materials, such as bone, cartilage, nacre or silk.[2,9,24,32,35]

This chapter aims a comprehensive review of the different microscopic and spectroscopic characterization of biomaterial-based polymer nanocomposites and engineering materials.

17.2 SCANNING ELECTRON MICROSCOPY

There are basically two important types of the electron microscopes: Scanning electron microscopes (SEM) and transmission electron microscopes (TEM). In an SEM, the secondary electrons generated by the specimen are identified to produce a picture, which has topological attributes of the specimen. The image in a TEM, alternatively, is produced by the electrons, which have passed throughout a slim specimen. We are going to understand precisely how both of these microscopes operate and also what type of data they are able to produce.[34]

Scanning electron microscopy is an essential tool for all experimental scientists and researchers. It has been widely used for imaging the various kinds of materials. The SEM has many advantages over traditional microscopes. The SEM has a large depth of field, which allows more of a specimen to be in focus at one time. The SEM also has much higher resolution, so closely spaced specimens can be magnified at much higher levels because electromagnets are employed in the SEM rather than lenses and instead of light electrons are used to form an image and the electron has much shorter wavelength than light (wavelengths of visible light vary from 400–700 nm, light diffraction limits resolution to 200–250 nm and magnification limited to around 1500x, whereas the wavelength of electrons in an SEM is 0.017 nm, resolution clarity to 1 nm, magnification up to 500,000×). For enabling better resolution, the electron microscope was developed in the year 1950 when wavelength became the limiting factor in light microscopes.[29] The main SEM signal depends on collecting secondary electrons, which typically form a large proportion of the emitted electrons.[8]

Electrons can be produced from thermionic heating at the top of the column, in which electron source [tungsten filament, solid state crystal (CeB6 or LaB6)], field emission gun) is mounted. To produce a focus on the electron beam on the surface of the sample, generated electrons (electron beam) are passed through a series of condenser lenses and apertures. When the electron beam hits and penetrate the surface of the sample, penetration depth can be few micron metres based on the applied voltage and density of samples, and thus, produce electrons, backscattered electrons and characteristic X-rays. The position of the electron beam on the sample is controlled by scanning coils which are situated above the objective lens. These coils allow the beam to be scanned over the surface of the sample by deflecting the beam in the X and Y axes (raster scanning pattern) and the chamber is totally evacuated. This beam rastering or scanning enables to collect information from a fixed area of the sample as the high-energy electrons interact with the sample in the rastering or scanning fashion.[7,8,29,38] Figure 17.1 demonstrates a simple schematic diagram of an SEM. The electrons generated by the electron gun are directed as well as targeted by the magnetic lenses on the specimen.

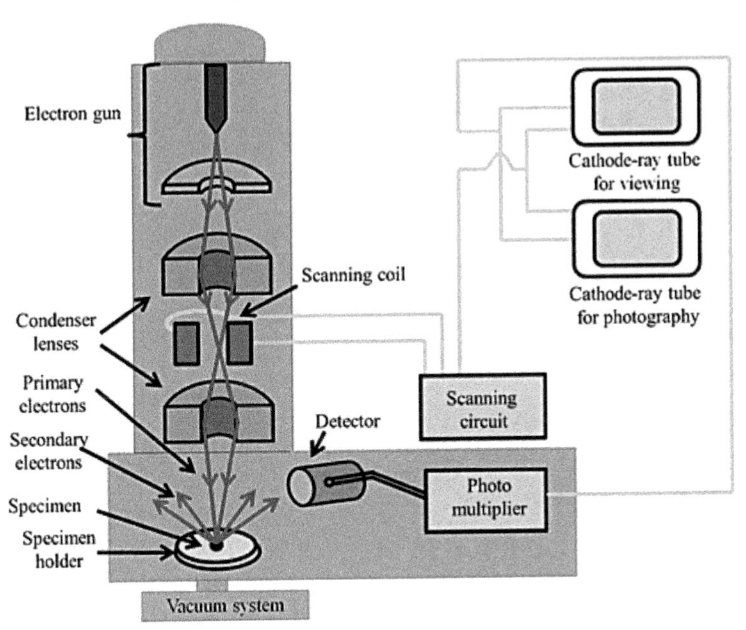

FIGURE 17.1 Schematic diagram for scanning electron microscope (SEM).
Source: Adapted from[7] with permissions.

17.2.1 EFFECT OF PARAMETER IN SCANNING ELECTRON MICROSCOPE ANALYSIS

1. Higher accelerating voltage (higher possibility of specimen damage, beam penetration and diffusion area become larger, unnecessary signals from backscattered electrons, reduces the image contrast and veils fine surface structures, high resolution).
2. Low accelerating voltage (clear surface structures, less damage, less charge-up, less edge effect, low resolution, difficult to obtain sharp micrographs at high magnifications).
3. Higher probe current (deteriorated resolution, more damage, but smooth image).
4. Low probe current (high resolution obtainable, less damage, sharper image, but grainy image, lost the surface smoothness).
5. Loss of electron beam energy is due to increase in temperature at an irradiated point and depends on electron beam accelerating voltage and dosage, scanning area, scanning time. Polymer materials and biological specimens are easily damaged by the electron beam, because of their low heat conductivity).
6. Avoid the sample damage by lowering accelerating voltage, decreasing electron beam intensity, shortening the exposure time, large scanning areas with low magnifications and metal coating on the specimen surface.

17.2.2 SCANNING ELECTRON MICROSCOPE MORPHOLOGY OF COMPOSITES AND NANOCOMPOSITES

17.2.2 1 NATURAL FIBRES

Petroleum-dependent polymers are being used as one of the primary materials for a wide variety of purposes, steadily displacing traditional materials. The development of plastic material is basically as a result of its fantastic features as well as handling opportunities. Despite this, the product packaging sector is dealing with leading problems, for example, reducing the prevalence of petrochemical sources, rises in their cost along with the determination of these materials in the natural environment beyond their functional life. Hence, this inspires product packaging companies to identify new and innovative choices in generating green and eco-friendly goods including fibre-reinforced recyclable and consequently biodegradable product packaging materials. Green composites, referred to as biodegradable biopolymers reinforced by natural fibres, have minimal effect on our environment; therefore, these are possibly the potency options to substitute traditional petroleum-dependent polymers and polymer composites. Because of environmental issues along with the requirement for practical properties, substitution of petrochemical-based thermoplastics by biopolymers is being extensively studied.

In recent times, biopolymers have been commonly used in form of a matrix in biocomposites with natural fibres as reinforcement. Even so, the pattern in many of these particular scientific studies is the introduction of reinforcement in structure of short or unidirectional natural fibres. This pattern has started to modify through the use of natural fabrics to reinforce biopolymer composites to be able to find out more effective composites with regard to stiffness and strength. The natural fibre fabric is developed in an effort to enhance particular boundaries in the biopolymer, for example, its impact strength, brittleness and also heat deflection temperature. Natural fibres are classified on the basis of their origin that is coming from plants, animals or minerals. All plant fibres are composed of cellulose; however, animal fibres are composed of proteins (hair, silk and wool). Bast (or stem or soft sclerenchyma) fibres, leaf or hard fibres, seed, fruit, wood, cereal straw and other grass fibres are examples of plant fibres. The increasing demand for natural fibres is basically a result of their lower density that is usually 1.25–1.50 g/cm^3 as compared to glass fibres that have a density of 2.6 g/cm^3 permitting natural fibres to impart greater specific strength as well as stiffness in plastic materials. Natural fibres also provide numerous features because they are recyclable, biodegradable, plentiful, manifest high-quality mechanical characteristics, offer a much better functioning environment and are significantly less abrasive to machine components in comparison to typical synthetic fibres, which often can lead to considerable less price. These kinds of features help to make their application extremely delightful for the production of polymer matrix composites. On the other hand, scientists who have dealt in the field of natural fibres and their composites concur that these types of environmentally friendly options possess certain downsides, such as inadequate wettability, incompatibility with certain polymeric matrices as well as excessive moisture absorption. Most of the important natural fibres are mentioned in Table 17.1.[16] Recently, the nanocellulose is becoming a vital theme for many areas as a result of its renewability, accessibility, biocompatibility and also various amazing features. Considering developed bioethanol, the research on nanocellulose is progressively intending in the direction of the indirect application of lignocellulosic biomass. The exploitation of various cellulosic residues as well as waste materials produced by agricultural and commercial daily activities and the extensive accessibility as well as renewable characteristics of these kinds of feedstocks with different chemical composition, structure and recalcitrance are listed in Table 17.2.[10] The lignocellulose recalcitrance or even treatability indicates that precisely how easier would be to treat, fractionate as well as alter the biomass, which is directly linked to the crystallinity as well as the polymerization degree of cellulose.

TABLE 17.1 List of Important Biofibres. (*Source:* Adapted from[16] with permissions.)

Fibre source	Species	Origin
Abaca	*Musa textilis*	Leaf
Bagasse	–	Grass
Bamboo	(>1250 species)	Grass
Banana	*Musa indica*	Leaf
Broom root	*Muhlenbergia macroura*	Root
Cantala	*Agave cantala*	Leaf
Caroa	*Neoglaziovia variegata*	Leaf
China jute	*Abutilon theophrasti*	Stem
Coir	*Cocos nucifera*	Fruit
Cotton	*Gossypium* sp.	Seed
Curaua	*Ananas erectifolius*	Leaf
Date palm	*Phoenix dactylifera*	Leaf
Flax	*Linum usitatissimum*	Stem
Hemp	*Cannabis sativa*	Stem
Henequen	*Agave fourcroydes*	Leaf
Isora	*Helicteres isora*	Stem
Istle	*Samuela carnerosana*	Leaf

TABLE 17.1 *(Continued)*

Fibre source	Species	Origin
Jute	*Corchorus* capsularis	Stem
Kapok	*Ceiba pentranda*	Fruit
Kenaf	*Hibiscus cannabinus*	Stem
Kudzu	*Pueraria thunbergiana*	Stem
Mauritius hemp	*Furcraea gigantea*	Leaf
Nettle	*Urtica dioica*	Stem
Oil palm	*Elaeis guineensis*	Fruit
Piassava	*Attalea funifera*	Leaf
Pineapple	*Ananas comosus*	Leaf
Phormium	*Phormium tenas*	Leaf
Roselle	*Hibiscus sabdariffa*	Stem
Ramie	*Boehmeria nivea*	Stem
Sansevieria (Bowstring hemp)	*Sansevieria*	Leaf
Sisal	*Agave sisalana*	Leaf
Sponge gourd	*Luffa cylindrica*	Fruit
Straw (Cereal)	–	Stalk
Sun hemp	*Crotalaria juncea*	Stem
Cadillo/urena	*Urena lobata*	Stem
Wood	(>10,000 species)	Stem

(Reprinted/adapted with permission from John, M. J.; Thomas, S. Biofibres and Biocomposites. *Carbohydr.Polym.* **2008,** *71*(3), 343–364. © 2008 Elsevier.)

TABLE 17.2 Main Composition (Cellulose, Hemicelluloses and Lignin in % of Raw Material, Wt. Dry Basis) of Some Lignocelluloses Feed Stocks.

Category	Cellulosic biomass	Cellulose	Hemicellulose	Lignin
RAW MATERIALS	Cotton	87.5	17.1	0.0
	Flax fibres	75.9	20.7	3.4
	Eucalyptus	52.7	15.4	31.9
	Switchgrass	48.7	38.4	12.9
	Pine	48.1	23.5	28.4
	Elephant grass	31.5	34.3	34.2

TABLE 17.2 *(Continued)*

Category	Cellulosic biomass	Cellulose	Hemicellulose	Lignin
NON-PROCESSED WASTES	Cotton stalk	66.2	18.4	15.4
	Sunflower shells	56.5	28.0	15.5
	Rice straw	52.3	32.8	14.9
	Barley straw	48.6	29.7	21.7
	Coconut coirs	52.2	28.4	19.4
	Corn stover	47.4	30.3	22.3
	Tobacco stalks	44.6	30.2	25.2
	Corn cob	48.1	37.2	14.7
	Wheat straw	44.5	33.2	22.3
	Flax shives	39.9	26.8	24.2
	Legume straw	29.2	35.5	35.3
	Oilseed rape	27.3	20.5	14.2
	Wood bark	25.2	30.3	44.5
	Olive husks	25.0	24.6	50.4
PROCESSED WASTES	Newspaper	45.6	31.3	23.1
	Sugarcane bagasse	47.4	29.1	23.5
	Tea residue	33.3	23.2	43.5

(Reprinted with permission from García, A et al. Industrial and Crop Wastes: A New Source for Nanocellulose Biorefinery. *Ind. Crops Prod.* **2016**, 93, 26–38. © 2016 Elsevier.)

The various kinds of biodegradable polymer families, which are derived from various sources and are indicated in Figure 17.2; most polymers are derived from renewable resources (biomass). The first kind is agro-polymers (e.g. polysaccharides, protein and lipids) which are derived from biomass by fractionation. The second and third kinds are polyesters, produced, respectively, by fermentation from biomass or from usually modified plants (e.g. polyhydroxyalkanoate) and by fabrication from monomers obtained from biomass (e.g. PLA, polylactic acid). The last kinds are polyesters, completely obtained by the petrochemical process (e.g. polycaprolactone, polyesteramide, aliphatic or aromatic co-polyesters).[16]

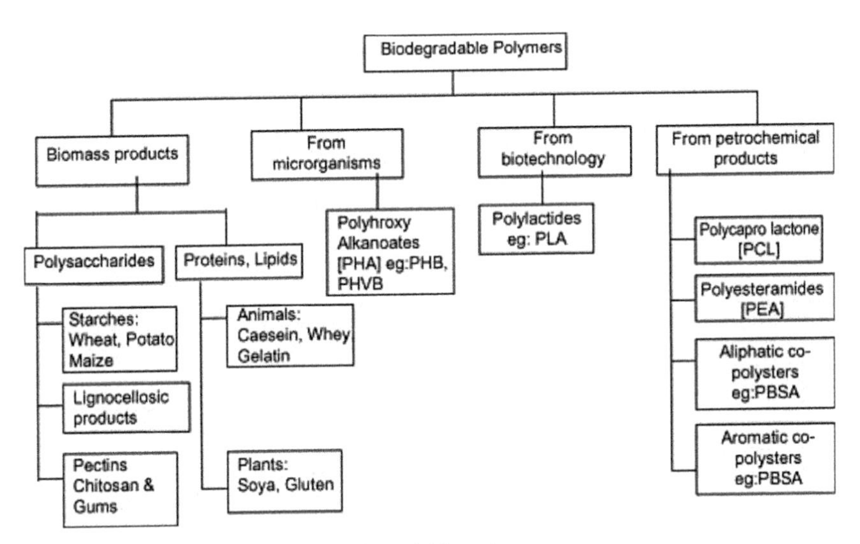

FIGURE 17.2 Classification of biodegradable polymers.
(Reprinted with permission from John, M. J.; Thomas, S. Biofibres and Biocomposites. *Carbohydr.Polym.* **2008**, *71*(3), 343–364. © 2008 Elsevier.)

Authors have produced nanocoated polypyrrole on natural cellulose fibre surface, it has been done by polymerization-induced adsorption; this process is well suitable for producing poly pyrrolenano coating on natural cellulose fibres without altering the hierarchical network structures of individual cellulose fibres.[16] The SEM morphology is shown in Figure 17.3. Another fundamental bio-composites group is dependent on agro-polymers matrixes, primarily focussed on starchy materials. Plasticized starch, known as 'thermoplastic starch' (TPS) is achieved after disturbances and also plasticization of natural starch, with water together with a plasticizer by using thermomechanical energy in a consecutive extrusion process. However, TPS displays certain disadvantages, for example, a substantial hydrophilic nature (water responsive), relatively inadequate mechanical characteristics in comparison to traditional polymers and a significant post-processing difference of the characteristics. TPS characteristics achieve equilibrium merely after few weeks. To enhance these types of material some of these factors, TPS is normally connected with supplementary compounds. The structure and life cycle evaluation of environmentally friendly composites are being primarily dealt. Green composites are being employed successfully in several applications, for example, mass-produced personal items with small life cycles or items meant for the single time or limited time use before disposal. Green composites can also be used for interior purposes with an effective life of many years. The reinforcement of bio fibres in environmentally friendly composites has been recommended. Cellulose is probably the most plentiful polysaccharide on the planet with around 109 t generated yearly. It is really an extremely arranged polymer of cellobiose (D-glucopyranosyl-β-1,4-D-glucopyranose) chains aggregated by many robust intermolecular hydrogen bonds between hydroxyl groups of surrounding macromolecules, developing cellulose microfibrils. Cellulosic materials exhibit crystalline domains (Fig. 17.4) split up by significantly less arranged types known as an amorphous domain which are possible reasons for chemical as well as biochemical attacks. Crystalline and amorphous domains are seen in natural cellulose fibres in varying quantities, because of the feature of the plant type, with the planting situations or even the section of the plant play the major role. Therefore, the features of cellulosic nanocrystals really rely mostly on the cellulose origin.[22]

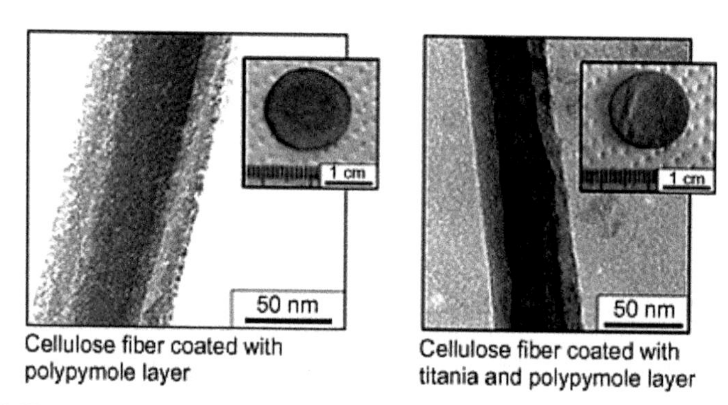

FIGURE 17.3 Scanning electron micrographs of nanocoated cellulose fibres. (Reprinted with permission from John, M. J.; Thomas, S. Biofibres and Biocomposites. *Carbohydr.Polym.* **2008**, *71*(3), 343–364. © 2008 Elsevier.)

An additional crucial parameter that played a major part in make the generation of nanofibrillated cellulose through a high-speed mixer feasible is the chemical constitution of the fiber after lignin extraction, and particularly, the remaining amount of hemicellulose left in the fibers. The higher amount of hemicelluloses is anticipated to additional take care of the fibrillation operation by avoiding fibrils aggregation throughout lignin extraction.

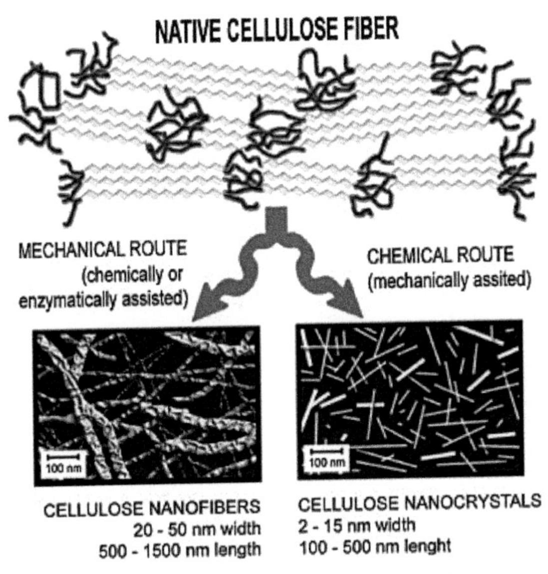

FIGURE 17.4 Crystalline and amorphous domain of cellulose, processing routes of nanocellulose from cellulose, and cellulose nanofibres and nanocrystals.
(Reprinted/adapted with permission from García, A et al. Industrial and Crop Wastes: A New Source for Nanocellulose Biorefinery. *Ind. Crops Prod.* **2016**, *93*, 26–38. © 2016 Elsevier.)

Hemicelluloses are closely restricted to the cellulose fibrils via hydrogen and covalent bonding in order to fill up space as amorphous elements between cellulose fibrils. Consequently the existence of hemicelluloses as a defensive layer encompassing the microfibrils prevents the fibrils from approaching near to aggregate as well as breakdown under the influence of interfibrillar hydrogen bonding interaction. Additionally facts of the crucial responsibility of hemicelluloses amount may be pointed out by evaluating the fibrillation yield of pulp from corn stalk with hemicelluloses amount of 15%, received via NaOH pulping method. If pretreated by TEMPO-mediated oxidation at a carboxyl amount of around 500 mol/g, the fibrillation yield failed to surpass 38% after disintegration for greater than 30 min. A schematic demonstration showing the nanoscale arrangement of the cellulose fibrils based on the hemicelluloses amount is presented in Figure 17.5. In that case if the substantial proportion of hemicelluloses is still left in the fiber after the delignification procedure, the split up of the cell wall throughout the mechanical disintegration is more convenient to apply, as well as the produced fibrils are much less aggregated with the lateral size near to the primary crystallites (roughly 3–4 nm).

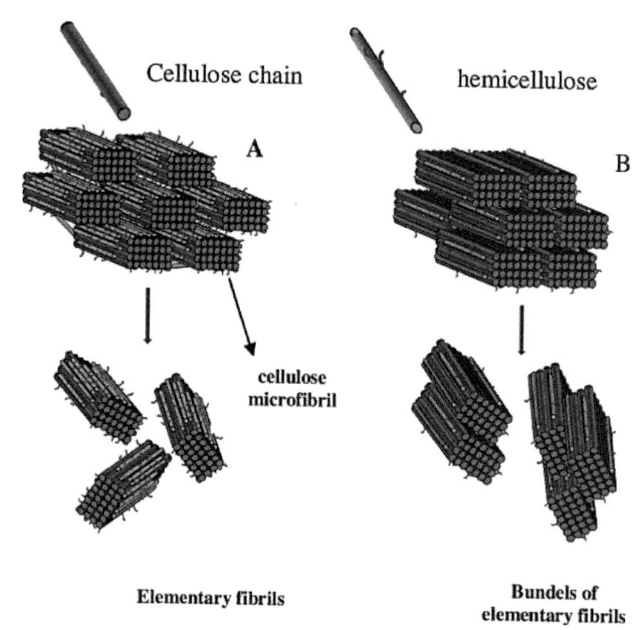

FIGURE 17.5 Schematic illustration of the aspect of the cellulose nanofibrils (CNF) according to their hemicelluloses content: (A) hemicelluloses content 30%, (B) Hemicellulose content (<20 %).

(Reprinted with permission from Boufi, S.; Chaker A.: Easy Production of Cellulose Nanofibrils from Corn Stalk by a Conventional High Speed Blender. *Ind. Crops Prod.* **2016**, *93,* 39–47. © 2016 Elsevier.)

Scanning electron microscope morphology of cellulose nanofibrils (CNF) from corn stalk by a conventional high-speed blender are shown in Figure 17.6. The nanofibrillated cellulose (NFC) was contained of individual fibrils 4–5 nm in width and length in the microscale. With increasing time of disintegration, the content of nanosized fibrils are increased by breaking down cell walls of fibres, and nanosized fibrils can be obtained up to 3 nm in width with increasing time of disintegration. This is because high-speed blender process produced the individual elementary cellulose fibrils, whereas full conversion of fibres from corn stalk into NFC by disintegration during 30 min in a high-speed blender would be favourable, when the hemicelluloses quantity of the fibres was over 30% and the carboxyl content more than the 500 μmol/g.[4] Figure 17.7 displays the variety of type cellulose sources provide nanocellulose with various morphology.

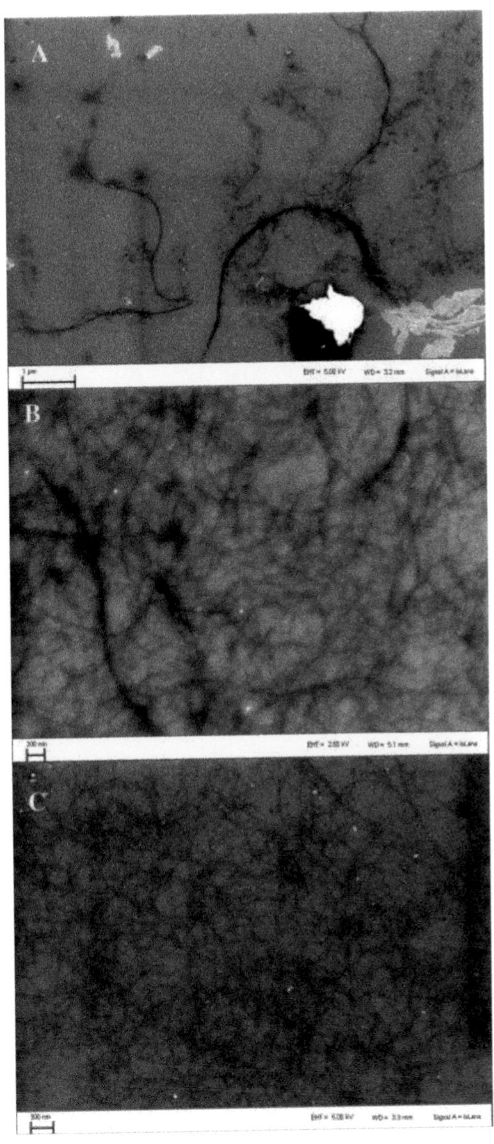

FIGURE 17.6 Field emission scanning electron microscope images of nanofibrillated cellulose (NFC) from oxidized corn pulp with carboxyl content 480 μmol/g at different time of disintegration via high-speed blender (A) after 10 min, (B) after 20 min and (C) after 30 min (prior to analysis, micro-sized fragment of fibres were extracted by centrifugation). (Reprinted with permission from Boufi, S.; Chaker A. Easy Production of Cellulose Nanofibrils from Corn Stalk by a Conventional High Speed Blender. *Ind.Crops Prod.* **2016**, *93*, 39–47. © 2016 Elsevier.)

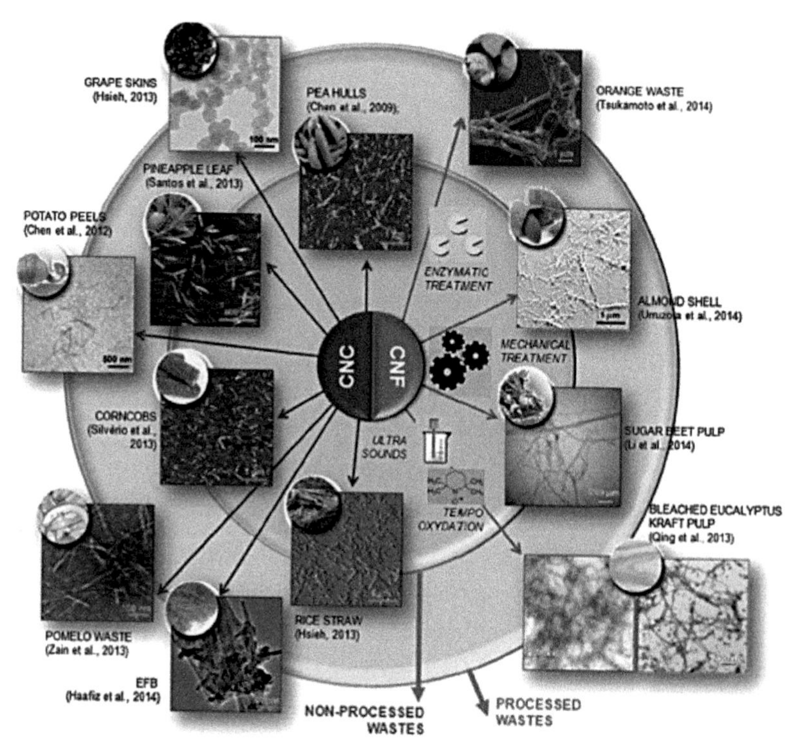

FIGURE 17.7 Microscopy images of cellulose nanofibres from crystals of non-processed and processed biowastes within the wide shape range.

(Reprinted/adapted with permission from García, A et al. Industrial and Crop Wastes: A New Source for Nanocellulose Biorefinery. *Ind.Crops Prod.***2016**, *93*, 26–38. © 2016 Elsevier.)

Scanning electron microscopic analysis on pineapple leaf fibres (PALF) are shown in Figure 17.8. Result showed the effect of processing routes on the microstructure of PALF. Author found from the environmental SEM images that fibrillation trends influenced along with the acid treatment and high pressure drop. The drop in pressure enhance the increase in the fibrillation process of the PALF whose size ranges in nanometres, and PALF fibres are in the entangled fibril form and the length to diameter ratios are in the range of 1–15 nm (Fig. 17.9).[6] Well dispersed cellulose nanofibres (15 nm in diameter) in the polyurethane matrix are shown in Figure 17.10.

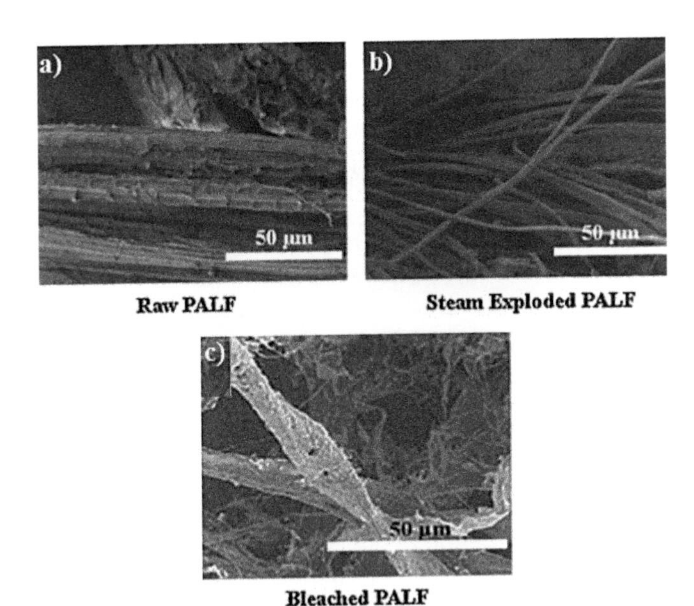

FIGURE 17.8 SEM morphology of raw, steam-exploded and bleached pineapple leaf fibres (PALF), and morphology of PALF morphology from the micro- to the nanoscale. (a) Fibrils with the diameter 50–125 μm, (b) small fibrils with the diameter 10–15 μm after steam treatment, (c) small fibrils with the diameter less than 1 μm was found with high pressure steaming coupled with acid. All the samples were coated with platinum for protecting the samples from electron beam.

(Reprinted with permission from Cherian, B. M. et al. Cellulose Nanocomposites with Nanofibres Isolated from Pineapple Leaf Fibers for Medical Applications. *Carbohydr.Polym.* **2011**, *86*(4), 1790–1798. © 2011 Elsevier.)

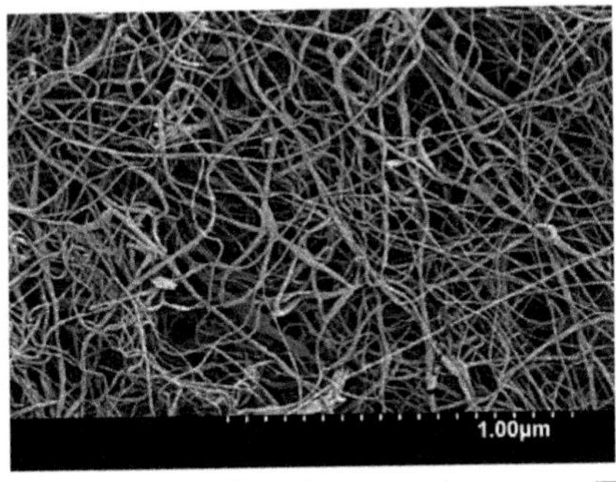

FIGURE 17.9 Environmental scanning electron microscope (ESEM) image of acid-treated PALF.

(Reprinted with permission from Cherian, B. M. et al. Cellulose Nanocomposites with Nanofibres Isolated from Pineapple Leaf Fibers for Medical Applications. *Carbohydr. Polym.* **2011**, *86*(4), 1790–1798. © 2011 Elsevier.)

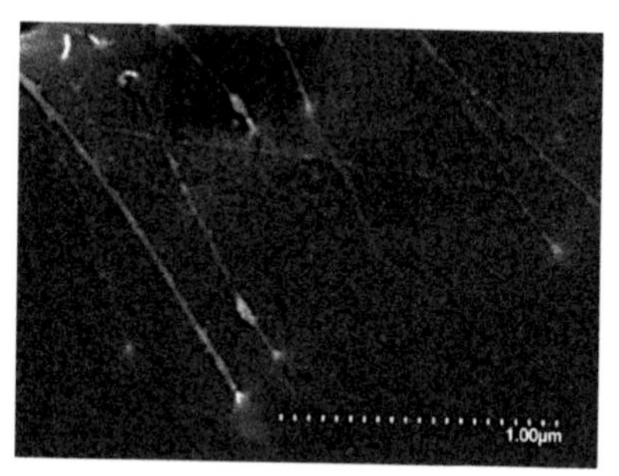

FIGURE 17.10 ESEM image of polyurethane–nanocellulose (5%) nanocomposite.
(Reprinted with permission from Cherian, B. M. et al. Cellulose Nanocomposites with Nanofibres Isolated from Pineapple Leaf Fibers for Medical Applications. *Carbohydr.Polym.* **2011**, *86*(4), 1790–1798. © 2011 Elsevier.)

Scanning electron microscope images of microfibrillated cellulose from agricultural residues and interfibrillar hydrogen bonding of cellulose nanofibre network are displayed in Figure 17.11.

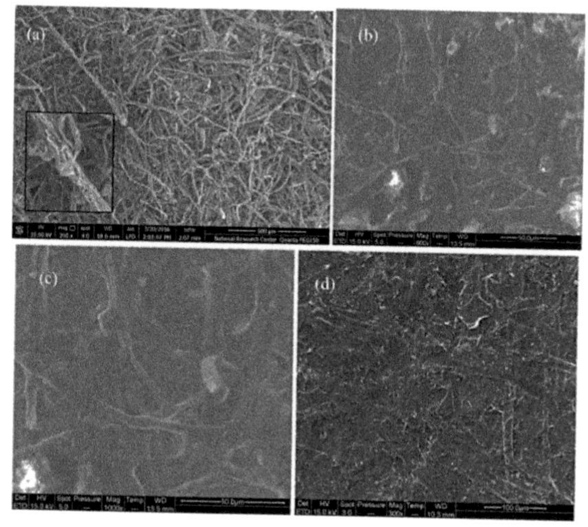

FIGURE 17.11 SEM images of (a) softwood pulp (200×) fibres tend to split in longitudinal direction as flatten fibres and they have quite compact structure (surface defect in the cellulose fibres is analysed at higher magnification), (b) MFCRE30 from rice straw and interfibrillar hydrogen bonding of cellulose nanofibre network(600×), (c) MFCBE30 from bagasse and interfibrillar hydrogen bonding of cellulose nanofibre network (1000×), (d) MFCCsE30 from cotton stalk and interfibrillar hydrogen bonding of cellulose nanofibre network (300×).

(Reprinted with permission from Adel, A. M. et al. Microfibrillated Cellulose from Agricultural Residues. Part II: Strategic Evaluation and Market Analysis for MFCE30. *Ind. Crops Prod.* **2016**, *93*, 175–185. © 2016 Elsevier.)

17.3 SCANNING PROBE MICROSCOPY

Scanning probe microscopy (SPM) is a tool that produces images of surfaces at nanoscale. A physical probe is used to scan over the surface of the samples in this tool. The SPM was developed early in the year 1981.[3,25] It has a resolution in the nanometre range and sometimes, less than a nanometre. Scanning probe microscopy imaging is based on the contact between the probe tip and the sample surface. The probe tip is mounted on the cantilever and the tip can be moved over the surface in the upward and backward (scanning) manner. Many forces and contacts such as mechanical contact, electrostatic forces, magnetic forces, chemical bonding, van der Waals forces and capillary forces deflect the cantilever during the scanning process. The distance of the deflection of the tip of the cantilever is measured by a laser and an array of photodiodes (such as the devices used in digital cameras) during scanning (movement) of the tip.

There are several varieties of SPMs (refer Fig. 17.12) as named[25] in the following:

1. Atomic force microscopes (AFMs): Measure the electrostatic forces between the cantilever tip and the sample (van der Waals forces or contact forces between a tip and the sample for measuring the sample topography or mechanical properties).
2. Magnetic force microscopes: Measure magnetic forces.
3. Scanning tunnelling microscopes (STMs): Measure the electrical current flowing between the cantilever tip and the sample (used for measuring the sample topography and electrical properties).
4. Near field scanning optical microscope (NSOM): It is used to analyse optical materials (scattered light through a sub-wavelength aperture to form an image).

For compression between these different types of SPMs, STM tip should be conducting, can be simply cut freshly by a normal wire cutter. AFM tip should be sharp enough to get a good resolution (fat-tip effect) and stiff enough to sense the atomic interaction with sample surface (the distance). AFM tip need not be conducting. Near field scanning optical microscope tip should be sharp enough to get good topography resolution; the aperture of NSOM tip should be small for better optical resolution, and to avoid artificial effects from the scanning, the outer surface of NSOM tip should be flat.

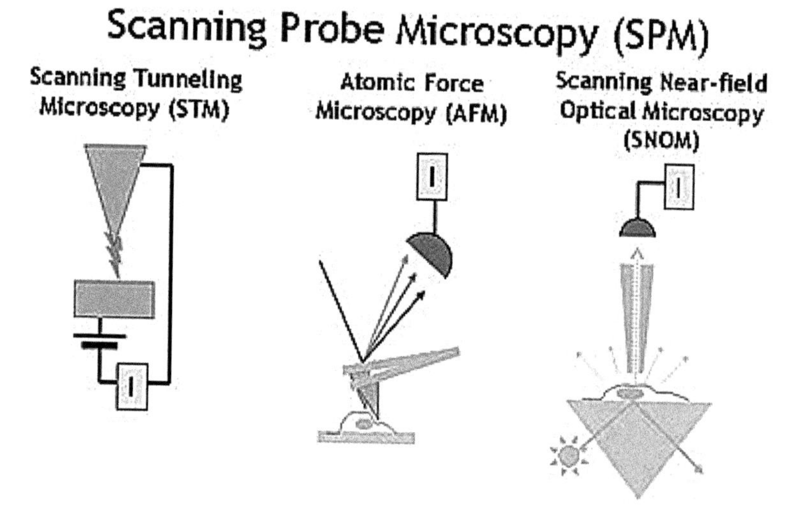

FIGURE 17.12 A schematic overview of the main types of scanning probe microscopy.
Source: (Reprinted from Nanotechnology/Scanning Probe Microscopy. https://en.wikibooks.org/wiki/Nanotechnology/Scanning_probe_microscopy (Creative Commons Attribution-Share-Alike License: https://creativecommons.org/licenses/by-sa/3.0/) (accessed Nov 24, 2016).

17.3.1 CHRONICLE ASPECT OF SPM

1981: Dr. Binnig and Dr. Rohrer invented the STM; 1982: Dr. Binnig and Dr. Rohrer are the first persons to 'see' atoms when they used an STM to create an image of a silicon sample; 1985: Dr. Binnig, Dr. Christoph Gerber (IBM Zurich Research Center) and Dr. C. F. Quate (Stanford University) developed the AFM; 1986: Dr. Binnig and Dr. Rohrer were awarded the Nobel Prize in physics for the invention of the STM; 1987: Tom Albrecht of Stanford University is the first person to use an AFM to create images of individual atoms; 1988: AFM became available commercially.[25] The SPM cantilever behaviour for quantitative measurements of microscopic tribological properties is estimated. The mentioned approach showed that a combination of finite element analysis calculations and the resonant frequency is used to measure spring constants of the SPM cantilevers.[31]

The detection of surface micromechanical properties is quite easy at submicron level resolution after the discovery of AFM and scanning force microscopy (SFM). The SFM have the capability to offer information about local surface topography, elastic surface properties, adhesive forces and shear stresses, but the only challenge is the analysis of quantitative characterization of the micromechanical with SFM. However, recently, it has been analysed that by combining optimal cantilever parameters and experimental conditions, it is possible to obtain reliable force–distance data which can correlate for further contact mechanics analysis for wide kinds of polymeric materials. In this case, Sneddon's and Hertzian models of elastic contact is useful, and a wide range of elastic polymer properties can be probed if a selection of cantilever stiffness from 0.1 to 50 N/m is there. It is also possible to analyse the dynamic behaviour of materials with correlate their microstructural properties.[14] Assembly of microcantilevers in AFM are important techniques, which can be used as force sensors for nanomaterials; however, Young's modulus is one of the key physical properties of the cantilevers. A method has been adopted to major Young's modulus of microcantilevers, which is not well suitable for normal testing techniques. The method based on the vibration model of the microcantilever studied immersed in air; study vibration behaviour of the corresponding cantilevers with the same geometry but different Young's modulus and finally estimating the resonate frequencies of the microcantilevers immersed in viscous fluid, making a correlation between experimental data with the numerical results to obtain Young's modulus of the cantilever. The surface energy data from AFM are demonstrated, in which the surface energy of various functional group and polymeric film have been measured. The result showed the low energy domains in heterogeneous surface.[36] High-resolution studies of dental tissues can be considered a sensational area of interest for biomedical applications. The piezoresponse force microscopy can be used for analysing the biological structure without special surface modification. The analysis is performed by monitoring their electromechanical behaviour because piezoelectricity is exhibited by a number of biological systems.[20] Nowadays, cellulose nanocrystals (CNCs) and fibrils are gaining more attention due to their 'green' and eco-friendly nanomaterial with superior mechanical, chemical characteristics for various polymer matrices nanocomposite and application. Scanning probe microscopy can be used for analysing the surface chemistry and mechanical properties of CNC and fibrils. However, bulk CNC and nanofibrils properties can be easily evaluated, but evaluation of individual CNC and nanofibrils is a difficult task. AFM can be used for estimating the topography, elastic and adhesive properties of individual nanocrystal and nanofibrils.[11] Collagen is one of the structural protein materials for various applications such as bone, cartilage and tendon, however, the physico-chemical properties of collagen refers to collagen hierarchical structure and length scale from micro- to nanorange. Scanning probe microscopy has been used for measuring the nanoscale structural and functional imaging of the individual collagen fibrils with nanoscale resolution and piezoelectric activity of collagen fibrils.[19] With the help of various forms of SPM, many types of nanostructured materials have been studied. Scanning probe microscopy allows new directions to researchers for evaluating the properties of individual nanostructured materials.

The surface morphology of the nanofibres surface obtained from AFM in tapping mode is displayed in Figure 17.13. The 69% yield nanofibres is received after the acid-treated PALF; acid-treated PALF is excellent mechanism of reduction in size to submicron level and fibres of certain diameters were obtained between 5 and 15 nm as shown in Figure 17.13.[6]

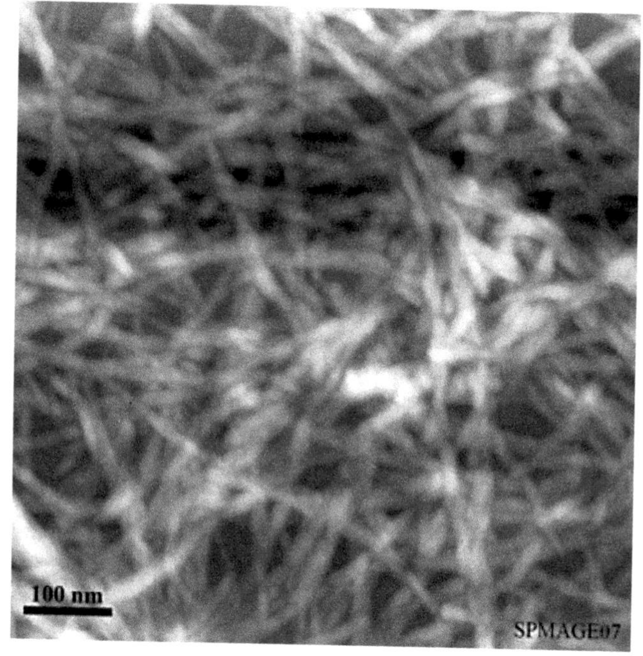

FIGURE 17.13 AFM micrographs of acid-treated PALF.

(Reprinted with permission from Cherian, B. M. et al. Cellulose Nanocomposites with Nanofibres Isolated from Pineapple Leaf Fibers for Medical Applications. *Carbohydr. Polym.* **2011**, *86*(4), 1790–1798. © 2011 Elsevier.)

Figure 17.14 displays AFM height sensor images of CNF, produced using various pretreatments, followed by homogenization in a microfluidizer. Cellulose nanofibrils diameter is estimated through the height profile of the images. Non-pretreated CNF (Fig. 17.14(a)) have a higher diameter compared with that of the chemically pretreated at nanoscale.

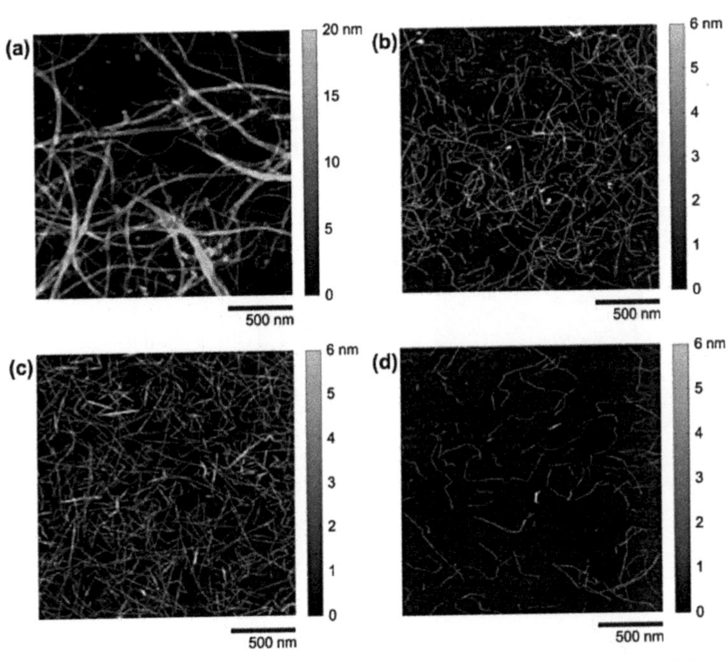

FIGURE 17.14 AFM height images of CNF, produced from hardwood kraft pulp using homogenization of: (a) non-pretreated (enzymatically hydrolysed CNF), (b) carboxymethylated (chemically pretreated), (c) carboxylated via TEMPO-mediated oxidation (chemically pretreated) or (d) quaternized cellulose (chemically pretreated).

(Reprinted with permission from Nechyporchuk, O.; Belgacem, M. N.; Bras, J. Production of Cellulose Nanofibrils: A Review of Recent Advances. *Ind. Crops Prod.* **2016**, *93*, 2–25. © 2016 Elsevier.)

The morphology of the dispersed nanofibrils was analysed by AFM imaging as shown in Figure 17.15. For AFM analysis, samples were prepared by the dilute NFC suspension and deposition on a wafer substrate, finally, dried at room temperature. All of the NFC samples contained different carboxyl groups; NFC samples exhibited nanosized thin fibrils forming entangled network. The fibrils were highly flexible, and the formation of bended zone and their length was in the range of 200 nm up to several microns. The width of the fibril was measured from the height profile and authors have found that width of individual non-aggregated fibrils exist between 3 and 5 nm. However, authors reported that the larger width distribution around 5 nm was obtained for the NFC produced from fibres with the lowest carboxyl content of 320 and over 500 μmol/g, the width level off around 3 nm, which was near to the average thickness of cellulose crystallites (around 3.5 nm), which was calculated on the basis of the Scherrer's equation. Many authors have remarked that the NFC produced through high-speed blender is composed by individual elementary cellulose fibrils.

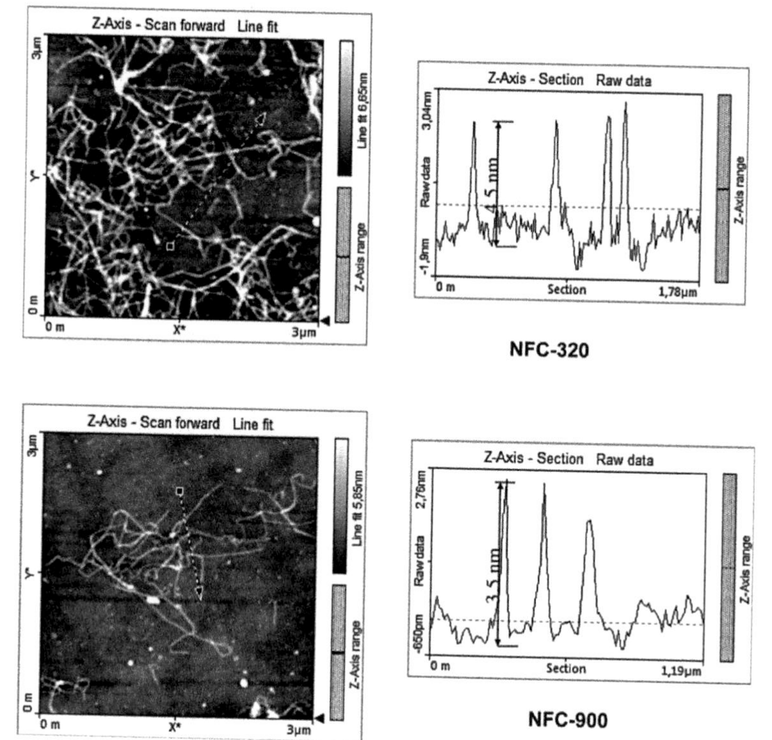

FIGURE 17.15 AFM height images of NFC prepared via high-speed blender with carboxyl content 300 and 900 μmol/g and via high pressure homogenization and the corresponding height profile analysis (the arrow marks the points used for the measurements of the height profile).

(Reprinted with permission from Boufi, S.; Chaker A. Easy Production of Cellulose Nanofibrils from Corn Stalk by a Conventional High Speed Blender. *Ind.Crops Prod.* **2016**, *93*, 39–47. © 2016 Elsevier.)

17.4 WIDE-ANGLE X-RAY DIFFRACTION

Wide-angle X-ray scattering (WAXS) or wide-angle X-ray diffraction (WAXD) is a fundamental technique that is often used for analysing the crystalline structure (atomic arrangement) of materials. Wide-angle X-ray scattering is the same technique as small-angle X-ray scattering (SAXS). However, the distance from the sample to the detector is shorter and thus diffraction maxima at larger angles are observed. It is possible to do WAXS and SAXS in a single run (small- and wide-angle scattering) depending on the measurement instrument used. The technique is time-honoured but is somewhat out-of-favour technique for the determination of the degree of crystallinity of polymer samples. The diffraction pattern generated allows determination of the chemical composition or phase composition of the film, the texture of the film (preferred alignment of crystallites), the crystallite size and presence of film stress. This method is a non-destructive method of characterization of solid materials.

This technique specifically refers to the Bragg's Law of diffraction: $2d\sin\theta = n\lambda$, where n is an integer, d is the spacing between the adjacent crystal planes, θ is the angle between incident X-ray beam and scattering plane and λ is the wavelength of incident X-ray. X-rays that are reflected from the adjacent crystal planes will undergo constructive interference only when the path difference between them is an integer multiple of the X-ray's wavelength.[12].

When X-rays incident on a sample, they are bent or diffracted in different directions depending on the locations of the atoms in the sample, and the final direction of the X-rays can be recorded. It is important that the atoms always occur in the same crystalline arrangement as the X-rays must travel through many layers of atoms. If they don't, the X-rays are bent into overlapping patterns, leaving the results fuzzy, indistinct and blur. However, if the structure has a repeating arrangement of atoms, they leave a pattern of sharp, clear spots. Different structures scatter the X-rays into different characteristic patterns. Per this method, the sample is scanned in a wide-angle X-ray goniometer and the scattering intensity is plotted as a function of the 2θ angle. When X-rays are directed in solids they will scatter in predictable patterns based on the internal structure of the solid. A crystalline solid consists of regularly spaced atoms (electrons) that can be described by imaginary planes. The distance between these planes is called the d-spacing. The intensity of the d-space pattern is directly proportional to the number of electrons (atoms) that are present in the imaginary planes. Every crystalline solid will have a particular pattern of d-spacing, which offer a 'fingerprint' of crystalline solid. In fact, solids with the same chemical contain but different phases can be identified by their pattern of d-spacing.[14] Analysis of Bragg peaks scattered to wide angles, which (by Bragg's law) implies that they are caused by sub-nanometre-sized structures.

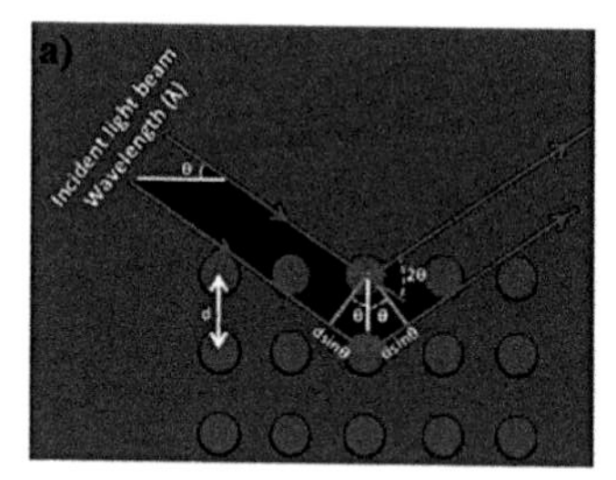

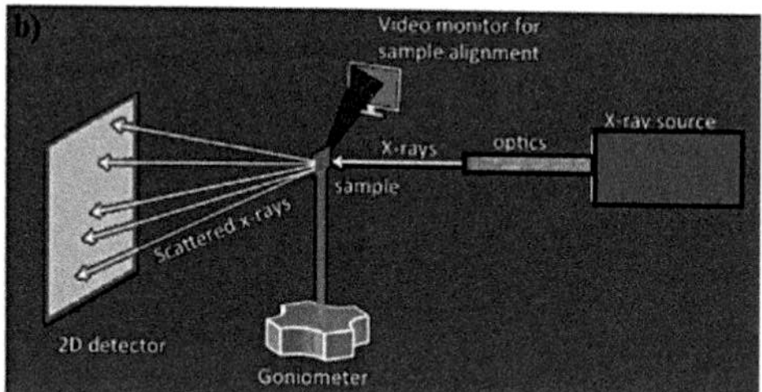

FIGURE 17.16 (a) Schematic diagram for X-ray diffraction (XRD) analysis (b) Schematic illustration of basic two-dimensional wide angle X-rays diffraction setup.

As mentioned above in the X-ray diffraction principle, the crystalline structure of materials can be analysed by this technique; X-ray diffraction characterization method gives more detailed data about the crystalline and less about the non-crystalline structure of materials compared with other techniques such as solid-state nuclear magnetic resonance, Fourier transform-infrared spectroscopy and so on. Therefore, X-ray diffraction (XRD) has been used to estimate the crystallinity and regular molecular arrangements of individual materials and composites system. If materials are not perfectly crystalline, its X-ray diffraction data shows low intensity, broad diffraction peaks, noise in the intensity scale and a large amorphous scattering distribution. The estimated X-ray powder diffraction will be continuous in appearance if we are considering the non-crystalline materials. Glassy materials are one of the solid state systems that show X-ray amorphous powder patterns. Generally, single phase non-crystalline materials have short range molecular order and show X-ray amorphous powder patterns, but they have local molecular order. To understand the chemical and physical stability of non-crystalline materials, it is essential to study the local molecular order of non-crystalline materials. The total diffraction analysis is mainly used to measures the local molecular order and structure of non-crystalline materials. The diffraction peaks show the crystalline form in the individual material and composites. In the case of polymers, polymers are never perfectly crystalline, so XRD is employed for estimating the degree of crystallinity in polymers and polymer-based system, and in the case of polymers, generally, X-ray follows the transmission mechanism because the polymer is not able to absorb X-rays in large amount.

Native maize starch (MS) granules with different amylose contents, the waxy maize starch, MSA, MSB, MSC, MSD and MSE and potato starch was represent according to the apparent amylose contents 0, 28, 40, 56, 65 and 84%, respectively. The X-ray diffractograms of native MS samples with differing amylose concentration are displayed in Figure 17.17. The corresponding X-ray diffraction parameters and crystallinity level estimated from the ratio of diffraction peak area and total diffraction area are demonstrated in Table 17.3. Amylose content reduces the degree of starch crystallinity.[5]

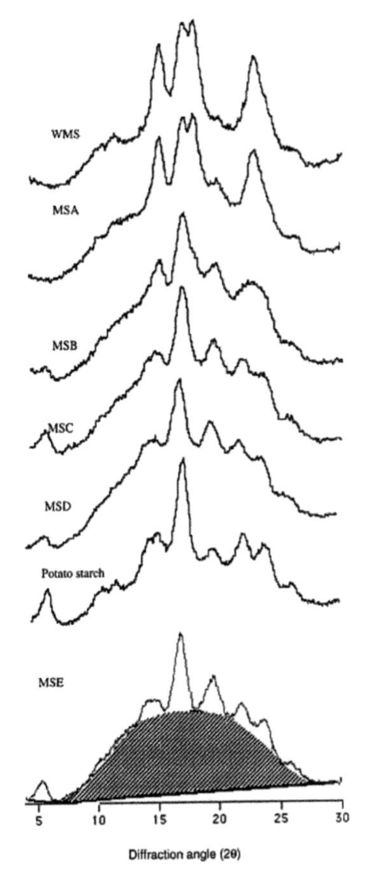

FIGURE 17.17 Wide-angle X-ray powder diffraction spectra for maize starches with different amylose contents.

Source: Adapted from[5] with permissions.

TABLE 17.3 X-ray Powder Diffraction Data of Various Maize Starch Samples Corresponding Figure 17.17.

Samples	Amylose (%)	Diffraction peaks at 2θ value (° angle)							Degree of crystallinity (%)	Crystal pattern
		5°	15°	17°	18°	20°	22°	23°		
Waxy maize starch	0	–	14.86[a] (5.96 Å)[b]	16.70 (5.3 Å)	17.84 (5.02 Å)	19.70 (4.50 Å)	–	22.86 (3.89 Å)	41.8	A
MSA	28	–	14.96 (5.91 Å)	16.96 (5.22 Å)	17.78 (4.98 Å)	19.70 (4.50 Å)	–	22.86 (3.89 Å)	30.3	A
MSB	40	5.24 (16.85 Å)	14.66 (6.04 Å)	16.80 (5.27 Å)	–	19.46 (4.56 Å)	–	22.60 (3.90 Å)	21.8	C
MSC	56	5.30 (16.66 Å)	14.46 (6.12 Å)	16.74 (5.29 Å)	–	19.50 (4.55 Å)	21.84 (4.07 Å)	23.60 (3.77 Å)	19.5	B
MSD	65	5.60 (15.77 Å)	14.68 (6.03 Å)	16.96 (5.22 Å)	–	19.60 (4.53 Å)	21.98 (4.04 Å)	23.78 (3.74 Å)	17.6	B
MSE	84	5.40 (16.35 Å)	14.42 (6.14 Å)	16.94 (5.23 Å)	–	19.60 (4.53 Å)	21.90 (4.05 Å)	23.80 (3.74 Å)	17.2	B

[a]Relative intensity.

[b]The figures in parentheses represent interplanar spacings.

(Reprinted with permission from Cheetham, N. W. H.; Tao, L. Variation in Crystalline Type with Amylose Content in Maize Starch Granules: An X-Ray Powder Diffraction Study. *Carbohydr. Polym.* **1998**, *36* (4), 277–284. © 1998 Elsevier.)

The synthetic polymers exist in oriented or polycrystalline form. Therefore, diffraction pattern refers to the powder pattern in the case of polycrystalline and show the fibre pattern in the case of oriented polycrystalline form. The size of crystallites domain in the polymer are determined by the Scherrer equation and size of crystallites domain are varied in the nanoscale range, and due to the giant structure of the polymer, Hermans orientation function is used to determine the crystalline orientation. If polymer crystals have many defects, the diffraction pattern will be in broad in nature.[18] The polymer crystallites are present in the nanoscale range. They also have high surface to volume ratio, which also take part in diffraction pattern. Considering diffraction and scattering, it may be said that both are two observation methods. But in both cases, the interference of electromagnetic radiation gives the nature of materials structure. According to Bragg's law, the structural size is inversely proportional to a reduced scattering angle, therefore high angle refers to smaller structure, whereas low angle for large structure. Considering the small-angle scattering is used to measure colloidal to nanoscale sizes. There is no large-scale limit to diffraction, whereas Bragg's law has certain size limitation; the smallest measurable size should be ($\lambda/2$). X-ray scattering from amorphous material formed a 'halo' of intensity, which when integrated obtains a broad, low-intensity 'hump'. However, X-ray scattering from a crystalline material produces well-defined spots or rings which integrate to sharp, higher-intensity peaks. Per cent crystallinity is the ratio of the intensity of crystalline peaks to the total intensity (sum of the crystalline and amorphous intensities). The lamellar architecture of semi-crystalline growth rings is only detected by small-angle scattering in the case of semi-crystalline polymers/composites system. However, amorphous polymers and amorphous phase in semi-crystalline polymers or polymer-based composites show the diffuse X-ray scattering. Intensity versus θ graph of XRD shows absence of a sharp peak for amorphous nature. However, crystalline structure show sharp and intense peak. Therefore, fibre-reinforced composites show better peak and intensity than amorphous (Fig. 17.18).

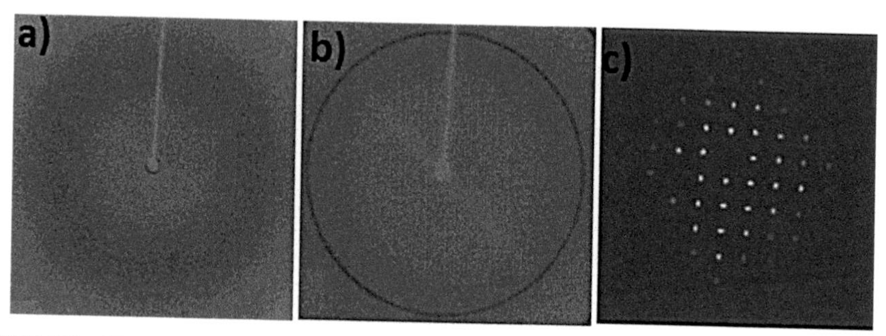

FIGURE 17.18 Scattering pattern of (a) amorphous polymers (b) polycrystalline biopolymers fibres, (c) single crystalline materials.

In Figures 17.19–17.21, X-ray results indicate that the electrospun Poly L Lactic Acid (PLLA) nanofibre is completely non-crystalline, while having highly oriented chains. Figure 17.19 indicates the 2D WAXD pattern of an electrospun non-woven PLLA membrane. No crystalline peak is displayed, which offer the evidence that electrospinning almost completely decelerate the crystallization process of PLLA (the fully crystallized PLLA usually contain about 30% crystallinity with distinct crystalline reflections in WAXD). This suggests that polymer chains are highly oriented, but polymer chains are non-crystalline in the electrospun nanofibres (it is due to lack of time during solidification, means, stretched chains do not have enough time to organize into suitable crystal prior they are solidified). Figure 17.20 displays the integrated intensity profile of the 2D WAXD pattern in Figure 17.19. The two dotted-line peaks express the Gaussian fits of the amorphous background using the programme Grams/32 Spectral Notebase™ (it is also noted that a single Gaussian or Lorentzian peak cannot fit the amorphous profile). Authors pointed out that the observed amorphous scattering profile from the electrospun PLLA membrane is not the similar as that from the quenched amorphous PLLA film. It has been observed that in electrospun fibre, the chains are highly oriented, but in the quenched amorphous film, the chains are in the random-coil state.[5]

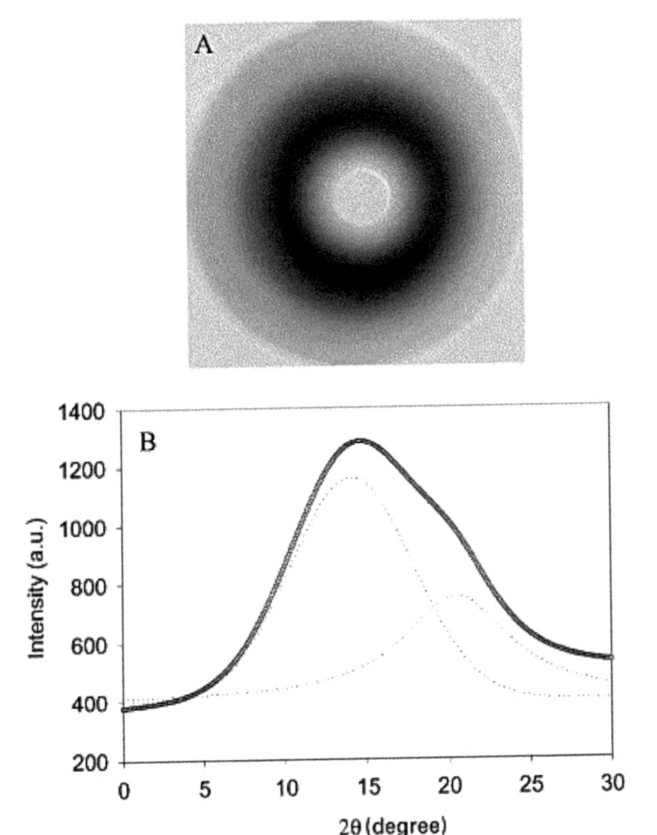

FIGURE 17.19 (A) Wide-angle X-ray diffraction (WAXD) pattern of electrospun Poly L Lactic Acid membranes dried at room temperature; (B) the integrated intensity profile from the WAXD pattern (the two dotted-line peaks represent the Gaussian fits for the amorphous background using the programme Grams/32 Spectral Notebase™).

(Reprinted with permission from Zong, X. et al. Structure and Process Relationship of Electrospun Bioabsorbable Nanofiber Membranes. *Polymer*. **2002**, *43*(16), 4403–4412. © 2002 Elsevier.)

After annealing at 55°C for 24 h, the electrospun PLLA membrane shows two strong crystalline reflection peaks (Fig. 17.20). The integrated intensity profile of the 2D WAXD pattern is also displayed in Figure 17.18. It can be notified as 110 and 131 reflections, which is two peaks at 2θ values of 16.4 and 18.7°. Author has mentioned that the α form of the PLLA crystal having a pseudo-orthorhombic unit cell with dimensions

($a=1.07$, $b=0.595$ and $c=2.78$ nm). This unit cell structure include of two 10_3 helices. In addition, the electrospun PLLA fibres were in a metastable state, it was confirmed by the diffraction peaks of the annealed samples, as shown in Figure 17.20.

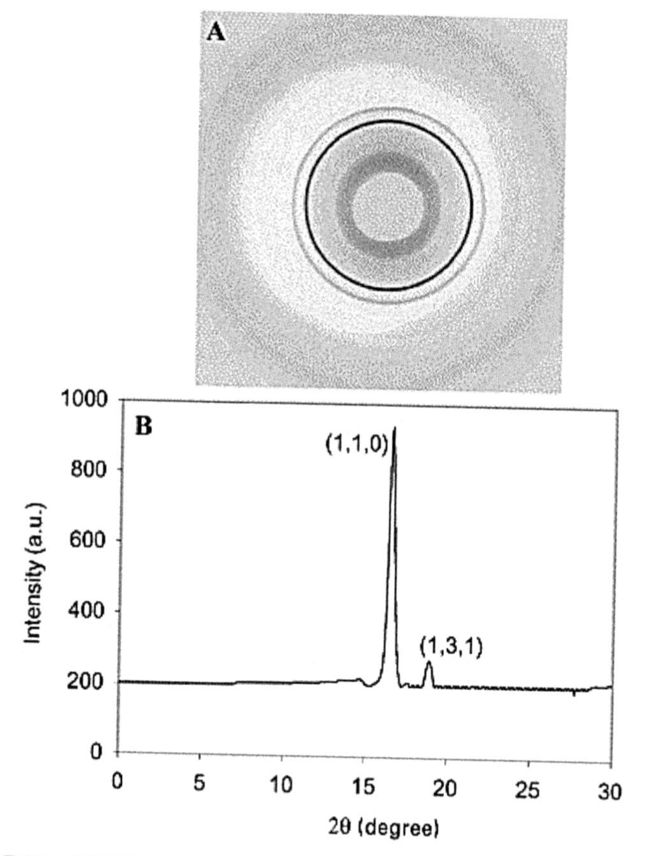

FIGURE 17.20 (A) WAXD pattern of electrospun PLLA membranes annealed at 55 8C for 24 h; (B) the integrated intensity profile from the WAXD pattern.

(Reprinted with permission from Zong, X. et al. Structure and Process Relationship of Electrospun Bioabsorbable Nanofiber Membranes. *Polymer*. **2002**, *43*(16), 4403–4412. © 2002 Elsevier.)

Figure 17.21(a) is the SAXS pattern of the as-spun PLLA membrane. The SAXS pattern of the annealed sample at 55°C for 24 h is mentioned in Figure 17.21(b). Both patterns indicate strong diffuse scattering near the beam stop. As the electrospun fibres are randomly oriented in the membrane, it is easy to understand that the pattern show the scattering from fibres of all directions. The authors came to the conclusion that the scatterers in the electrospun PLLA fibres have a microfibrillar structure, but they do not contain shish–kebab microstructure. Electrospinning was found to significantly retard the crystallization of PLLA. The amorphous phase in electrospun PLLA membrane is probably not in a pure amorphous phase. Some metastable states, such as oriented chains with no helical structures, may exist in between the amorphous and crystalline states of electrospun PLLA fibres.[40]

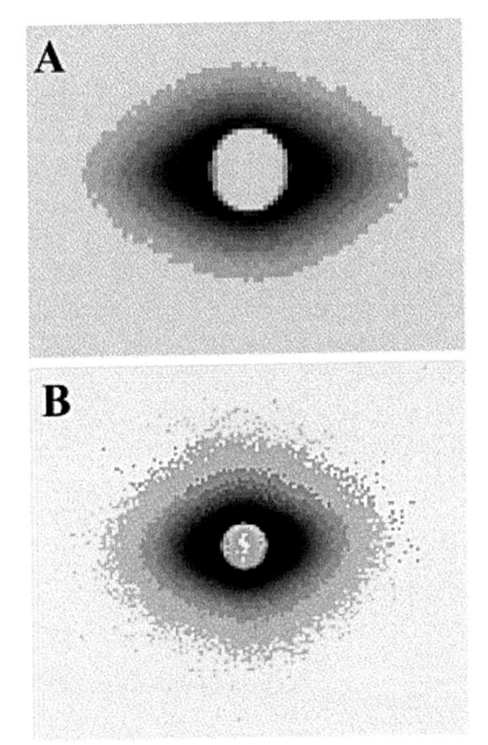

FIGURE 17.21 SAXS patterns of (A) electrospun PLLA membranes dried at room temperature; (B) membrane annealed at 55°C for 24 h.

(Reprinted with permission from Zong, X. et al. Structure and Process Relationship of Electrospun Bioabsorbable Nanofiber Membranes. *Polymer*. **2002**, *43*(16), 4403–4412. © 2002 Elsevier.)

The drawing process of polymer always enhances the intensity, and a sharp peak of polymeric system is obtained. XRD is also useful to characterize exfoliated nanocomposite based on d-spacing of nanoclay intercalated biodegradable polymer composites systems. But, this analysis is only possible with sufficient clay loadings. In addition, the d-spacing of clay in polymer materials has been used to describe the nanoscale dispersion of the clay. The types of dispersion are immiscible, intercalated and exfoliated in nature. No biopolymer chain enters the gallery spacing between clay layers for an immiscible system. Therefore, changes in d-spacing of clay between layers are not observed. But in the case of intercalated nanocomposites, the d-spacing is increased; it attributes to fill the polymer chain in between clays galleries. If d-spacing is increased tremendously, there will be no peak in the WAXD pattern, which suggests that the exfoliated types are composites present in the system and if clay is present in the highly-disordered state, then also XRD peak will not be found. In Figure 17.22 shows the transmission electron microscopy (TEM) microstructure of Cloisite 15A (intercalant: quaternary ammonium salt; intercalant content: 125 meq/100 g; interlayer spacing: d001 = 3.15 nm) and Cloisite 30 B (intercalant: quaternary ammonium salt; intercalant content: 90 meq/100 g; interlayer spacing: d001 = 1.85 nm) and polypropylene (PP) nanocomposites. The interlayer spacing of the less-ordered particles in Figure 17.4(c), (d) is significantly greater than in ordered particles. Irregularity in the interlayer spacing to larger distance is mentioned to be the effect of the penetration of PP into the interlayer spacing of the particles that occurs and pronounced to the formation of small 2–3 platelets thick particles.[28]

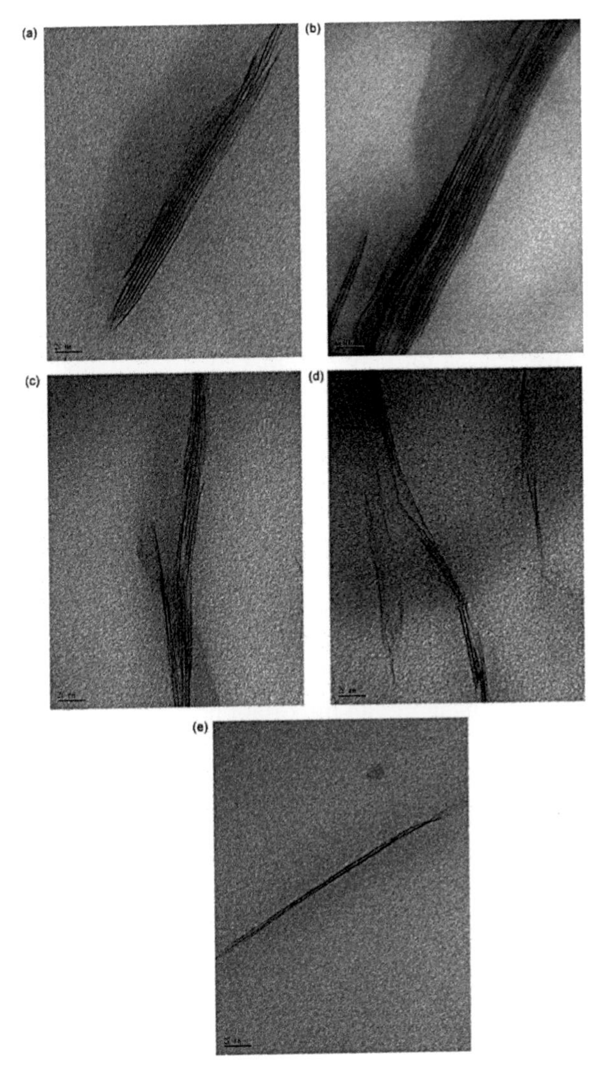

FIGURE 17.22 High magnification transmission electron microscopy micrographs of clay intercalated polymer nanocomposites. Clay particles consist of well-aligned and regularly-spaced platelets: (a) PP/15A (2 w %) /MA9k (coupling agent 4 w %), (b) PP/15A (2 w %) /MA9k (coupling agent 4 w %), wider range of alignment and order, from well-aligned ordered platelets to disordered platelets, (c) PP/15A (2 w %)/MA330k (coupling agent 4 w %), (d) PP/15A (2 w %) /MA330k (coupling agent 4 w %), small particles of 2–3 platelets, (e) PP/15A (2 w %) /MA330k (coupling agent 4 w %).

(Reprinted with permission Perrin-Sarazin, F. et al. Micro- and Nano-Structure in Polypropylene/Clay Nanocomposites. *Polymer*. **2005**, *46*(25), 11624–11634. © 2005 Elsevier.)

In Figure 17.23(a), natural montmorillonite clay gives a diffraction peak at $2\theta = 7.3°$, which writes to an interlayer spacing of 1.2 nm. Peak fitting study for organo-modified clays (clays 15A and 30B) reveals the existence of a low intensity peak at the similar angle (peak 3 in Table 17.4), as well as displays the behaviour of two higher intensity peaks at lower angles (peaks 1 and 2 in Table 17.4). The organic modification therefore contributes to, as expected, an increase of the interlayer spacing adequate to the appearance of the intercalant between the clay platelets but also to a wider distribution of the interlayer spacing. Authors explained that the highest intensity peak is for clay 15A placed at lower angle than for clay 30B, hence suggesting that higher clay intercalation is achieved in 15A ($d_{001} = 3.3$ nm) compared with 30B ($d_{001} = 1.8$ nm).[28]

In PP/15A, the maximum intensity peak is displaced to higher angle estimated relatively to that of clay 15A indicating that the interlayer spacing of clay 15A leads to depression once disposed in the polymer. Low thermal stability of the clay intercalant leading to clay collapse during mixturing may be responsible for this interlayer spacing reduction. When a coupling agent is applied in the nanocomposites, such experience is greatly decreased demonstrating that despite the probable intercalant degradation, the existence of coupling agent shows an important part on the intercalation. This is the case for PP/15A/MA9k, PP/15A/MA330k and m (PP/15A)/MA330k for which the interlayer spacing measured from the highest intensity peak is above 3.0 nm, higher than for PP/15A (Table 17.4). Nevertheless, their peak shape is relatively distinct as demonstrated in Figure 17.23(a). An intense and sharp peak (FWHM=0.6°) is seen at relatively small angles for PP/15A/MA9k, whereas a smaller and broader peak (FWHM=1.4°) displaced between PP/15A, and the latter is found for both PP/15A/MA330k and m (PP/15A)/MA330k. The appearance of this broad peak represents a wider distribution (i.e. range) of interlayer spacing probably referred to a larger heterogeneous intercalation. The relative high intensity of peak 3 achieved for these nanocomposites (Table 17.4) infers that less clay appears to support intercalation. However, for PP/15A/MA330k, the presence of nano-level particles detected by TEM (Fig. 17.22(c)) implies that intercalation possibly happened to a different term.[28]

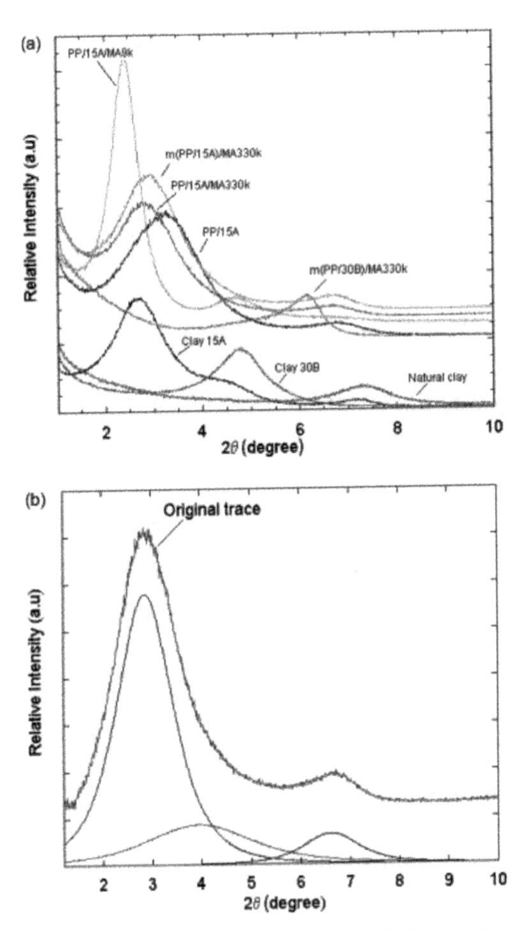

FIGURE 17.23 XRD patterns (a) original traces for all clays and nanocomposites and (b) example of peak fitting.

(Reprinted with permission Perrin-Sarazin, F. et al. Micro- and Nano-Structure in Polypropylene/Clay Nanocomposites. *Polymer*. **2005**, *46*(25), 11624–11634. © 2005 Elsevier.)

TABLE 17.4 X-ray Diffraction Result.

	From peak fitting					
	Interlayer d_{001} spacing (nm) (intensity %)			Full width at half maximum (deg) of the 100%-intensity peak	Mean particle size t (nm) (from Eq. (1))	Number of platelets per particle (from Eq. (2))
	From peak 1	From peak 2	From peak 3			
Natural montmorillonite			1.2 (100)	1.3		
Clay 15A	3.3 (100)	2.0 (23)	1.2 (8)	1.2		
Clay 30B	2.3 (20)	1.8 (100)	1.5 (8)	1.0		
PP/15A	3.3 (23)	2.6 (100)	1.3 (11)	1.4	5.5	2.7
PP/15A/MA9k	3.6 (100)	1.9 (10)	1.3 (2)	0.6	12.7	4.2
PP/15A/MA330k	3.1 (100)	2.2 (15)	1.3 (11)	1.3	6.0	2.6
m(PP/15A)/MA330k	3.0 (100)	2.1 (11)	1.3 (12)	1.4	5.5	2.5
m(PP/30B)/MA330k	1.8 (41)	1.5 (44)	1.4 (100)	0.5	15	10.8

(Reprinted with permission Perrin-Sarazin, F. et al. Micro- and Nano-Structure in Polypropylene/Clay Nanocomposites. *Polymer*. **2005**, *46*(25), 11624–11634. © 2005 Elsevier.)

17.5 X-RAY PHOTOELECTRON SPECTROSCOPY

To analyse the composition and electronic state of the surface section of a sample, photoelectron spectroscopy uses photo-ionization and examination of the kinetic energy (KE) distribution of the emitted photoelectrons. In general, when the procedure has been employed for surface investigations, it has been divided per the source of exciting radiation into X-ray photoelectron spectroscopy (XPS), a soft X-rays (with a photon energy of 200–2000 eV) is applied to analyse the core-levels, and Ultraviolet photoelectron spectroscopy (UPS) can be used by vacuum UV radiation (with a photon energy of 10^{-45} eV) to look at valence levels. Photoelectron spectroscopy is established upon a single photon in/electron out process. This controlling of this process is a much simpler experience than the Auger spectroscopy. According to the Einstein relation:

$$E = h\nu$$

where, h is Planck constant (6.62×10^{-34} J s) and ν is frequency (Hz) of the radiation.

Photoelectron spectroscopy is based on the monochromatic sources of radiation (i.e. photons of fixed energy). In XPS, the photon is received by an atom in a molecule or solid, contributing to ionization and the ejection of a core (inner shell) electron. In UPS, the photon interacts with valence levels of the molecule or solid, leading to ionization by removal of one of these valence electrons. The KE distribution of the emitted photoelectrons (i.e. the number of emitted photoelectrons as a function of their KE) can be measured using any appropriate electron energy analyser, and a photoelectron spectrum can be recorded. The process of photoionization can be regarded in several ways: one way is to look at the overall process as follows:

$$A + h\nu \rightarrow A^{+} + e^{-}$$

Energy balance: $E(A) + h\nu = E(A^{+}) + E(e^{-})$

As the electron's energy is present solely as KE, this can be rearranged to give the following expression for the KE of the photoelectron:

$$KE = h\nu - (E(A^{+}) - E(A))$$

The final term in brackets, representing the difference in energy between the ionized and neutral atoms, is generally called the binding energy (BE) of the electron—that leads to the following commonly quoted equation:

$$KE = h\nu - BE$$

The BEs of energy levels in solids are conventionally determined with regard to the Fermi level of the solid. This suggests a slight change to the equation that is presented above; this correction is called as the work function (φ) of a solid, but this correction will be omitted for the means of the study.

The essential conditions for a photoemission analysis (XPS or UPS) are as follows:

1. Origin of fixed-energy radiation (an x-ray source for XPS or, generally, a He discharge lamp for UPS)
2. Electron energy analyser (which can spread the expelled electrons corresponding to their KE and thoroughly estimate the flux of expelled electrons of a specific energy)
3. High vacuum environment (to make possible the ejected photo-electrons to be examined without interruption from gas phase collisions).
4. For each element, there will be a characteristic BE compared with each core atomic orbital that is each element will produce to a characteristic set of peaks in the photoelectron spectrum, and kinetic energies is figured out by the photon energy and the corresponding binding energies.

The intensity of the peaks is described as the concentration of the individual element within the sample. Thus, the method gives a quantitative evaluation of the surface composition and is occasionally known as the Electron Spectroscopy for Chemical Analysis.

The most commonly employed X-ray sources are those giving rise to

$$\text{Mg K}\alpha \text{ radiation: } h\nu = 1253.6 \text{ eV}$$

$$\text{Al K}\alpha \text{ radiation: } h\nu = 1486.6 \text{ eV}$$

In polymer science and technology field, the objective of X-ray photoemission spectroscopy analysis is to demonstrate the chemical constitution of polymer blends and composites in few nanometre scale. But it is essential to accomplish accurate information from XPS. These results consistently depend on several aspects such a photo ionization cross sections, inelastic electron mean free paths, influence of elastic electron scattering. The peak intensity of XPS depends on the in-depth distribution of atoms, and in-depth distribution of atoms has unknown parameter in XPS.

Nowadays, XPS have achieved remarkable interest to receive the above-quoted information; XPS investigation has been used to analyse the chemical bonding of the plasma polymerized nanocomposites polyethylene oxide (PEO) film, which have been built for high-voltage insulation. The O/C proportions of the PEO-tapes are mentioned the 0.27 for 5 W plasma, 0.31 for 10 W, 0.2 for 15 and 0.23 for 20 W. It has been published that in these kinds of film, 10 W plasma polymerizations are fairly satisfactory than other plasma process. It was stated that the peak is at 286.5 eV conforming to the ether bonds, which is supposed to be the characteristic peak of general PEO polymers that was much (45.97%) at 10 W among all the plasma state. However, change in hydrophilicity have been described in terms of polar sites, which are produced by the carboxyl composition, and hydrophilicity was the completion to existence of hydrocarbon content, which is apt to the nonpolar groups, as observed in Figure 17.24.[40] C1s Binding Energy (BE) and Surface Concentration from the X-Ray Photoelectron Spectroscopy (XPS) is mentioned in the Table 17.5. Figure 17.25 show the 1s peaks of the deconvoluted XPS spectra of the pure epoxy resin, the nanocomposite with as-received nanoparticles, and the nanocomposite with plasma polymer coated nanoparticles.

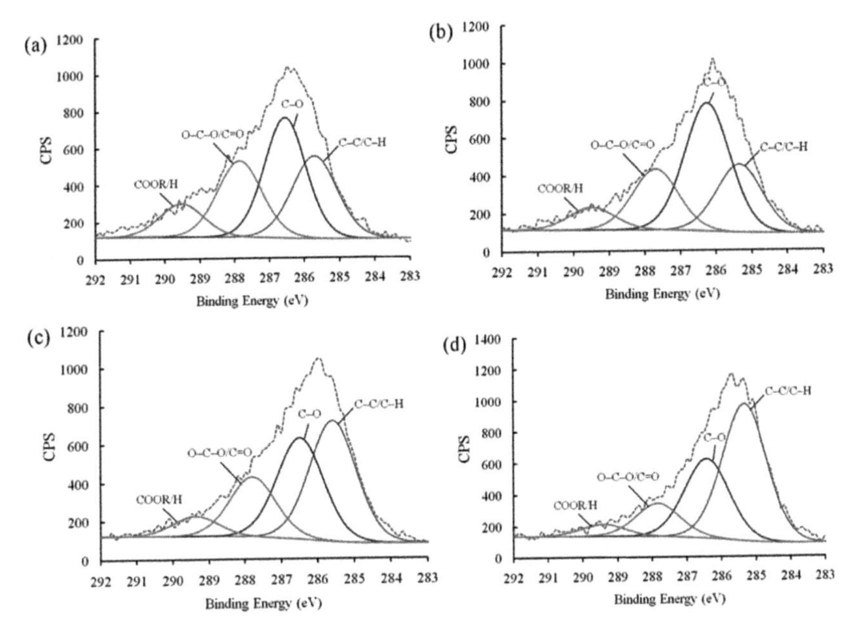

FIGURE 17.24 X-ray photoelectron spectroscopy (XPS) spectra of C 1s peaks of plasma polymerized polyethylene oxide-like film at (a) 5 W, (b) 10 W, (c) 15 W and (d) 20 W for 4 min.

(Reprinted from Yan, W.; Han, Z. J.; Phung, B. T.; Faupel, F.; Ostrikov, K. High-Voltage Insulation Organic-Inorganic Nanocomposites by Plasma Polymerization. *Materials*. **2014**, 7(1), 563–575. http://creativecommons.org/licenses/by/3.0/)

TABLE 17.5 C Spectra of the Plasma Polymerized Polyethylene Oxide-Like Film at 5, 10, 15, and 20 W.

BE (eV)	Chemical bonds	Concentration (%)			
		5 W	10 W	15 W	20 W
289.2	COOR/H	11.08	8.07	6.66	4.51
288	O–C–O/C=O	25.16	21.78	19.91	12.62
286.5	C–O	36.67	45.97	33.46	30.36
285	C–C/C–H	27.08	24.19	39.96	52.52

(Reprinted from Yan, W.; Han, Z. J.; Phung, B. T.; Faupel, F.; Ostrikov, K. High-Voltage Insulation Organic-Inorganic Nanocomposites by Plasma Polymerization. *Materials.* **2014**, 7(1), 563–575. http://creativecommons.org/licenses/by/3.0/)

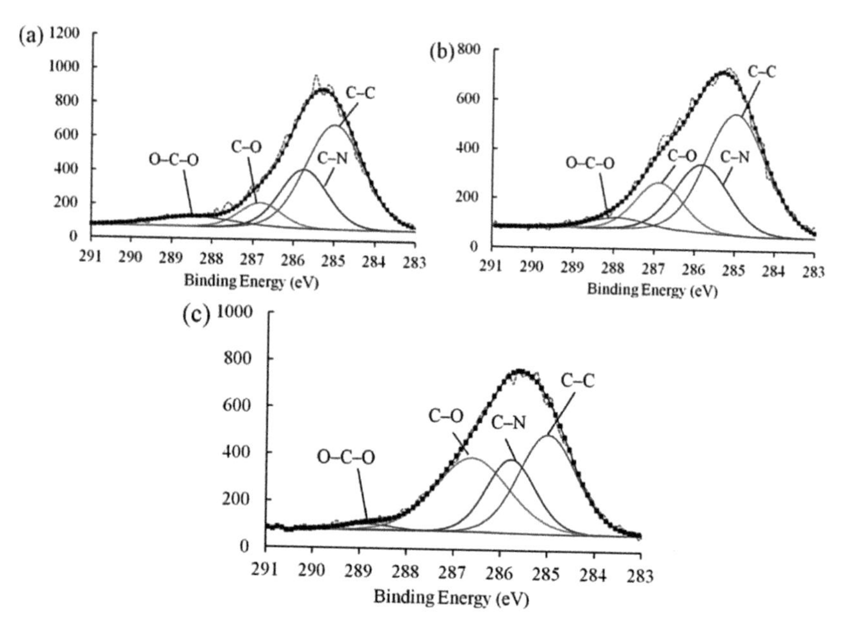

FIGURE 17.25 C 1s peaks of the deconvoluted XPS spectra of (a) the pure epoxy resin; (b) the nanocomposite with as-received nanoparticles; and (c) the nanocomposite with plasma polymer coated nanoparticles.

(Reprinted from Yan, W.; Han, Z. J.; Phung, B. T.; Faupel, F.; Ostrikov, K. High-Voltage Insulation Organic-Inorganic Nanocomposites by Plasma Polymerization. *Materials.* **2014**, 7(1), 563–575. http://creativecommons.org/licenses/by/3.0/)

TABLE 17.6 Characteristics of the Peaks in C 1s Spectra for Pure Epoxy Resin, Nanocomposite with as-Received Nanoparticles, and Nanocomposite with Plasma Polymer Coated Nanoparticles. (*Source:* Adopted from[39] with permission.)

Sample	BE (eV)	Concentration (%)	Groups
Pure epoxy resin	285.00	56.86	C–C/C–H
	285.76	26.47	C–N
	286.82	9.49	C–O
	288.49	7.18	C–O–C
Nanocomposite with as-received silica	285.00	53.08	C–C/C–H
	285.83	26.67	C–N
	286.89	16.14	C–O
	287.93	4.11	C–O–C
Nanocomposite with plasma polymer coated silica	285.00	37.96	C–C/C–H
	285.78	24.02	C–N
	286.61	36.15	C–O
	288.84	1.87	C–O–C

For determining the qualifications of MWNT-C_{60} composites, X-ray photoelectron spectroscopy study has been carried out.[30] They commented about two kinds of composites, one is the immersion of nanotubes into fullerene solution in CS2 and other is the modification of these composites by laser. Authors have described the full-width half maxima and symmetry for both MWCNT and C_{60}. In addition, it has been indicated that the C1s spectra of composites are dissimilar from the simple addition of MWNT and C_{60} characteristics.[18] Authors presented that the XPS is very useful techniques to investigate the polymer and inorganic interface. It is also a very useful procedure to analyse the content and interphase structure at the molecular level.[17]

X-ray photoelectron spectroscopy can be applied for explaining the chemistry of waterborne polyurethane film. The XPS results indicated that appropriate structure for the interaction of silane ingredient on the waterborne polyurethane film surface and cross-link ability of chain during film formation process can be analysed. The silane is quite useful for promoting the surface contact angle to water and water adsorption of the waterborne polyurethane film coating films, as shown in Table 17.7 and Figure 17.26.[21] The XPS can be applied to evaluate the O/C ratios and C1 percentage, lignin, extractives, hydrophilic groups and carbohydrate on the surface of pulp fibre. The fungal treated eucalyptus chemo-thermomechanical pulp fibres exhibit higher O/C ratios, lower C1 percentage, lower lignin and extractives and more hydrophilic groups and carbohydrate on the surface, which is appropriate for producing hydrogen bond than unmodified fibres.[15]

TABLE 17.7 XPS Results of Hybrid WPU Films.

Sample	Concentration (%)			Si Concentration (%)	
	C	N	O	Theoretical	Experimental
S2	66.47	0.62	29.99	0.10	2.92
S3	65.81	0.77	29.49	0.27	3.93
S4	65.15	0.51	29.97	0.39	4.37
S5	66.99	1.32	25.48	0.79	6.21
S6	68.74	0.91	22.93	1.07	7.42
S7	71.03	1.79	19.09	1.31	8.09

(Reprinted with permission from Li, Q.; Guo, L.; Qiu, T.; Xiao, W.; Du, D.; Li, X. Synthesis of Waterborne Polyurethane Containing Alkoxysilane Side Groups and the Properties of the Hybrid Coating Films. *Appl.Surf.Sci.* **2016**, *377*, 66–74. © 2016 Elsevier.)

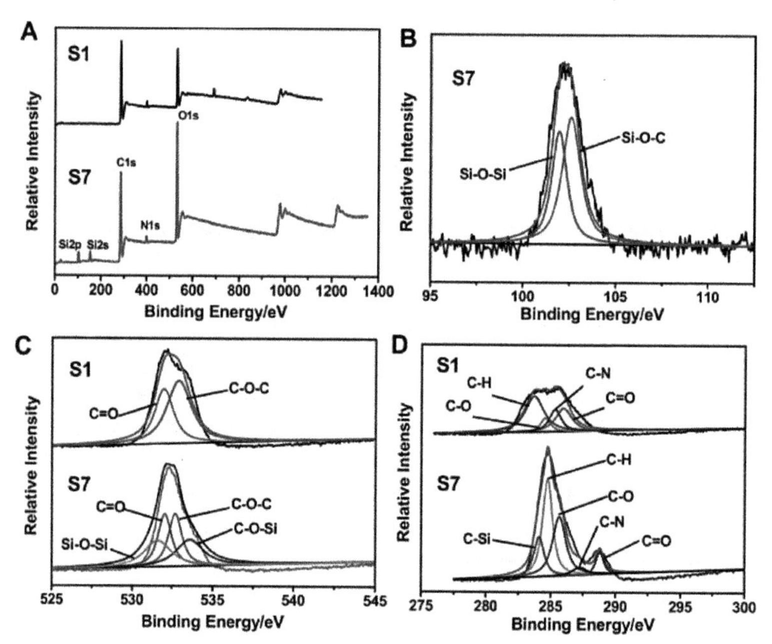

FIGURE 17.26 XPS spectra of (A) survey spectrum (S1) reveals the characteristic signal of carbon (C 1s), nitrogen (N 1s) and oxygen (O 1s) at about 285, 400 and 533 eV, respectively, comparatively, there are additional signals of silicon (Si 2p at about 102 eV and Si 2s at about 153 eV in sample S7); (B) Si 2p (peaks at 101.95 and 102.6 eV are the binding energy of Si–O–Si and Si–O–C, respectively); (C) O 1s; (D) C 1s of neat polyurethane film (S1) and hybrid polyurethane film (S7).

(Reprinted with permission from Li, Q.; Guo, L.; Qiu, T.; Xiao, W.; Du, D.; Li, X. Synthesis of Waterborne Polyurethane Containing Alkoxysilane Side Groups and the Properties of the Hybrid Coating Films. *Appl.Surf.Sci.* **2016**, *377*, 66–74. © 2016 Elsevier.)

X-ray photoelectron spectroscopy can provide the clue about the existence of functional group on the surface of natural fibre in composite system. X-ray photoelectron spectroscopy based study has illustrated the larger polarity of flax fibres corresponding to coir fibres, and alkali treatment can improve the fibre surface energies and polarity of the flax fibres.[37] The surface morphology of silver ions based electrospun polylactide PLA porous membrane can be examined by the XPS. These electrospun membranes are well suited for antibacterial wound dressing materials.[33] The above mentioned considerations confirm the usefulness of XPS, and consequently, it could be accomplished that the XPS is a sophisticated analysis approach to evaluating several surface chemistry of materials.

17.6 CONCLUSION

The current chapter aims is to a comprehensive review of the different microscopic and spectroscopic characterization of eco-friendly polymer composites and nanocomposites. The most important goal of microscopy and spectroscopy is to observe and understand the composite materials with respect to their sizes, shapes and compositions at different length scale.

KEYWORDS

- **natural fibre**
- **biopolymer**
- **spectroscopy**
- **microscopy**
- **functional properties**

REFERENCES

1. Adel, A. M. et al. Microfibrillated Cellulose from Agricultural Residues. Part II: Strategic Evaluation and Market Analysis for MFCE30. *Ind. Crops Prod.* **2016,** *93*, 175–185.

2. Alhadlaq, A.; Tang, M.; Mao, J. J. Engineered Adipose Tissue from Human Mesenchymal Stem Cells Maintains Predefined Shape and Dimension: Implications in Soft Tissue Augmentation and Reconstruction. *Tissue Eng.* **2005,** *11*(3–4), 556–566.
3. Arshad, M.; Kaur, M.; Ullah, A. Green Biocomposites from Nanoengineered Hybrid Natural Fiber and Biopolymer. *ACS Sustainable Chem. Eng.* **2016,** *4*(3), 1785–1793.
4. Boufi, S.; Chaker A. Easy Production of Cellulose Nanofibrils from Corn Stalk by a Conventional High Speed Blender. *Ind. Crops Prod.* **2016,** *93*, 39–47.
5. Cheetham, N. W. H.; Tao, L. Variation in Crystalline Type with Amylose Content in Maize Starch Granules: An X-Ray Powder Diffraction Study. *Carbohydr. Polym.* **1998,** *36*(4), 277–284.
6. Cherian, B. M. et al. Cellulose Nanocomposites with Nanofibres Isolated from Pineapple Leaf Fibers for Medical Applications. *Carbohydr. Polym.* **2011,** *86*(4), 1790–1798.
7. Electron Microscopy. National Programme on Technology Enhanced Learning (NPTEL). http://nptel.ac.in/courses/102103044/18# (Distributed under Creative Commons Attribution-Share. https://creativecommons.org/licenses/by-sa/4.0) (accessed Nov 23, 2016).
8. Futamoto, M.; Hanbücken, M.; Harland, C. J.; Jones, G. W.; Venables, J. A. Visualization of Submonolayers and Surface Topography by Biassed Secondary Electron Imaging: Application to Ag Layers on Si and W Surfaces. *Surf. Sci.* **1985,** *150*(2), 430–450.
9. Gao, R.; Brigstock, D. R. Low Density Lipoprotein Receptor-Related Protein (LRP) is a Heparin-Dependent Adhesion Receptor for Connective Tissue Growth Factor (CTGF) in Rat Activated Hepatic Stellate Cells. *Hepatol. Res.* **2003,** *27*(3), 214–220.
10. García, A et al. Industrial and Crop Wastes: A New Source for Nanocellulose Biorefinery. *Ind. Crops Prod.* **2016**, *93*, 26–38.
11. Gruverman, A.; Wu, D.; Rodriguez, B. J.; Kalinin, S. V.; Habelitz, S. High-Resolution Imaging of Proteins in Human Teeth by Scanning Probe Microscopy. *Biochem. Biophys. Res. Commun.* **2007,** *352*(1), 142–146.
12. Harnagea, C.; Vallières, M.; Pfeffer, C. P.; Wu, D.; Olsen, B. R.; Pignolet, A.; Légaré, F.; Gruverman, A. Two-Dimensional Nanoscale Structural and Functional Imaging in Individual Collagen Type I Fibrils. *Biophys. J.* **2010,** *98*(12), 3070–3077.
13. Hay, J. N.; Shaw, S. J. Nanocomposites—Properties and Applications. http://www.azom.com/Details.asp?ArticleID=921 (accessed Nov 24, 2016).
14. Hazel, J. L.; Tsukruk, V. V. Spring Constants of Composite Ceramic/Gold Cantilevers for Scanning Probe Microscopy. *Thin Solid Films.* **1999,** *339*(1), 249–257.
15. Huang, C.; Yang, Q.; Wang, S. XPS Characterization of Fiber Surface of Chemithermomechanical Pulp Fibers Modified by White-Rot Fungi. *Asian J. Chem.* **2012,** *24*(12), 5476.
16. John, M. J.; Thomas, S. Biofibres and Biocomposites. *Carbohydr. Polym.* **2008,** *71*(3), 343–364.
17. Kalina, L.; Másilko, J.; Koplík, J.; Šoukal, F. XPS Characterization of Polymer–Monocalcium Aluminate interface. *Cem. Concr. Res.* **2014,** *66*, 110–114.
18. Khokhlov, A. R. Statistical Physics of Macromolecules. American Institute of Physics, 1994.

19. Lahiji, R. R.; Xu, X.; Reifenberger, R.; Raman, A.; Rudie, A.; Moon, R. J. Atomic Force Microscopy Characterization of Cellulose Nanocrystals. *Langmuir.* **2010,** *26*(6), 4480–4488.
20. Lamprou, D. A.; Smith, J. R.; Nevell, T. G.; Barbu, E.; Stone, C.; Willis, C. R.; Tsibouklis, J. A Comparative Study of Surface Energy Data from Atomic Force Microscopy and from Contact Angle Goniometry. *Appl. Surf. Sci.* **2010,** *256*(16), 5082–5087.
21. Li, Q.; Guo, L.; Qiu, T.; Xiao, W.; Du, D.; Li, X. Synthesis of Waterborne Polyurethane Containing Alkoxysilane Side Groups and the Properties of the Hybrid Coating Films. *Appl. Surf. Sci.* **2016,** *377*, 66–74.
22. Mishra, R. K.; Sravanthi, L. Fabrication Techniques of Micro/Nano Fibres based Nonwoven Composites: A Review. *Mod. Chem. Appl.* **2017,** *5*. DOI: 10.4172/2329-6798.1000206.
23. Mishra, R. K.; Thomas, S.; Karikal, N. *Micro and Nano Fibrillar Composites (MFCs and NFCs) from Polymer Blends;* Elsevier, 2017. DOI: 10.1016/B978-0-08-101991-7.09989-1.
24. Murphy, W. L.; Dennis, R. G.; Kileny, J. L.; Mooney, D. J. Salt Fusion: An Approach to Improve Pore Interconnectivity within Tissue Engineering Scaffolds. *Tissue Eng.* **2002,** *8*(1), 43–52.
25. Nanotechnology/Scanning Probe Microscopy. https://en.wikibooks.org/wiki/Nanotechnology/Scanning_probe_microscopy (Creative Commons Attribution-Share-Alike License: https://creativecommons.org/licenses/by-sa/3.0/) (accessed Nov 24, 2016).
26. Nechyporchuk, O.; Belgacem, M. N.; Bras, J. Production of Cellulose Nanofibrils: A Review of Recent Advances. *Ind. Crops Prod.* **2016,** *93*, 2–25.
27. Njuguna, J.; Pielichowski, K.; Desai, S. Nanofiller-Reinforced Polymer Nanocomposites. *Polym. Adv. Technol.* **2008,** *19*, 947–959.
28. Perrin-Sarazin, F. et al. Micro-and Nano-Structure in Polypropylene/Clay Nanocomposites. *Polymer.* **2005**, *46*(25), 11624–11634.
29. Purdue University, West Lafayette. https://www.purdue.edu/ehps/rem/rs/sem.htm (accessed Nov 24, 2016).
30. Skryleva, E. A.; Parkhomenko, Yu N.; Karnaukh, I. M.; Zhukova, E. A.; Karaeva, A. R.; Mordkovich, V. Z. XPS Characterization of MWCNT and C60–Based Composites. *Fullerenes, Nanotubes Carbon Nanostruct. (just-accepted)* **2016,** 00–00.
31. Stanford University. Stanford, CA 94305. (650) 723–2300. http://web.stanford.edu/group/cpn/index.html (accessed Nov 24, 2016).
32. Stevens, M. M.; Mayer, M.; Anderson, D. G.; Weibel, D. B.; Whitesides, G. M.; Langer, R. Direct Patterning of Mammalian Cells onto Porous Tissue Engineering Substrates Using Agarose Stamps. *Biomaterials* **2005,** *26*(36), 7636–7641.
33. Sun, Z.; Fan, C.; Tang, X.; Zhao, J.; Song, Y.; Shao, Z.; Xu, L. Characterization and Antibacterial Properties of Porous Fibers Containing Silver Ions. *Appl. Surf. Sci.* **2016,** *387*, 828–838.
34. Thomas, S.; Thomas, R.; Zachariah, A. K.; Mishra, R. K. Microscopy Methods in Nanomaterials Characterization. In *Microscopy Methods in Nanomaterials*

Characterization, 1st ed.; Thomas, S., Thomas, R., Zachariah, A. K., Mishra, R. K., Eds.; Elsevier, 2017; p 432. DOI: 10.1016/B978-0-323-46141-2.01001-4.

35. Tibbitt, M. W.; Anseth, K. S. Hydrogels as Extracellular Matrix Mimics for 3D Cell Culture. *Biotechnol. Bioeng.* **2009,** *103*(4), 655–663.
36. Wang, F.; Zhao, X. *Determination of Young's Modulus for Microcantilevers in Atomic Force Microscopy*. International Conference on Experimental Mechnics 2008 and Seventh Asian Conference on Experimental Mechanics, International Society for Optics and Photonics, 2008, pp 737538–737538.
37. Wan-Gyu, K.; Jae-Ryung C.; Myoung-Seon G. Surface Modification of Polyester Fibers by Thermal Reduction with Silver Carbamate Complexes. *Fibers Polym.* **2016,** *17*(8), 1146–1153.
38. What is the SEM? Iowa State University of Science and Technology. http://www.mse.iastate.edu/research/laboratories/sem/microscopy/what-is-the-sem/ (accessed Nov 23, 2016).
39. Yan, W.; Han, Z. J.; Phung, B. T.; Faupel, F.; Ostrikov, K. High-Voltage Insulation Organic-Inorganic Nanocomposites by Plasma Polymerization. *Materials.* **2014,** *7*(1), 563–575. http://creativecommons.org/licenses/by/3.0/
40. Zong, X. et al. Structure and Process Relationship of Electrospun Bioabsorbable Nanofiber Membranes. *Polymer.* **2002,** *43*(16), 4403–4412.

CHAPTER 18

NANOTECHNOLOGY FOR SMART AND INTELLIGENT FOOD PACKAGING

AJITH JAMES JOSE[1] and MUTHUKARUPPAN ALAGAR[2]

[1]Post graduate & Research Department of Chemistry, St Berchman's College (Autonomous), Changanassery 686101, Kerala, India

[2]Centre of Excellence for Advanced Materials Manufacturing Processing and Characterisation (CoExAMMPC), VFSTR University, Vadlamudi-522 213, Guntur, India

CONTENTS

ABSTRACT

Currently, food packaging and monitoring are a major focus of food industry-related nanotechnology research. It is hoped that nanotechnology has a great potential in food industries as it may be used to manufacture about 25% of all food packaging in near future. Nanotechnology is an emerging novel food packaging technique which can increase the shelf life of foods, minimize the spoilage, ensure the food safety, repair the tears in packaging, reduce the problem of food shortage and finally improve the health of the people. In this chapter, several applications of nanomaterials in food packaging and food safety are reviewed, including polymer/clay nanocomposites as high-barrier packaging materials, silver nanoparticles as potent antimicrobial agents and nanosensors.

18.1 INTRODUCTION

In food industries, there has been a gradual increase in the importance and popularity of food packaging, in which more and more attention is given to the advances in functionality such as convenience and portioning. Modern packaging, however, should serve not only as an efficient tool for keeping quality of foodstuffs but also for increasing product values, promoting sales and imparting information. In today's world, the safety and quality assurance of packaged food products have become vital concerns in worldwide-integrated food supply chains.[1] The growing demand of minimally or unprocessed packaged foods has further aggravated the safety concerns which in turn has fuelled in extensive research to develop novel techniques of food processing, preservation and packaging as well as for rapid, accurate and early detection of contaminant products/microbes.[2]

The packaging industry is coming up with new ways to entertain, add functionality and to create an experience that consumers have never encountered before. In addition, all food companies are aiming for better packaging to make sure that it is a part of their brand story. Consumers want to be assured that these packaging are good for me, my family, my community and the world. Innovative active and intelligent packaging concepts are being developed to provide shelf life extension and to improve the quality, safety and integrity of the packaged food. Active packaging technology is a relatively novel concept designed to provide interaction between food and packaging material, while sustaining the microenvironment contained within. It is aimed at extending the product shelf life, maintaining its nutritional and sensory quality, as well as providing microbial safety. Intelligent packaging systems monitor the condition of packaged foods to give information about the quality of the packaged food during transport and storage.[3]

Traditional food packaging materials include metal, ceramic (glass) and paper (cardboard). Although these materials are still in use, the light weight, low cost, ease of processing and formability, and remarkable diversity in physical properties of organic polymeric materials causes plastics to be attractive alternatives for food packaging. Some of the properties that are offered by polymer packaging are strength and stiffness, barrier to oxygen and moisture, resistance to food component attack and flexibility. They can be flexible or rigid, transparent or opaque, thermosetting or thermoplastic (heat sealable), fairly crystalline or practically amorphous and are less expensive than metal or glass. They can be produced as films or as containers of any shape and size.[4] Novel and efficient polymer materials for food packaging based on nanotechnology can provide the polymer a global market, solutions to increase the performance of the polymers further by adding safety, economic and environmental advantages, reduction of energy input for production, transport and storage, increase of biodegradability and protection from gases and light, reduction in volume of waste material to be disposed in landfills and decrease the contributions to CO_2 emissions.

18.2 NANOTECHNOLOGY IN FOOD PACKAGING

The nanofood market has increased from a value of USD 2.6 billion in 2003 to USD 5.3 billion in 2005; and a surge of USD 20.4 billion is expected in 2020. This trend is a clear indication that nanotechnology will progress within the food and drink packaging industry, and that all companies, should they wish not to lose out, need to stay on top of this dynamic development due to the increasing consumer demand for high quality and microbiologically safer foods, together with a longer product shelf life. 'Nanotechnology' is crossing many technology boundaries as the scientists from disciplines such as chemistry, physics and other pure sciences to medical, materials, sensors and food to name a few, interact to link their researches together to improve current products.[5] The application of nanotechnology to the food sector offers various benefits, including enhancements in production and processing—suggested improvements being consonant with the following preferences: superior food contact materials, quality and freshness monitoring, traceability and product security, sensation, consistency, fat content and nutrient absorption.

A polymer nanocomposite is a combination of polymers with inorganic or organic fillers of certain geometries (fibres, flakes, spheres, particulates). The use of fillers which have at least one dimension in the nanometric range (nanoparticles) produces polymer nanocomposites. Depending on how many dimensions are in the nanometric range, nanofillers can be divided into three types. Isodimensional nanoparticles, such as spherical silica nanoparticles or semiconductor nanoclusters, have three nanometric dimensions. Nanotubes or whiskers are elongated structures in which two dimensions are in the nanometre scale. The composites are known as polymer-layered crystal nanocomposites when only one dimension is in the nanometre range, almost exclusively obtained by the intercalation of the polymer (or a monomer subsequently polymerized) inside the galleries of layered host crystals.[6]

Polymer matrices used in nanocomposites include polyamides, nylons, polyolefins, ethylene-vinyl acetate copolymer, polystyrene, epoxy resins, polyurethane and polyethylene terephthalate (PET). The efficacy of the active components in the film depends upon the choice of polymer matrix. Density is an important factor as it determines the rate of release of bioactives, which may be required to either be bound in the matrix or be released over time. The shelf life of the food and the contact area between the packaging and food must also be taken into consideration. By reinforcing appropriate nanoparticles, it is possible to produce packages with stronger mechanical, barrier and thermal performance. To take advantage of their substantially enhanced properties, polymer nanocomposites have also been studied for food packaging applications including injection moulded bottles for beverage or beer, coatings for paperboard juice cartons and cast and blown films. It is also possible that combining two or more nanoparticulate components in one film can optimize a food packaging system. Due to very large aspect ratios, relatively low levels of nanoparticles are sufficient to change the properties of packaging materials without significant changes in density, transparency and processing characteristics. Food safety will be the main focus of new packaging solutions, and this will be done by controlling microbial growth, delaying oxidation, improving tamper visibility and convenience. Application of polymer nanotechnology can provide new food packaging materials with improved mechanical, barrier and antimicrobial properties, together with nanosensors for tracing and monitoring the condition of food during transport and storage. The use of nanosilver is an example for the packaging industry. On account of its antimicrobial properties, nanosilver is used to coat packaging materials and inner surfaces of fridges and dishwashers, as well as is being incorporated into plastic food containers. Another example is the use of nanoclays, which can be incorporated into plastic bottles for drinks. It will prevent oxygen from migrating through the plastic bottle walls and destabilize the drink. Therefore, it extends the product shelf life. In short, nanotechnology is quickly moving from the laboratory onto supermarket shelves and our kitchen tables and has the potential to revolutionize food systems.[7]

18.3 NANOREINFORCEMENTS

Nanoparticles have larger surface area than their microscale counterparts, which favours the filler–matrix interactions and the performance of the resulting material. High aspect ratio fillers are of particular interest because of their high specific surface area, which provides better reinforcing effects. A uniform dispersion of nanoparticles leads to a very large matrix/filler interfacial area, which changes the molecular mobility, the relaxation behaviour and the consequent thermal and mechanical properties of the material. Active or 'smart' properties are sometimes provided to the packaging system by these reinforcing nanoparticles such as antimicrobial activity, enzyme immobilization and biosensing.

The incorporation of nanoparticles into packaging offers the following:

1. **Reduction in raw materials**. Improved stiffness enables the use of fewer raw materials and down-gauging by 20% can be achieved. Lighter packaging may lead to savings in the cost of transportation, storage and recycling.
2. **Less dependence on speciality products.** Polymer nanocomposites can be an alternative to the expensive specialty materials.

3. **Elimination of secondary processes.** High-cost incurred during operations such as laminations for barrier packaging or mechanical surface finishing can be eliminated.
4. **Less complex structures.** Nanocomposites may have, for example, less complex structures than multi-laminates and this can lead to easier recycling.
5. **Reduction in machine cycle time.** By changing the physical and thermal properties of polymers it is possible to reduce pack production times.

The main kinds of nanoparticles as well as their effects and applications which have been studied for the use in food packaging systems are discussed below. Some particles can have multiple applications, and sometimes, these applications can overlap, such as some immobilized enzymes which can act as antimicrobial components, oxygen scavengers and/or biosensors.[8]

18.3.1 CLAYS AND LAYERED SILICATES

Although several nanoparticles have been recognized as possible additives to enhance polymer performance, the packaging industry has focused its attention mainly on layered inorganic solids such as clays and silicates, due to their easy availability, low cost, significant enhancements and relative simple processing. The layered silicates commonly used in nanocomposites consist of two-dimensional layers, which are 1 nm thick and several microns long depending on the particular silicate. In general, there are three possible arrangements for layered silicate clay nanocomposite materials: non-intercalated, intercalated, and exfoliated or delaminated. In non-intercalated materials, the polymer does not fit between the layered clay. This leads to a microphase separated final structure. The polymer is located between clay layers in intercalated systems, and this helps in increasing the interlayer spacing. Some degree of order is retained in parallel clay layers, which are separated by alternating polymer layers with a repeated distance of few nanometres. Exfoliated systems achieve complete separation of clay platelets in random arrangements.

Layered silicates in polymer formulations increase the tortuosity of the diffusive path for a penetrant molecule, providing excellent barrier properties. The most widely studied type of clay fillers is montmorillonite, hydrated alumina-silicate layered clay consisting of an edge-shared octahedral sheet of aluminium hydroxide between two silica tetrahedral layers. Due to the hydrophilic character of its surface, the homogeneous dispersion of most of clays in organic polymers is not easy. Surfactants can be used to improve the dispersing power of the clay.

Large amounts of silicate nanoparticles are interspersed in polymer films. These nanoparticles block oxygen, carbon dioxide and moisture from reaching fresh meats and other food products. The nanoclay particles act as an impermeable obstacle in the path of the diffusion process, which helps in extending the shelf life of foods while improving their quality. The final package is also considerably lighter, stronger and more heat-resistant.

18.3.2 CELLULOSE-BASED NANOREINFORCEMENTS

Cellulose, the building material of long fibrous cells, is a highly strong naturally occurring polymer. Cellulose nanofibres are inherently a low cost and widely available material. Moreover, they are environment friendly and easy to recycle by combustion, and energy consumption in manufacturing is low. All of this makes cellulose nanofibres an attractive class of nanomaterials for the production of low cost, lightweight and high-strength nanocomposites.

Basically two types of nanoreinforcements can be obtained from cellulose—microfibrils and whiskers. In plants and animals, the cellulose chains are synthesized to form microfibrils (or nanofibres), which are bundles of molecules that are elongated and stabilized through hydrogen bonding. The microfibrils have diameters in nanometre range (2–20 nm, depending on the origin), and their lengths are in the micrometre range. Each microfibril is formed by the aggregation of elementary fibrils, which are made up of crystalline and amorphous parts. The crystalline parts are the whiskers, which can be isolated by several treatments, and also known as nanocrystals, nanorods or rod-like cellulose microcrystals, with lengths ranging from 500 nm up to 1–2 μm, and about 8–20 nm or less in diameter, resulting in high aspect ratios.

18.3.3 CARBON NANOTUBES

Carbon nanotubes (CNTs) can consist of a one-atom thick single-wall nanotube or a number of concentric tubes called multiwalled nanotubes, having extraordinarily high aspect ratios and elastic modulus. It has been reported that CNTs have theoretical elastic modulus and tensile strength values as high as 1 TPa and 200 GPa, respectively. Carbon nanotubes are modified by introducing carboxylic acid groups on their surfaces to enhance their intermolecular interactions with the polymer matrix. Carbon nanotubes greatly improved the thermal stability, tensile strength and modulus of polymer nanocomposite, even in concentrations as low as 0.1 wt%.

18.3.4 STARCH NANOCRYSTALS

Native starch granules can be submitted to an extended time hydrolysis at temperatures below the gelatinization temperature. When the amorphous regions are hydrolysed separation of crystalline lamellae occurs, which are more resistant to hydrolysis. The starch crystalline particles show platelet morphology with thicknesses of 6–8 nm. The tensile strength and modulus of polymer films is improved by the addition of starch nanocrystals, but decreases their elongation. The T_g values shifted to higher temperatures with increasing nanocrystal content, and this was attributed to a restricted mobility of polymer chains due to the formation of strong interactions between nanocrystals and as well as between filler and matrix. The water vapour permeability of polymer films was found to decrease by addition of starch granules.

18.3.5 CHITOSAN/CHITIN NANOPARTICLES

Chitosan is a partially deacetylated polymer of N-acetylglucosamine that can be obtained through alkaline deacetylation of chitin. It consists of a β-(1,4)-linked-D-glucosamine residue with the amine groups randomly acetylated. Nanoparticles can be prepared by adding drop-wise tripoly-phosphate-pentasodium solution to chitosan solutions under stirring. Trehalose, mannitol and polyethylene-glycol are used as bioprotectants to prevent particle aggregation and to reduce mechanical stress during freezing and drying processes. Addition of chitosan nanoparticles significantly improved mechanical and barrier properties of the hydroxypropyl methylcellulose films. Acid hydrolysis of chitin helps in preparing chitin whiskers. The average dimensions of the whiskers are 500 nm (length) and 50 nm (diameter). Addition of chitin whiskers to soy protein isolate thermoplastics greatly improved not only the tensile properties (tensile strength and elastic modulus) of the matrix but also its water resistance. Addition of chitin whiskers to chitosan films showed an improvement of chitosan tensile strength until a whisker content of 2.96%, whereas higher increases of whiskers contents resulted in decreasing strength. The elongation of the films was impaired by addition of whiskers up to 2.96%, and then it levelled off at higher whiskers contents. The addition of a-chitin whiskers improved water resistance of the films.

18.3.6 GRAPHENE

Graphene is a monolayer of sp^2-hybridized carbon atoms arranged in a two-dimensional lattice and has successfully attracted tremendous attention in recent years owing to its exceptional thermal, mechanical and electrical properties. Graphene can be synthesized in various ways and on different substrates. One of the most promising applications of this material is in polymer nanocomposites, polymer matrix composites which incorporate nanoscale filler materials.

For gas-barrier applications, functional graphene-based polymer nanocomposites have been studied. Recently, the trustees of Princeton University received a patent for graphene–elastomer nanocomposites in which functionalized graphene sheets had been dispersed in vulcanized natural rubber and styrene butadiene rubber. The patented work can find a wide range of industrial applications, and this also includes food packaging.

However, there are many safety concerns about nanomaterials, as their size may allow them to penetrate into cells and eventually remain in the system. There is no consensus about categorizing nanomaterials as new (or unnatural) materials. The properties and safety of the materials in its bulk form are usually well known, but different properties are exhibited by the nanosized counterparts frequently and are different from those that are found at the macro-scale. There is limited scientific data about migration of most types of nanoparticles from the packaging material into food, as well as their eventual toxicological effects. It is reasonable to assume that migration may occur; hence, the need for accurate information on the effects of nanoparticles to human health following chronic exposure is imperative.

18.4 NANOCOMPOSITES PACKAGING

In the near future, nanocomposite packages are predicted to make up a significant portion of the food packaging market.[9] Many nanocomposite food packages are either already in the marketplace or are being developed. The majority of these are being targeted at beverage packaging. By using nanocomposite materials, fundamental characteristics of food packaging materials such as strength, barrier properties, antimicrobial properties and stability to heat and cold are improved. Three common methods are used to process nanocomposites: solution method, in situ or interlamellar polymerization technique and melt processing.

Earlier, packaging of beer in plastic bottles was not possible due to oxidation and flavour problems. However, this challenge has been overcome with the help of nanotechnology. For example, Nanocor, a subsidiary of Amcol International Corp., is producing nanocomposites for using in plastic beer bottles that facilitate a 6-months shelf life. A new family of barrier nylons was recently developed by combining the nanocomposite and oxygen scavenger technologies for use in multilayer, co-injection blow-moulded PET bottles. In the near future, nanocrystals embedded in plastic bottles may increase beer shelf life up to 18 months by minimizing loss of carbon dioxide from and entrance of oxygen into bottles. Similar materials are being developed to extend the shelf life of soft drinks. Another advantage of these nanocomposite bottles is that they are lightweight, thereby aiding in the reduction of transportation costs. A considerable amount of research is also taking place in the area of biodegradable nanocomposite food packages. These films act as effective moisture barriers by increasing the tortuosity of the path that water must take to penetrate the films.[10]

18.5 BIONANOCOMPOSITES FOR FOOD PACKAGING

Nowadays, most of the materials used in packaging industries are produced from fossil fuels and are practically nondegradable causing serious environmental problems.[11] Biodegradable films from renewable resources have been encouraged in an effort to extend the shelf life and to enhance food quality at the time of reducing packaging waste in the exploration of new bio-based packaging materials. The materials which can be degraded by the enzymatic action of living organisms, such as bacteria, yeasts, fungi and the ultimate end products of the degradation process, these being CO_2, H_2O and biomass under aerobic conditions and hydrocarbons, methane and biomass under anaerobic conditions are known as 'biodegradable' materials. The long polymer molecules are reduced to shorter and shorter lengths first and undergo oxidation during the process of biodegradation. This process may be triggered by heat, UV light and mechanical stress. Oxidation causes the molecules to become hydrophilic and small enough to be ingestible by microorganisms, setting the stage for biodegradation to begin. As microorganisms consume this degraded plastic, carbon dioxide, water and biomass are produced and is returned to nature through biocycle.

Biodegradable polymers can be classified according to their source:

1. Polymers extracted or removed directly from biomass (i.e. polysaccharides, proteins, polypeptides, polynucleotides).
2. Polymers produced by classical chemical synthesis using renewable bio-based monomers or mixed sources of biomass and petroleum (i.e. polylactic acid or bio-polyester)
3. Polymers produced by microorganism or genetically modified bacteria [polyhydroxybutyrate (PHB), bacterial cellulose, xanthan, curdian, pullan].

Thus, the development of biodegradable polymer solves the menace of discarding the plastics leading to complete bioassimilation of the plastics. However, the use of biodegradable polymers is limited due to issues related to performance (such as brittleness, poor gas and moisture barrier), processing (such as low heat distortion temperature) and cost. For this reason, blends of biopolymers with other biodegradable polymers have been considered as a promising avenue for preparing polymers with 'tailor-made' properties (functional physical properties and biodegradability). A way to reduce the overall cost of the material and to offer a method of modifying both properties and degradation rates is provided with the incorporation of relatively low cost natural biopolymers into biodegradable synthetic polymers, with the aim of extending their applications in more special or severe circumstances.[12]

The extraordinary success of the nanocomposite concept in the area of synthetic polymer has stimulated the new research on nanocomposites based on biodegradable polymers as matrix. The incorporation of inorganic particles into biopolymeric matrix not only shows significant improvement of mechanical, thermal and barrier properties of the virgin polymers but is also suitable for enhancing their rate of biodegradation. Thus, the application of nanotechnology plays an important role to improve the feasibility of use of biopolymers that can reduce the packaging waste associated with processed foods. This supports the preservation of foods by extending their shelf life and cost-price efficiency. The most studied biopolymer nanocomposites are starch, cellulose, pectin, gluten, gelatine, soya protein, polylactide, PHB and so on.

Bionanocomposites are a promising way to further improve material properties while maintaining their biodegradability. The major challenge is to reduce production and material cost of these bionanocomposites to make them cost-effective against synthetic polymers. There is a need to improve bionanocomposite formulation and the processing method to produce these bionanocomposites at a lower cost to overcome this challenge. There has been some success in commercializing bionanocomposites, whereas more products are still under research stage. More research is also needed to utilize different types of nanoparticles such as CNTs, graphene and nanocelluloses for producing new nanocomposite materials with improved properties.[13]

18.6 EDIBLE POLYMER NANOCOMPOSITE FILMS AND COATINGS

Edible coating is an environment-friendly technology that is applied on many products to control moisture transfer, gas exchange or oxidation processes, which can be easily consumed by human beings or lower animals in whole or part via the oral cavity and has harmless effect to the health. Edible films or coatings are biopolymers with numerous desirable properties and may be made from a variety of materials, alone or in combination with other components. Edible biopolymers have also been developed from sources such as fungal exopolysaccharide (pullan) or fermentation byproducts (polylactic acid) and applied to foods. Suitable polysaccharides include cellulose derivatives, alginates, pectins, starches, chitosan and other polysaccharides. To make edible films and coatings, many lipid compounds such as animal and vegetable fats including waxes, acylglycerols and fatty acids have been used. Lipid films possess excellent moisture barrier properties and can be used as coating agents for adding gloss to confectionery products. In general, waxes are used for coating fruits and vegetables to retard respiration and lessen moisture loss. As edible films and coatings are both a packaging and a food component, they have to fulfil requirements such as good sensory qualities, high barrier and mechanical efficiencies, enough biochemical, physicochemical and microbial stability, free of toxins and safe for health, simple technology, nonpolluting and low cost of raw materials and process.

Currently, edible films and coatings are used as casings for sausages and chocolate coatings for nuts and fruits. Food products are usually coated by dipping or spraying, forming a thin film on the food surface. This thin film acts as a semipermeable membrane, which in turn control the moisture loss or/and suppress the gas transfer.

Production of edible polymer causes less waste and pollution; however, their commercial uses have been limited because of their poor mechanical and barrier properties when compared with synthetic polymers. The functionality and performance of edible polymer mainly depend on their barrier, mechanical and colour properties, which in turn depend on film composition and its formation process. Adding reinforcing compounds (fillers) to polymers to enhance their thermal, mechanical and barrier properties, several nanocomposites have been developed. Plasticizers are often added to film-forming solutions to enhance properties of the final film. Common food-grade plasticizers such as sorbitol, glycerol, mannitol, sucrose and polyethylene glycol decrease brittleness and increase flexibility of the film, which is important in packaging applications. Plasticizers used for protein-based edible films decrease protein interactions and increase both polymer chain mobility and intermolecular spacing.[14]

Edible coatings and films are a viable means for incorporating food additives and other substances to enhance the product colour, flavour and texture and to control microbial growth. Edible polymers can be applied directly on to the surface as additional protection to preserve product quality and stability. Mechanical, sensory and even functional properties may be affected drastically when active ingredients (antimicrobials, antioxidants and nutrients) are added to edible coatings. The development of new technologies to improve the delivery properties of edible polymer is a major issue for future research.

Nanocomposites perception represents a motivating route for creating new and innovative materials, in area of edible polymers as well. Materials with a large variety of properties have been realized, and even more are due to be realized. Micro- and nanoencapsulation of active compounds with edible polymer coatings may help to control their release under specific conditions, thus protecting them from moisture, heat or other extreme conditions and thus enhancing their stability and viability. Nanocomposite edible films have been developed by the addition of cellulose nanofibres in different concentrations to mango puree edible films. They act as a nanoreinforcing component on tensile properties, water vapour permeability and glass transition temperature. Recently, most studies on food applications are being conducted at a laboratory scale. However, further research should be focused on a commercial scale with the purpose of providing more accurate information that can be used to commercialize fresh-cut products coated with edible polymers.

18.6.1 ANTIMICROBIAL NANOCOMPOSITES

Antimicrobial packaging encompasses any packaging technique that can be used to control microbial growth in a food product. In recent years, from the food packaging industry, antimicrobial packaging is gaining more attention, as the use of preservative packaging films has several advantages compared with the direct addition of preservatives into food products. Reflecting this demand, the preservative agents must be applied to packaging in such a way that only low levels of preservatives comes in contact with the food. These active food contacting materials can extend the product shelf life, enhance food quality and safety and ultimately lead to less food waste.[15]

The incorporation of antimicrobial nanoparticles directly into polymers, coating antimicrobials onto polymer surfaces, immobilizing antimicrobials by chemical grafting or using polymers that are antimicrobial by themselves allows industry to combine the preservative functions of antimicrobials with the protective functions of the pre-existing packaging concepts. In controlling the growth of pathogenic and spoilage microorganisms, films with antimicrobial activity helps greatly. An antimicrobial nanocomposite film is particularly desirable due to its acceptable structural integrity and barrier properties imparted by the nanocomposite matrix and the antimicrobial properties contributed by the natural antimicrobial agents impregnated within. Materials in the nanoscale range have a higher surface-to-volume ratio when compared with their microscale counterparts. This allows nanomaterials to be able to attach more copies of biological molecules, which imparts it with greater efficiency. Investigation on nanoscale materials have been carried out for antimicrobial activity so that they can be used as growth inhibitors, killing agents or antibiotic carriers.

Nanosilver, nanochitosan, nanomagnesium oxide, nanocopper oxide, nanotitanium dioxide and CNTs are also predicted for future use in antimicrobial food packaging.[16] The biggest advantage of inorganic nanoparticles over molecular antimicrobials is the ease with which the former can be incorporated into polymers to form functional antimicrobial materials. Silver is a well-known antimicrobial agent and is being infused into storage containers to retard bacterial growth and also allows longer storage of foods. In a case study, the 24-h growth of bacteria was reduced by over 98% because of the silver nanoparticles. Thus, silver/polymer nanocomposites are attractive materials for use in both medical devices as well as food packaging materials to preserve shelf life. Titanium dioxide (TiO_2) is widely used as a photocatalytic disinfecting material for surface coatings. TiO_2 photocatalysis, which promotes peroxidation of the polyunsaturated phospholipids of microbial cell membranes, has been used to inactivate several food-related pathogenic bacteria. Studies suggest that TiO_2 powder-coated packaging film has the ability to reduce E. coli contamination on food surfaces, suggesting that the film could be used for fresh-cut produce. Moreover, TiO_2-coated films when exposed to sunlight can inactivate faecal coliforms in water. Antimicrobial mechanism of nanoscale chitosan involves interactions between positively charged chitosan and negatively charged cell membranes, increasing membrane permeability and eventually causing rupture and leakage of intracellular material. This is consistent with the observation that both raw chitosan and engineered nanoparticles are ineffective at pH values above 6, which would be due to the absence of protonated amino groups. Another two antimicrobial mechanisms proposed were chelation of trace metals by chitosan, inhibiting enzyme activities and in fungal cells, penetration through the cell wall and membranes to bind DNA and inhibit RNA synthesis.

There is a probability that new applications such as the use of nanoparticles as antimicrobial agents in edible films and coatings will arise in the future and will bring up new issues on the ingestion, accumulation and safety of such products.

18.6.2 CONDUCTING POLYMER NANOCOMPOSITE SENSORS

Food spoilage is caused by microorganisms whose metabolism produces gases which may be detected by several types of gas sensors which have been developed to translate chemical interactions between particles on a surface into response signals. Nanosensors, which are able to quantify and/or identify microorganisms based on their gas emissions, are usually conducting polymer nanocomposites and are fabricated by embedding conducting particles into an insulating polymer matrix packaging. A specific response pattern for each microorganism was developed in nanosensors containing carbon black and polyaniline. This was also used to detect and identify food-borne pathogens. Sometimes, the embedded sensors in a packaging film may be able to detect food-spoilage organisms and trigger a colour change to alert the consumer that the shelf life is ending or has ended.

On the account of the conducting polymers' electrical, electronic, magnetic and optical properties, which are related to their conjugated π electron backbones, they are also very important. Polyene and polyaromatic conducting polymers such as polyaniline, polyacetylene and polypyrrole have been widely studied. Electrochemically polymerized conducting polymers have a remarkable ability to switch between conducting oxidized (doped) and insulating reduced (undoped), which is the basis of several applications.

18.6.3 NANOSENSORS

In addition to this, nanomaterials can be used to make packages that have the ability to keep the product inside fresher for longer period of time. Intelligent food packaging, incorporating nanosensors, could be used to detect chemicals, pathogens and toxins in foods. Food packages embedded with nanoparticles can alert consumers when a product is no longer safe to eat. Sensors can warn before the food goes rotten or can inform us the exact nutritional status contained in the contents.

Numerous research reports delineate detection methods for bacteria, viruses, toxins and allergens using nanotechnology. For example, adhering antibodies to *Staphylococcus* enterotoxin B onto poly (dimethyl-siloxane) chips formed biosensors that have a detection limit of 0.5 ng/ml. Nanovesicles have been developed to simultaneously detect *E. coli* 0157:H7, *Salmonella* spp. and *Listeria monocytogenes*. Liposome nanovesicles have been devised to detect peanut allergen proteins. In addition, AgroMicron has developed a Nanobioluminescence detection spray containing a luminescent protein that has been engineered to bind to the surface of microbes such as *Salmonella* and *E. coli*. It emits a visible glow that varies in intensity according to the amount of bacterial contamination when bound to the surface of these microbes.

DNA biochips that can detect pathogens are being developed. Nano-sized carbon tubes coated with strands of DNA were fabricated to create nanosensors with abilities to detect odours and tastes. A single strand of DNA serves as the sensor and a CNT functions as the transmitter. By using similar technologies, electronic tongue nanosensors are being developed to detect substances in parts per trillion, which could be used to trigger colour changes in food packages to alert consumers when food is spoiled. A unique aspect of these biochips is that the DNA is self-assembled onto the chips and repairs itself if damaged. Another colour-changing film that could find its way into food packages is polymer opal films. Polymer opal films belonging to a class of materials known as photonic crystals. The crystals are built of tiny repeating units of carbon nanoparticles wedged between spheres, leading to intense colours that mimic the colours associated with the photonic crystals found on butterfly wings and peacock feathers.

18.7 CONCLUSION

Since its inception in the 18th century, the food industry has seen great advances in the packaging sector with most active and intelligent of the innovations occurring during the past century. These advances have led to improved food quality and safety. The packaging sector is greatly influenced by nanotechnology. Nanoscale innovations in the forms of pathogen detection, active packaging and barrier formation have the potential to elevate food packaging to new heights. Nanocomposites concept represents an inspiring route for creating new and innovative materials, in the area of polymers as well. Materials with a large variety of properties have been realized, and even more are due to be realized. The nanocomposites materials obtained by mixing polymers and or nanoparticles materials offer great improvements over conventional composites in mechanical, thermal, electrical and barrier properties. Furthermore, they significantly reduce flammability and maintain the transparency of the polymer matrix. They are even able to compete in price, performance and in applications particularly in packaging. Even so, food packing nanotechnology is still young, and it is also true that still now great uncertainty prevails in the future of this exciting field. Regardless of how applications of nanotechnology in the food sector are ultimately marketed, governed or perceived by the public, it is clear that the manipulation of matter on the nanoscale will continue to yield exciting and unforeseen products. In fact, nanotechnology is going to change the fabrication of the entire food packaging industry.

18.8 ACKNOWLEDGEMENTS

The authors would like to thank *Ms Christy Mathew* and *Ms Annu John,* postgraduate students, Department of Chemistry, St Berchmans College, Changanassery for their precious help.

KEYWORDS

- **nanotechnology**
- **food packaging**
- **nanocomposites**
- **edible packaging**
- **antimicrobials**
- **sensors**

REFERENCES

1. Aaron L.B.; Betty, B.; Jung, H. H.; Claire, K. S.; Tara H. M. Innovative Food Packaging Solutions. *J. Food Sci.* **2008,** *73*, R108–R116.
2. Qasim, C.; Laurence, C. Food Applications of Nanotechnologies: An Overview of Opportunities and Challenges for Developing Countries. *Trends Food Sci. Technol.* **2011,** *22*, 595–603.
3. Ahmed, M. Y. Polymer Nanocomposites as a New Trend for Packaging Applications. *Polym. Plastic Technol.* **2013,** *52*, 635–660.
4. Cushena, M.; Kerryb, J.; Morrisc, M.; Cruz-Romerob M.; Cummins, E. Nanotechnologies in the Food Industry—Recent Developments, Risks and Regulation. *Trends Food Sci. Technol.* **2012,** *24*, 30–46.
5. Ravichandran. R. Nanotechnology Applications in Food and Food Processing: Innovative Green Approaches, Opportunities and Uncertainties for Global Market. *Int. J. Green Nanotechnol. Phys. Chem.* **2010,** *1*, 72–96.
6. Henriette, M. C. A. Nanocomposites for Food Packaging Applications. *Food Res. Int.* **2009,** *42*, 1240–1253.
7. John Wesley, S.; Raja, P.; Allwyn Sundar R. A.; Tiroutchelvamae, D. Review on Nanotechnology Applications in Food Packaging and Safety. *Int. J. Eng. Res.* **2014,** *3*, 645–651.
8. Malathi, A. N.; Santhosh, K. S.; Udaykumar, N. Recent Trends of Biodegradable Polymer: Biodegradable Films for Food Packaging and Application of Nanotechnology in Biodegradable Food Packaging. *Curr. Trends Technol. Sci.* **2014,** *2*, 2279–0535.
9. Timothy V. D. Applications of Nanotechnology in Food Packaging and Food Safety: Barrier Materials, Antimicrobials and Sensors. *J. Colloid Interface Sci.* **2011,** *361*, 1–24.
10. Sudip, R.; Siew, Y. Q.; Allan, E.; Xiao, D. C. The Potential Use of Polymer-Clay Nanocomposites in Food Packaging. *Int. J. Food Eng.* **2006**, *2*, 1–15.
11. Andrea, S.; Giuliana, G.; Vittoria V. Potential Perspectives of Bio-Nanocomposites for Food Packaging Applications. *Trends Food Sci. Technol.* **2007,** *18*, 84–95.
12. Daniela, G.; Ovidiu, T. Biopolymers Used in Food Packaging: A Review. *Acta Univ. Cibiniensis, Ser. E: Food Technol.* **2012,** *XVI*, 1–19.
13. Marcia, R. M.; Fauze A. A.; Roberto J. A. B.; Tara H. H.; John, M. K.; Luiz, H. C. M. Improved Barrier and Mechanical Properties of Novel Hydroxypropyl Methylcellulose Edible Films with Chitosan/Tripolyphosphate Nanoparticles. *J. Food Eng.* **2009,** *92*, 448–453.
14. Henriette, M. C. A.; Luiz Henrique, C. M.; Delilah, W.; Tina, G. W.; Roberto J. A. B.; Tara H. M. Nanocomposite Edible Films from Mango Puree Reinforced with Cellulose Nanofibers. *J. Food Sci.* **2009,** *72*, R31–R35.
15. Metak, A. M.; Ajaal T. T. Investigation on Polymer Based Nano-Silver as Food Packaging Materials. *Int. J. Biol. Vet. Agric. Food Eng.* **2013,** *12*, 772–778.
16. Paola, P.; Veronica, A.; Cosimo, C.; Pierfrancesco, C.; Ilario, F.; Gianluigi M. Nanocomposite Polymer Films Containing Carvacrol for Antimicrobial Active Packaging. *Polym. Eng. Sci.* **2009,** 1447–1455.

INDEX

****

B

C

D

E

H

I

J

K

L

O

P

S

X

Y

Z